Molecular Biology of Cancer

Second edition

Molecular Biology of Cancer

Second edition

F. Macdonald
Regional Genetics Service, Birmingham Women's Hospital, Queen Elizabeth Hospital, Birmingham, UK

C.H.J. Ford
Department of Surgery, Kuwait University, Kuwait

A.G. Casson
Department of Surgery, Dalhousie University, Halifax, Nova Scotia, Canada

BIOS Scientific Publishers
Taylor & Francis Group

LONDON AND NEW YORK

A CIP catalogue record for this book is available from the British Library.

ISBN 1 85996 247 5

Garland Science/BIOS Scientific Publishers
4 Park Square, Milton Park, Abingdon, Oxon, OX14 4RN, UK and
29 West 35th Street, New York, NY 10001-2299, USA
World Wide Web home page: www.garlandscience.com

Garland Science/BIOS Scientific Publishers is a member of the Taylor & Francis Group.

Distributed in the USA by
Fulfilment Center
Taylor & Francis
10650 Toebben Drive
Independence, KY 41051, USA
Toll Free Tel.: +1 800 634 7064; E-mail: taylorandfrancis@thomsonlearning.com

Distributed in Canada by
Taylor & Francis
74 Rolark Drive
Scarborough, Ontario M1R 4G2, Canada
Toll Free Tel.: +1 877 226 2237; E-mail: tal_fran@istar.ca

Distributed in the rest of the world by
Thomson Publishing Services
Cheriton House
North Way
Andover, Hampshire SP10 5BE, UK
Tel.: +44 (0)1264 332424; E-mail: salesorder.tandf@thomsonpublishingservices.co.uk

Library of Congress Cataloging-in-Publication Data

Macdonald, F.
Molecular biology of cancer / Macdonald, Ford & Casson.-- [2nd ed.]
        p. cm.
Includes bibliographical references and index.
ISBN 1-85996-247-5 (pbk.)
1. Cancer-Molecular aspects. 2. Cancer-Genetic aspects. 3. Cancer-Molecular diagnosis.
I. Ford, C. H. J. II. Casson, Alan G. III. Title.

RC268.4.M23 2004
616.99'4042--dc22
        2004001336

Production Editors: Georgia Bushell and Harriet Milles
Typeset by Charon Tec Pvt. Ltd, Chennai, India
Printed and bound by Cromwell Press, Trowbridge, UK

Contents

Abbreviations

2D-PAGE	two-dimensional polyacrylamide gel electrophoresis
AC	Amsterdam criteria
ACF	aberrant crypt foci
ADCC	antibody-dependent cellular cytotoxicity
ADEPT	antibody-directed enzyme prodrug therapy
AGENT	antibody-guided enzyme nitrile therapy
AJCC	American Joint Committee on Cancer
ALL	acute lymphocytic/lymphoblastic leukemia
AML	acute myeloid/myelogenous leukemia
APC	adenomatous polyposis coli
APC	anaphase-promoting complex
APL	acute promyelocytic leukemia
AR	androgen receptor
BAC	bronchioloalveolar carcinoma
BCNS	basal cell nevus syndrome
BER	base excision repair
BIC	breast cancer information care
BL	Burkitt's lymphoma
bp	basepair
BPH	benign prostate hyperplasia
BWS	Beckwith–Wiedemann syndrome
CAK	CDK-activating complex
CBF	core-binding factor
CDK	cyclin-dependent kinase
cDNA	complementary DNA
CEA	carcinoembryonic antigen
CIMP	CpG island methylator phenotype
CHRPE	congenital hypertrophy of the retinal pigment epithelium
CIS	carcinoma *in situ*
CK	cytokeratin
CKI	cyclin-dependent kinase inhibitors
CLL	chronic lymphocytic leukemia
CML	chronic myeloid/myelogenous leukemia
CMML	chronic myelomonocytic leukemia
CNTF	ciliary neurotrophic factor
CS	Cowden syndrome
CTL	cytotoxic T cell
DAM	DNA adenosine methylase
DD	dimerization domain
DDS	Denys–Drash syndrome
DHT	dihydrotestosterone
DLBCL	diffuse large B cell lymphoma
DLC	delocalized lipophilic cation
DM	double-minute chromosomes

DMR	differentially methylated region
DSB	double-strand break
EGF	epidermal growth factor
EGFR	epidermal growth factor receptor
EOC	epithelial ovarian cancer
ERE	estrogen-responsive element
ERT	estrogen replacement therapy
ETS	external transcribed spacer
FACS	fluorescence-activated cell sorter
FAP	familial adenomatous polyposis
FGF	fibroblast growth factor
FMTC	familial medullary thyroid cancer
FRET	fluorescence resonance energy transfer
GAP	GTPase activating protein
GDEPT	gene-directed enzyme prodrug therapy
GEF	guanine-nucleotide exchange factor
GERD	gastroesophageal reflux disease
HDAC	histone deacetylase
HIF	hypoxia-inducible factor
HIV	human immunodeficiency virus
HNPCC	hereditary nonpolyposis colon cancer
HPV	human papillomavirus
HSR	homogeneous-straining repeat
HSV	herpes simplex virus
HTLV	human T-cell leukemia virus
IASLC	International Association for the Study of Lung Cancer
Ig	immunoglobulin
IGF	insulin-like growth factor
JPS	juvenile polyposis syndrome
LEF	lymphoid enhancer factor
LFS	Li–Fraumeni syndrome
LOH	loss of heterozygosity
LPD	lymphoproliferative disease
LTR	long terminal repeat
MALT	mucosa-associated lymphoid tissue
MAPK	mitogen-activated protein kinase
MCL	mantle cell lymphoma
MCP	modified citrus pectin
MCR	mutation cluster region
MDS	myelodysplastic syndrome
MEK	MAPK/ERK kinase
MEN	multiple endocrine neoplasia
MLL	mixed-lineage leukemia
MMCT	microcell-mediated chromosome transfer
MMP	matrix metalloproteinase
MMR	mismatch repair
MLPA	multiplex ligation-dependent probe amplification
mRNA	messenger RNA
MS	mass spectrometry
MSI	microsatellite instability
MTD	maximum tolerated dose

NER	nucleotide excision repair
NLS	nuclear localization signal
NSAID	nonsteroidal anti-inflammatory drug
NSCLC	non-small cell lung cancer
ORF	open reading frame
PCNA	proliferating cell nuclear antigen
PCR	polymerase chain reaction
PDGF	platelet-derived growth factor
PDGF-R	platelet-derived growth factor receptor
PI3K	phosphatidylinositol 3-kinase
PIG	p53-inducible gene
PIP	phosphatidylinositolphosphate
PLC	phospholipase C
PlGF	placenta growth factor
PLGF	placenta-like growth factor
PPARγ	peroxisome proliferator-activated receptor γ
PSA	prostate-specific antigen
PSCA	prostate stem cell antigen
PTK	protein tyrosine kinase
PTT	protein truncation test
RAR	retinoic acid receptor
RCC	renal cell carcinoma
RFC	replication factor C
RFLP	restriction fragment length polymorphism
RNAi	RNA interference/inhibition
RPA	replication protein A
RTK	receptor tyrosine kinase
RT-PCR	reverse transcriptase-*PCR*
RXR	retinoid X receptor
SAGE	serial analysis of gene expression
SCC	squamous cell carcinoma
SCE	sister chromatid exchanges
SCLC	small cell lung cancer
SERM	selective estrogen receptor modulator
SH2	src homology-2
SIN	squamous intraepithelial neoplasia
SNP	single nucleotide polymorphism
SSCP	single-strand conformation polymorphism
STAT	signal transduction and activation of transcription
TAA	tumor-associated antigen
TCC	transitional cell carcinoma
TCF	T cell factor
TCR	T cell receptor
TGF	transforming growth factor
TIL	tumor-infiltrating lymphocytes
TNFR	tumor necrosis factor receptor
TS	tuberous sclerosis
UICC	International Union Against Cancer
VEGF	vascular endothelial growth factor
VHL	Von Hippel Lindau
WHO	World Health Organization

Preface

This second edition of *The Molecular Biology of Cancer* is directed at graduate students embarking on cancer research studies, in addition to undergraduate and postgraduate medical trainees. The book may also be of interest to busy clinicians who want to update their knowledge of recent advances in cancer biology and how these may apply to future practice. The aim is to provide a readable text that summarizes the scientific aspects of the four main groups of genes involved in the development of human tumors and then deals with the ways in which these genes may be used to influence patient management in each of the commonest cancers. As with the first edition of this book and it predecessor *Oncogenes and Tumor Suppressor Genes*, it is not intended to give definitive details of the biology of the genes – a huge topic which is covered in many excellent and more specialized books. Instead the aim is to provide the reader with an overview of this exciting topic and to provide information on how it may impinge on their medical and scientific practices. The book gives a more detailed account than previous editions, reflecting the momentous developments in molecular biology, genetics and oncology over the last five years. The first four chapters cover the scientific aspects of the genes. Chapter 5 has been introduced to cover the major developments in hereditary cancers in recent years and the clear clinical utility of molecular testing. Subsequent chapters have been expanded to give more detail on the use of molecular testing in specific cancers; the penultimate chapter has been updated to reflect the increased role of these genes as therapeutic targets. Finally, the last chapter on the techniques involved in analysis of the genes has also been updated.

We would like to thank our long-suffering families for their support during the writing of this book – particularly Dr Alan Cockayne for reading sections of the manuscript and Jo Ford for assistance with the illustrations. We would also like to thank the many colleagues who provided us with examples of their work and the staff at Garland/BIOS Scientific Publications for their help.

Fiona Macdonald, Chris Ford and Alan Casson

General principles

1

1.1 Introduction

It has been realized for many years that cancer has a genetic component and at the level of the cell it can be said to be a genetic disease. In 1914, Boveri suggested that an aberration in the genome might be responsible for the origins of cancer. This was subsequently supported by the evidence that cancer, or the risk of cancer, could be inherited; that mutagens could cause tumors in both animals and humans; and that tumors are monoclonal in origin, that is, the cells of a tumor all show the genetic characteristics of the original transformed cell. It is only in recent years that the involvement of specific genes has been demonstrated at the molecular level.

Cancer cells contain many alterations which accumulate as tumors develop. Over the last 25 years, considerable information has been gathered on the regulation of cell growth and proliferation leading to the identification of the proto-oncogenes and the tumor suppressor genes. The proto-oncogenes encode proteins which are important in the control of cell proliferation, differentiation, cell cycle control and apoptosis. Mutations in these genes act dominantly and lead to a gain in function. In contrast the tumor suppressor genes inhibit cell proliferation by arresting progression through the cell cycle and block differentiation. They are recessive at the level of the cell although they show a dominant mode of inheritance. In addition, other genes are also important in the development of tumors. Mutations leading to increased genomic instability suggest defects in mismatch and excision repair pathways. Genes involved in DNA repair, when mutated, also predispose the patient to developing cancer. This failure of DNA repair is seen in xeroderma pigmentosum, ataxia telangiectasia, Fanconi's anemia and Bloom's syndrome [1]. In addition, many other genes encoding proteins, such as proteinases or other enzymes capable of disrupting tissues, and vascular permeability factors, have been shown to be involved in carcinogenesis. Epigenetic events such as alterations in the degree of methylation of DNA have also been detected in tumors [2]. Genomic imprinting, a process in which expression of two alleles is dependent on the parent of origin and controlled by changes in methylation, is now widely recognized in cancers [3]. Any combination of these changes may be found in an individual tumor. The overall progression to malignancy is therefore a complex event.

1.2 What is cancer?

In normal cell growth there is a finely controlled balance between growth-promoting and growth-restraining signals such that proliferation occurs only when required. The balance is tilted when increased cell numbers are required, for example during wound healing and during normal tissue turnover. Differentiation of cells during this process occurs in an ordered manner and proliferation ceases when no longer required. In tumor cells this process is disrupted, continued cell proliferation occurs and loss of differentiation may be found. In addition the normal process of programmed cell death may no longer operate.

Cancers arise from a single cell which has undergone mutation. Mutations in genes such as those described in the next three chapters give the cell increased growth advantages compared to others and allows them to escape normal controls on proliferation. The initial mutation will cause cells to divide to produce a genetically homogeneous clone. In turn, additional mutations occur which further enhance the cells' growth potential. These mutations give rise to subclones within the tumor each with differing properties so that most tumors are heterogeneous. This multistep process is described further in Section 1.7.

Tumors can be divided into two main groups, benign or malignant. Benign tumors are rarely life threatening, grow within a well-defined capsule which limits their size and maintain the characteristics of the cell of origin and are thus usually well differentiated. Malignant tumors invade surrounding tissues and spread to different areas of the body to generate further growths or metastases. It is this process which is often the most life threatening. Different clones within a tumor will have differing abilities to metastasize, a property which is genetically determined. The process of metastasis is likely to involve several different steps and only a few clones within a tumor will have all of these properties. Some of the genes involved in the process of metastasis are described in subsequent chapters but additional factors such as proteinases, the cell adhesion molecules E-cadherin [4] and the integrin family have been implicated in the invasive process.

Tumor cells show a number of features which differentiate them from normal cells: (1) They are no longer as dependent on growth factors as normal cells either because they are capable of secreting their own growth factors to stimulate their own proliferation, a process termed autocrine stimulation, or because growth factor receptors on the surface are altered in such a way that binding of growth factors is no longer necessary to stimulate proliferation; (2) normal cells require contact with the surface in the extracellular environment to be able to grow whereas tumor cells are anchorage independent; (3) normal cells respond to the presence of other cells, and in culture will form a monolayer due to contact inhibition, whereas tumor cells lack this and often grow over or under each other; (4) tumor cells are less adhesive than normal cells; (5) normal cells stop proliferating once they reach a certain density but tumor cells continue to proliferate.

1.3 The cell cycle

Central to the complex process of proliferation is the control of the processes involved in driving the cell through the cell cycle. The cell cycle involves a series of events which result in DNA duplication and cell division. In normal cells this process is carefully controlled but in tumor cells, mutations in the genes associated with the cell cycle result in progression of cells with damaged DNA through the cycle. The phases of the cell cycle have been known for around 40 years but it is only in the last few years that the genes which are involved in control of the cycling process have been identified, as described further in Chapter 4. The cell cycle is divided into four distinct phases. The first gap phase (G_1), DNA replication (S) and the second gap phase (G_2) together make up interphase, the period from the end of one nuclear division to the start of the next. This is followed by mitosis or M (*Figure 1.1*). The length of the cell cycle varies considerably from one species and tissue to another but is typically around 16–24 h. In G_1, cells are preparing to synthesize DNA and biosynthesis of both RNA and proteins occurs. The length of this phase is the most variable. During S phase, DNA is replicated and histones are synthesized. At the end of S phase the DNA content of the cell has doubled and the chromosomes have replicated. In G_2, cells are preparing for cell division, the replicated DNA complexes with proteins and biosynthesis continues. The nucleus and cytoplasm finally divide during mitosis and two daughter cells are produced which can then begin interphase of a new cell cycle

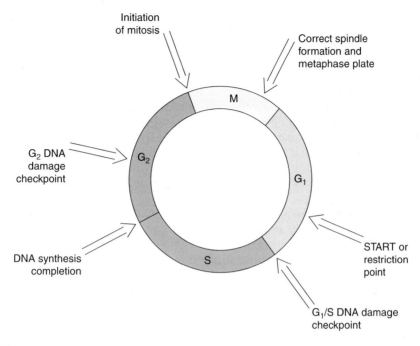

Initiation
of mitosis

Correct spindle
formation and
metaphase plate

M

G_2 DNA
damage
checkpoint

G_2

G_1

DNA synthesis
completion

S

START or
restriction
point

G_1/S DNA damage
checkpoint

Figure 1.1

The cell cycle showing checkpoints at which DNA is monitored before the next stage of the cycle is entered. Adapted from [5].

if conditions are suitable for further growth. In the absence of mitogens or when nutrients are depleted, cells can also enter a resting phase termed G_0. Similarly, cells which are terminally differentiated, for example neurons, are generally arrested in G_0. Transition from one stage to the next is regulated at a number of checkpoints which prevent premature entry into the next phase of the cycle (*Figure 1.1*). Mammalian cells initially respond to external stimuli such as growth factors to enter into G_1. Once cells have become committed to replication, that is towards the end of G_1, they become refractory to growth factor-induced signals and to growth factor inhibitors such as TGFβ and then rely on the action of the cell cycle proteins to regulate progression. This progression from one stage to the next is carefully controlled by the sequential formation, activation and subsequent degradation or modification of a series of cyclins and their partners, the cyclin-dependent kinases (CDKs; see Chapter 4) [6]. In addition it has been shown more recently that a further group of proteins, the cyclin-dependent kinase inhibitors (CKIs) are important for signal transduction and coordination of each stage of the cell cycle.

There are a number of recognized checkpoints at which the integrity of the DNA and the formation of the spindle and spindle pole are checked. Defects in the DNA itself can lead to chromosomal rearrangements and transmission of damaged DNA. Defects in the spindle lead to mitotic nondisjunction, a process which results in the gain or loss of whole chromosomes. Defects in the spindle pole can lead to alterations in ploidy [7]. The first checkpoint in mammalian cells is known as START or restriction point (R) and occurs late in G_1 [8]. It is at this stage that cells must commit themselves to a further round of DNA replication and at this point any damage to DNA can be monitored and if necessary the cycle halted until it is repaired. Loss of this checkpoint can lead to genomic instability and survival of damaged cells. Further checkpoints occur during S phase and at the G_2 to M transition to monitor the completion of DNA synthesis and the formation of a functional spindle. This last checkpoint prevents chromosome segregation if the chromosomes are not intact. The degradation of various cyclins occurs at each checkpoint and it is this mechanism together with interaction of the CKIs which allows the cell to enter the next phase.

Initially it was believed that the basic cell cycle machinery would not be involved in tumorigenesis but it is now clear that this is wrong. Checkpoints are deregulated in tumors and mutations occur in the genes involved in cell cycle components, leading to genomic instability and cancer. The mode of action of cyclins, CDKs and CKIs are discussed in Chapter 4 and abnormalities associated with specific cancers are detailed in subsequent chapters.

1.4 Apoptosis

Apoptosis or programmed cell death, first described in 1972, is a series of genetically controlled events which result in the removal of unwanted cells, for example during morphogenesis of the embryo, without disruption of tissues [9]. As with differentiation or proliferation, apoptosis is an important method of cellular control and any disruption of this process, not surprisingly, leads to abnormal growth. It plays an important role in tumorigenesis as well as in many other disease processes such as degenerative and autoimmune diseases. In recent years the field has expanded rapidly and many extensive reviews have been written [10–12], therefore only an overview of the process will be presented here.

Apoptosis differs from necrosis or accidental cell death in a number of important ways. First, it is an active process as opposed to an unplanned process induced by cell injury. Second, apoptotic cells are recognized by phagocytes and removed before they disintegrate. As a consequence, there is no surrounding tissue damage or induction of inflammatory responses. In contrast, in necrosis, cells become leaky, release macromolecules and rapidly disintegrate, thereby inducing inflammation.

Apoptotic cells show a very characteristic morphology. They show condensation of nuclear heterochromatin, cell shrinkage and loss of positional organization of organelles in the cytoplasm. Electrophoresis of DNA extracted from apoptotic cells shows a characteristic laddering pattern of oligonucleosomal fragments resulting from internucleosomal chromatin cleavage by endogenous endonucleases. This differs from the smear of degraded DNA usually seen in necrotic cells.

In contrast to the well-established morphological features, the genetic events leading to apoptosis are only now beginning to be recognized [11]. A number of genes and their proteins, some acting within the cells themselves and others acting extrinsically, have now been identified which either promote or inhibit apoptosis acting in pathways which eventually impinge on caspase activation [11,13].

Both proto-oncogenes and tumor suppressor genes have been shown to be involved in the control of apoptosis (Figure 1.2). The MYC proto-oncogene has a dual role in cells inducing both cell proliferation and apoptosis [14–16]. As described in Chapter 2, MYC, complexed with its protein partner MAX, has a role in driving cell proliferation. However, it also plays a role in committing cells to enter the apoptotic pathway again complexed to MAX. The choice of which pathway to follow appears to depend on the cell's microenvironment, particularly the availability of growth factors. In the absence of suitable growth factors the result is apoptosis. Overexpression or deregulation of MYC in tumors therefore has important implications for the control of apoptosis. A number of tumor suppressor genes such as p53, APC and PTEN all exert tumor suppressor activity by inducing apoptosis with loss of suppressor function leading to reduced apoptosis and tumor growth. p53 acts

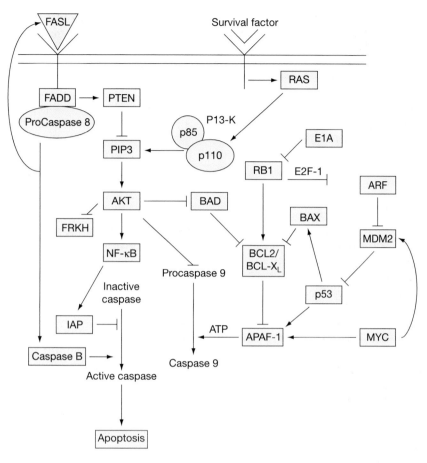

Figure 1.2

The role of proto-oncogenes and tumor suppressor genes in apoptosis. Adapted from (12). Oncogenes involved in the process include *MYC, E1a, AKT, RAS* and *REL*. The tumor suppressor genes, *PTEN, RB1, p53* and *ARF* regulate apoptosis through pathways which all impinge on caspase activation.

by driving cells with damaged DNA along the apoptotic pathway rather than allowing them to continue through the cell cycle [17,18]. Apoptotic regulators lying downstream of *p53*, such as *BAX, FAS/Apo1* and *BCL2*, have been shown to be critical to the function of *p53* in suppression of tumors [13,18].

Both *MYC* and *p53* play a role in the induction of apoptosis. Another proto-oncogene, the *BCL2* gene, acts to block apoptosis specifically [19]. Its deregulation will therefore remove suppression of programmed cell death and can promote tumor formation. The *BCL2* gene is widely expressed during embryonic development but in adults is confined to long-lived cells. It is now known to belong to a large family of related proteins, some of which (e.g. BCL2, BCLX$_L$, BCLW and MCL1) suppress apoptosis whereas others (e.g. BAX, BAK, BAD, BIK, BID and BCLX$_S$) act to antagonize the anti-apoptotic properties of the others.

Inhibition of cell death, either by suppression of those genes which induce cell death or by activation of those genes which cause cell survival, therefore contributes to the development of tumors. There is considerable interest in this area not just to gain a better understanding of tumorigenesis but also because they may be suitable targets for therapy. Recently strategies aimed at inducing apoptosis have begun to be investigated as alternatives to the more usual forms of chemotherapy [20].

1.5 Chromosomes and cancer

Normal human cells contain 46 chromosomes. Changes in this number as well as structural chromosomal abnormalities are common in the majority of tumors. The first consistent chromosome abnormality to be recognized was the Philadelphia chromosome seen in chronic myeloid leukemia (CML: see Sections 2.2.2 and 11.5). Since then many other changes have been found, including loss or gain of whole chromosomes or parts of chromosomes and chromosomal translocations, examples of which are given in Tables 1.1 and 1.2 [21,22]. These changes are nonrandom events and generally are somatically acquired alterations. Some may be primary events occurring early in the development of the tumor and are likely to be an important event in its development. Others are secondary events and may have a role in the subsequent behavior of the tumor. In addition, many other random changes in the chromosome complement are also found due to the instability of the tumor cell.

Loss of chromosomal material can often result in the deletion of a tumor suppressor gene (see Figure 3.4). Duplication of a region can lead to overexpression of an oncogene (see Figures 2.4 and 12.3). Molecular analysis of translocation breakpoint has led to the identification of a number of genes adjacent to the breakpoints which have been implicated in the initiation and progression of tumors.

Studies of chromosome translocations led to the identification of a number of oncogenes, for example, the 8;14 translocation found in Burkitt's lymphoma and shown to be associated with the MYC oncogene on chromosome 8 (see Section 2.2.2). Now that these chromosome abnormalities have been well studied, some, such as the presence of the t(9;22) translocation in CML, can be used diagnostically (see Section 11.5).

Constitutional chromosome abnormalities, present in all the cells of the body, have helped in the identification of a number of tumor suppressor genes. In general these changes are found in only one or a few patients but have been instrumental in the subsequent isolation of the gene. The classic examples, discussed more fully in Chapter 3, are the deletions found in retinoblastoma patients leading to the identification of the RB1 gene; the deletions associated with Wilms' tumor leading to the identification of WT1; and the deletion found on the long arm of chromosome 5 which led to the identification of the APC gene in familial adenomatous polyposis.

1.6 Inherited vs. sporadic cancer

Cancers can be classified into four main groups on the basis of the genetic defect:

(1) The majority of cancers are sporadic and are caused by environmental factors such as chemicals and radiation. Mutations in these tumors are found only in the cancer tissue itself.
(2) Some cancers, without a recognizable genetic basis, show clustering in families and may represent an underlying susceptibility to environmental carcinogens. Care has to be taken when looking at this group as it is possible to have apparent clustering of cancers in families due to the shared environment of family members rather than because of a genetic defect.

Table 1.1 Examples of primary chromosomal aberrations in hematological malignancies

Tumor	Abnormality	Involved gene(s)
AML	t(8;21)(q22;q22)	*AML1-ETO*
	t(15;17)(q22;q11.2)	*PML-RARα*
	inv(16)(p13;q22)	*CBFβ-MYH11*
	t(6;11)(q27;q23)	*MLL-AF6*
	t(9;11)(p21;q23)	*MLL-AF9*
	t(10;11)(p12;q23)	*MLL-AF10*
	t(11;17)(q23;q25)	*MLL-AF17*
	t(11;19)(q23;p13.3)	*MLL-ENL*
	t(11;19)(q23;p13.1)	*MLL-ELL*
	t(1;11)(q21;q23)	*MLL-AF1Q*
	t(2;11)(p21;q23)	*MLL*
	t(16;21)(p11;q22)	*FUS-ERG*
	t(6;9)(p23;q34)	*DEK-CAN*
	t(3;3)(q21;q26)	*EVI1*
	Trisomy 8	
	Monosomy 5	
	Monosomy 7	
	del/t(12p)	
CML	t(9;22)(q34;q11)	*BCR-ABL*
	t(7;11)(p15;p15)	*NUP98-HOXA9*
	t(3;21)(q26;q22)	*AML1-EVI1*
CLL	t(11;14)(q13;q32)	*BCL1-IGH*
	t(14;19)(q32;q13)	*IGH-BCL3*
	Trisomy 12	
ALL	t(1;19)(q23;p13)	*PBX1-E2A*
	t(5;14)((q31;q32)	*IL3-IGH*
	t(12;21)(p13;q22)	*TEL-AML1*
	t(17;19)(q22;p13)	*E2A-HLF*
	t(4;11)(q21;q23)	*AF4-MLL*
	t(9;22)(q34;q11)	*BCR-ABL*
	t(8;14)(q24;q32)	*MYC-IGH*
	t(8;22)(q24;q11)	*MYC-IGL*
	t(2;8)(p12;q24)	*IGK-MYC*
	t(7;19)(q35;p13)	*TCRB-LYL1*
	t(7;9)(q35;q34)	*TAL2*
	t(1;14)(p32;q11)	*TAL1*
	t(7;9)(q34;q34)	*TAN1*
	t(11;14)(p15;q11)	*RBTN2-TCRA*
	t(11;14)(p13;q11)	*RBTN1-TCRA*
	t(10;14)(q24;q11)	*HOX11-TCRA*
	t(7;10)(q35;q24)	*HOX11*
Lymphoma	t(8;14)(q24;q32)	*MYC-IGH*
	t(8;22)(q24;q11)	*MYC-IGL*
	t(2;8)(p12;q24)	*IGK-MYC*
	t(3;14)(q27;q32)	*BCL6-IGH*
	t(14;18)(q32;q21)	*IGH-BCL2*

Derived from [21] and [22].
This list shows examples of some of the rearrangements found in hematological malignancies. AML, acute myeloid leukemia; ALL, acute lymphocytic leukemia; CML, chronic myeloid leukemia; t, translocation; inv, inversion; del, deletion.

(3) A small proportion of cancers have a clearly defined genetic cause. This means that screening at-risk family members is immediately possible, thereby preventing unnecessary morbidity and mortality (see Chapter 5). In addition, the genes involved have a wider importance as the genes which cause the inherited form of the disease are often the same as those which are implicated in the sporadic form of the disease. Their study can therefore help in the understanding of the more common forms of the cancer. The classic example of this is the *APC* gene which is responsible for the inherited condition familial adenomatous polyposis (FAP) and which is also the earliest gene to be mutated in the development of sporadic colorectal cancer (see Sections 3.5 and 7.4.2). The main exception to this principle is familial breast cancer which is caused by defects in the *BRCA1* and *BRCA2* genes (see Sections 3.6.1 and 9.2.4). Defects in these genes have not so far been observed in sporadic breast cancer. In cancers with a true genetic basis, the causative mutation is found in all the cells of the body. This group can be

Table 1.2   Examples of primary chromosomal aberrations in solid tumors

Tumor (malignant)	Abnormality
Bladder carcinoma	+7, del(10)(q22-24), del(21)(q22),
Brain (rhabdoid tumor)	−22
Breast cancer	−17, i(1q), der(16)t(1:16)(q10;p10),
Colon	+7, +20,
Ewing's sarcoma	t(11;22)(q24;q12)
Glioma	+7, −10, −22, −X, +X, −Y
Kidney (renal carcinoma)	del(3)(p11-14), del(3)(p14-p21)
Liposarcoma	t of 12q13-q14
Lung	del(3)(p14;p23), +7
Meningioma	−22, +22, −Y, del(22)(q11-q13)
Neuroblastoma	del(1)(p32-p36)
Ovarian cancer	+12, +7, +8, −X
Prostate	del(10)(q24), +7, −Y
Retinoblastoma	i(6p), del(13)(q14)
Uterine carcinoma	+10
Wilms' tumor	del(11)(p13;p13), del(11)(p15;p15)

Tumor (benign)	Abnormality
Barrett's esophagus	−Y
Colorectal adenoma	+7
Giant cell tumor	+8
Lipoma	t(1;12)(p33-p34;q13-q15), t(2;12)(p22-p23;q13-q15), t(3;12)(q27-q28;q13-15), t(5;12)(q33;q13-q15), t(11;12)(q13;q13-15), del(12)(q13;q15), t(12;21) (q13-q15;q21), del(13)(q12-q22)
Neuroepithelioma	t(11;22)(q24;q12)
Neurinoma	−22, −Y
Ovary (adenoma)	+12
Ovary (fibroma)	+12
Mixed salivary gland	t(1;8)(p22;q22), t(3;8)(p21;q12), t(6;8)(p21-p22;q12), t(8;13)(p23;q13-14), del(8)(q12;q21-q22), t(8;9)(q12;p22), t(9;12)(p21-p24;q13-q15), inv(12)(p13;q13)
Uterus (leiomyoma)	r(1)(p11-p36;q11-q14), t(2;12)(q35-q27;q14)

Adapted from [21].
t, translocation; inv, inversion; del, deletion; I, isochromosome; r, ring.

recognized by an earlier age of onset, multiple cancers in individuals and by segregation of the disease through the family in a Mendelian manner.

(4) Individuals with some conditions, generally termed chromosome breakage syndromes
because of the increased chromosome fragility seen in cultured cells, for example those
from xeroderma pigmentosum and ataxia telangiectasia, have an increased risk of cancers although the incidence of cancers is not close to the levels seen in patients in
group 3.

1.7   The multistage nature of cancer development

In 1949, Berenblum and Shubik concluded that 'the recognition that carcinogenesis is at
least a two-stage process should invariably be borne in mind' [23]. Armitage and Doll
took this observation a step further; in 1954, they published age/incidence curves for 17
common types of cancer. From their figures they concluded that carcinogenesis was at
least a six- or seven-stage process [24]. Although each of these steps cannot usually be
clearly defined in an individual tumor, it is clear today that there is without doubt a
multistage progression to malignancy (*Figure 1.3*).

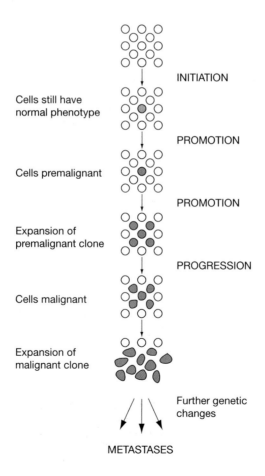

Figure 1.3

The multistage progression to carcinogenesis. Adapted from [5].

Tumors tend to acquire more aggressive characteristics as they develop, and in 1957 Foulds pointed out that tumor progression occurred in a stepwise fashion, each step determined by the activation, mutation or loss of specific genes [25]. Over the next two decades biochemical and cytogenetic studies demonstrated the sequential appearance of subpopulations of cells within a tumor, attributable, in part at least, to changes in the genes themselves.

The evidence suggests that, in the majority of cases, cancers arise from a single cell which has acquired some heritable form of growth advantage [26]. This initiation step is believed to be caused frequently by some form of genotoxic agent such as radiation or a chemical carcinogen. The cells at this stage, although altered at the DNA level, are phenotypically normal. Further mutational events involving genes responsible for control of cell growth lead to the emergence of clones with additional properties associated with tumor cell progression. Finally, additional changes allow the outgrowth of clones with metastatic potential. Each of these successive events is likely to make the cell more unstable so that the risk of subsequent changes increases. Animal models of carcinogenesis, primarily based on models of skin cancer development in mice, have enabled these steps to be divided into initiation events, promotion, malignant transformation and metastasis [26] (*Figure 1.3*).

Although it is clear that multiple changes are necessary for tumor development, it is not clear whether the order in which the changes occur is critical. Evidence suggests, however, that it is the accumulation of events that is important rather than the order in which they occur (see Section 3.7.2 for the classic example in colorectal cancer).

References

1. **Taylor, A.** (2001) *Best. Pract. Res. Clin. Haematol.*, **14**, 631.
2. **Jones, P.A. and Baylin, S.B.** (2002) *Nat. Rev. Genet.*, **3**, 415.
3. **Feinberg, A.P.** (2000) *Curr. Top. Microbiol. Immunol.*, **249**, 87.
4. **Hajra, K.M. and Fearon, E.R.** (2002) *Genes Chrom. and Cancer*, **34**, 255.
5. **Macdonald, F. and Ford, C.H.J.** (1997) *Molecular Biology of Cancer*, 1st Edition, BIOS Scientific Publishers, Oxford.
6. **Malumbres, M. and Barbacid, M.** (2001) *Nat. Rev. Cancer*, **1**, 222.
7. **Nigg, E.A.** (2001) *Nat. Rev. Mol. Cell. Biol.*, **2**, 21.
8. **Ho, A. and Dowdy, S.F.** (2002) *Curr. Opin. Genet. Dev.*, **12**, 47.
9. **Kerr, J.F.R., Wyllie, A.H. and Currie, A.R.** (1972) *Br. J. Cancer*, **26**, 239.
10. **Fisher, D.E.** (2001) *Hematol. Oncol. Clin. North Am.*, **15**, 931.
11. **Igney, F.H. and Krammer, P.H.** (2002) *Nat. Rev. Cancer*, **2**, 277.
12. **Eastman, A. and Rigas, J.R.** (1999) *Semin. Oncol.*, **26**, 7.
13. **Macleod, K.** (2000) *Curr. Opin. Genet. Dev.*, **10**, 81.
14. **Dang, C.V.** (1999) *Mol. Cell. Biol.*, **19**, 1.
15. **Evan, G.I. and Vousden, K.H.** (2001) *Nature*, **411**, 342.
16. **Prendergast, G.C.** (1999) *Oncogene*, **18**, 2967.
17. **Hickman, E.S., Moroni, M.C. and Helin, K.** (2002) *Curr. Opin. Genet. Dev.*, **12**, 60.
18. **Vousden, K.** (2000) *Cell*, **103**, 691.
19. **Newton, K. and Strasser, A.** (1998) *Curr. Opin. Genet. Dev.*, **8**, 68.
20. **Makin, G.** (2002) *Exp. Opin. Ther. Targets*, **6**, 73.
21. **Barch, M.J., Knutsen, T. and Spurbeck, J.L.** (1998) *The ACT Cytogenetics Laboratory Manual*, 3rd edition. Lippincott-Raven, Philadelphia and New York.
22. **Vogelstein, B. and Kinzler, K.W.** (1998) *The Genetic Basis of Human Cancer*. McGraw-Hill, New York.
23. **Berenblum, I. and Shubik, P.** (1949) *Br. J. Cancer*, **3**, 109.
24. **Armitage, P. and Doll, R.** (1954) *Br. J. Cancer*, **8**, 1.
25. **Foulds, L.** (1957) *Cancer Res.*, **17**, 355.
26. **Nowell, P.C.** (1976) *Science*, **194**, 23.

Oncogenes

2

2.1 Viruses and cancer

For over 30 years, it has been known that DNA viruses such as SV40 virus and RNA viruses such as the retrovirus, Rous sarcoma virus, are capable of transforming those cells they infect. These viruses are associated primarily with animals and are rarely implicated in human disease, although a few examples are known (*Table 2.1*). The viruses are particularly important because they have taught us a great deal about the molecular basis for transformation and have led to the identification of cellular oncogenes.

Most of this information has come from studies of the retroviruses. These RNA viruses encode three genes, *GAG*, *POL* and *ENV*, which produce a core protein, a reverse transcriptase and envelope glycoproteins, respectively. At either end of the viral genome are long terminal repeats (LTRs) which function as promoter or enhancer sequences (*Figure 2.1*) [1]. In those viruses capable of causing malignant transformation, a fourth gene, the oncogene, has been found (*Table 2.2*).

A clue as to how these genes might be involved in the pathogenesis of cancers came in 1976 when a group studying the Rous sarcoma virus, which causes malignancy in chickens, showed that the *SRC* oncogene carried by the virus was not viral in origin but had been picked up by the virus from the host genome, a process termed transduction [2]. This was possible because the retrovirus uses its reverse transcriptase enzyme to copy its RNA genome into duplex DNA. The DNA copy can integrate reversibly into the host genome (*Figure 2.2*). By recombination between viral and host DNA, the retroviruses 'kidnap' a host gene which then becomes part of the viral genome and is expressed under the control of viral promoters. The resulting virus is usually defective in replication, but can transform host cells following re-infection. These host genes are activated when transduced by the retrovirus either because the gene is altered, resulting in a protein with abnormal activity, or because the gene is brought under the control of a viral promoter, leading to aberrant, high levels of expression driven by retroviral transcriptional enhancers.

Table 2.1   Viruses associated with human cancers

Virus	Associated tumors
DNA viruses	
Epstein–Barr	Burkitt's lymphoma
	Nasopharyngeal cancer
Hepatitis B	Liver cancer
Papilloma virus	Benign warts
	Cervical cancer
RNA viruses	
Human immunodeficiency virus (HIV-1)	Kaposi's sarcoma
Human T-cell leukemia virus Type I (HTLV-1)	Adult T-cell leukemia
HTLV-2	Hairy cell leukemia
HTLV-5	Cutaneous T-cell leukemia

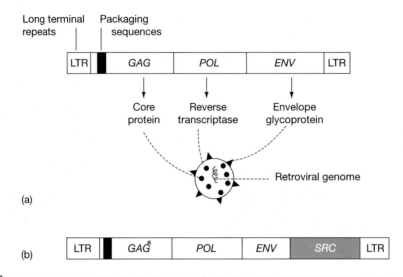

Figure 2.1

(a) The retroviral genome. Three genes are encoded by the viral genome. (b) In the transforming retroviruses, one mammalian gene, in this case *SRC*, is included and gives the virus its transforming properties. Adapted from [1].

Table 2.2 Retroviral oncogenes

Oncogene	Virus	Tumor
v-ABL	Abelson leukemia virus	Leukemia
v-ERBA	Avian erythroblastosis virus	Helps v-ERBB
v-ERBB	Avian erythroblastosis virus	Erythroleukemia
v-FMS	Feline sarcoma virus	Sarcoma
v-HRAS	Rat sarcoma virus (Harvey strain)	Sarcoma
v-KRAS	Rat sarcoma virus (Kirsten strain)	Sarcoma
v-JUN	Avian sarcoma virus	Fibrosarcoma
v-MYB	Avian myeloblastosis virus	Myeloblastosis
v-MYC	Avian myelocytomatosis virus	Leukemia
v-SIS	Simian sarcoma virus	Sarcoma
v-SRC	Rous sarcoma virus	Sarcoma

With reference to their potential function in tumor development, the original cellular genes were termed proto-oncogenes. This mechanism of activation is implicated in a wide range of animal tumors but it has never been shown convincingly to be associated with human cancers.

Retroviruses can also activate proto-oncogenes more directly by a process known as insertional mutagenesis. In this process, the insertion of a DNA copy of the retrovirus into the cellular genome close to a proto-oncogene causes abnormal activation of that gene by stimulation of gene expression via the promoter action of the LTRs. This has been demonstrated for the *INT1* gene which is activated in breast cancer in mice infected with the mouse mammary tumor virus. Most of the proto-oncogenes identified in this way are identical to those already found via transforming retroviruses, although a few additional genes such as *EVI1* have also been found.

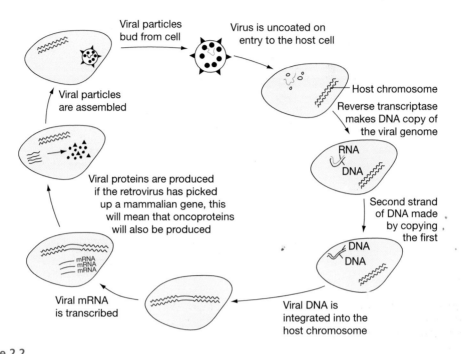

Figure 2.2

The retrovirus life cycle. Adapted from [1].

As few retroviruses had been shown to be the cause of human cancers, it was still not clear how these genes associated with the retroviruses might relate to the pathogenesis of cancer in man. It was only when it was shown that human tumors contained activated oncogenes homologous to those found in the retroviruses, but with no viral intermediary, that this whole area of research expanded rapidly.

2.2   Cellular oncogenes

Whilst research into the retroviral oncogenes continued, more direct methods of identifying oncogenic sequences in the human genome were examined. A DNA transfection assay was used as a method of identifying those sequences in tumor cells which were responsible for uncontrolled cell proliferation. DNA was extracted from human tumors and sheared into fragments. These were then transfected into a mouse-derived cell line called NIH-3T3 so that random fragments were incorporated into its genome. As a result, some cells were transformed and could be identified by their loss of contact inhibition which caused cells to pile up *in vitro* (*Figure 2.3*). These transformed cells were also capable of producing tumors when injected into athymic (nude) mice. The genome of the transformed cells was analyzed and shown to contain an oncogene which in many cases was similar to one which had been identified in the retroviruses [3] (*Table 2.3*). This transfection assay did not give a positive result with all tumors. Only about 20% of tumors contained oncogenes which could be identified in this way and about one quarter of these belonged to the *RAS* gene family. Additional oncogenes were identified by alternative strategies.

It had long been known that some tumors carry a consistent chromosome translocation. In the case of chronic myeloid leukemia (CML), this is a reciprocal translocation

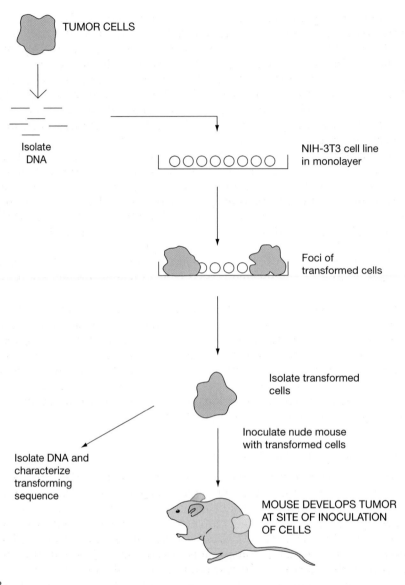

Figure 2.3

DNA transfection assay. Adapted from [1].

Table 2.3 Methods of oncogene identification in human tumors

Method of identification	Oncogene
Amplification	ERBB2, MYCL, MYCN
Chromosomal translocation	ABL, BCL1, BCL2, MYC
Homology to retroviruses	HRAS, KRAS, SRC
Insertional mutagenesis	EVI1, INT1
Transfection assay	MAS, MET, MYC, RAS, TRK

between chromosomes 9 and 22 (see Section 2.2.2). In Burkitt's lymphoma, there is a reciprocal translocation between chromosome 8 and, in the majority of cases, chromosome 14, although either chromosome 2 or chromosome 22 is occasionally involved. In both examples the breakpoints were shown to coincide with the location of oncogenes already identified from retroviral studies; *ABL* on chromosome 9 and *MYC* on chromosome 8. Other translocations in tumors identified the sites of further oncogenes (*Table 2.3 and Table 1.1*). In a comprehensive study of 5345 tumors [4], the distribution of cancer-specific breakpoints was compared with the site of 26 cellular oncogenes; of these, 19 were located at cancer-associated chromosome breakpoints.

Two other chromosome abnormalities observed in tumors have also identified the location of oncogenes (*Table 2.3*). Both the development of homogeneous-staining regions (HSRs) and formation of double-minute chromosomes (DMs, *Figure 2.4*) were associated with oncogene amplification. Some oncogenes identified in this way, for example *MYC*, had previously been detected by other techniques, but other genes such as *MYCN*, associated with neuroblastomas, and *MYCL*, found in small-cell lung carcinomas, were first revealed in this way [5].

Currently about 200 different cellular proto-oncogenes have been identified [6], which when activated by one of the mechanisms described below convert the gene into an oncogene.

The importance of these genes to the cell is clear as there is sequence conservation from organisms such as yeast, through the invertebrates and vertebrates to man and indeed homologs of the proto-oncogenes have been found in all multicellular organisms so far studied. In the normal cell, the expression of these proto-oncogenes is tightly controlled and they are transcribed at the appropriate stages of growth and development of cells. However, alterations in these genes or their control sequences lead to inappropriate expression. What therefore are the functions of these genes and how are they converted into genes capable of contributing to malignant progression?

2.2.1 Function of the proto-oncogenes

Many studies have supported the prediction that the proto-oncogenes would be involved in the basic essential functions of the cell related to control of cell proliferation and differentiation. Cells are stimulated by external signals such as growth factors, which bind to cell surface receptors, for example the transmembrane protein tyrosine kinases, thereby activating their function as signal transducers. In turn this stimulates intracellular signaling pathways, eventually leading to alterations in gene expression. The proto-oncogenes function at each of these steps and are found therefore at all levels of the cell (*Figure 2.5*). Mutations in any of the genes result in their abnormal activation promoting cell growth in the absence of external stimuli and leads to malignant transformation.

Growth factors

Growth factors constitute the products of the first group of proto-oncogenes and exist as polypeptides, oligopeptides or steroid hormones (*Table 2.4*). They bind to their own specific receptors or, in a few cases, cross react with a number of receptors, to stimulate or inhibit cell growth acting via a number of pathways culminating in alterations in the expression of genes including those involved in cell cycle regulation and differentiation. Growth factors play a major role in the development of tissues and if limited, result in cell death. Growth factors stimulate cells in the stationary stage to enter the cell cycle. This occurs in a two-stage process – the stimulation to proceed into G_1 is provided by competence factors such as platelet-derived growth factor (PDGF) or epidermal growth factor

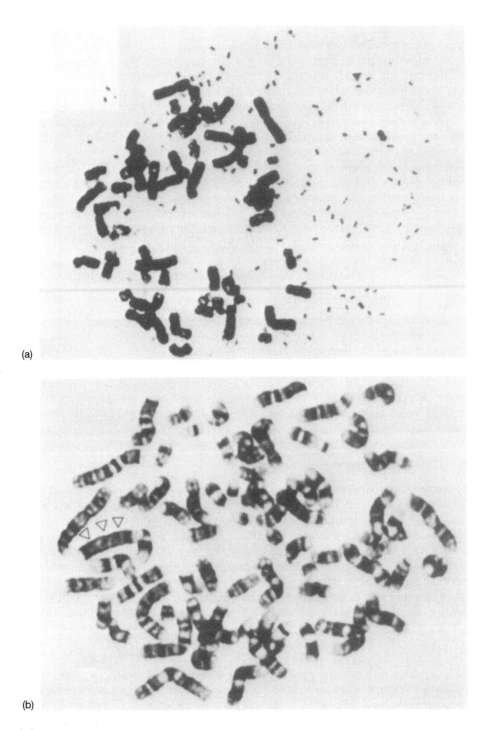

(a)

(b)

Figure 2.4

Examples of (a) DMs and (b) HSRs (arrowed). Figure courtesy of Dr J. Waters, NE London Regional Cytogenetics Laboratory, Gt Ormond St. Hospital. Adapted from [1].

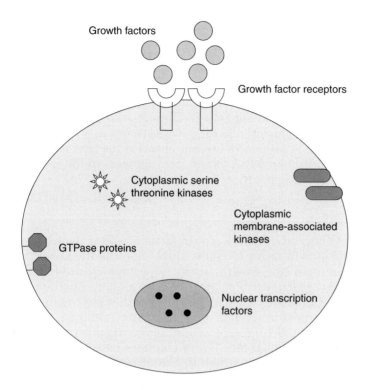

Growth factors

Growth factor receptors

Cytoplasmic serine
threonine kinases

Cytoplasmic
membrane-associated
kinases

GTPase proteins

Nuclear transcription
factors

Figure 2.5

Cellular functions of
proto-oncogene
products.

Table 2.4   Role of proto-oncogene products in
the cell: growth factors

Platelet-derived growth factor (PDGF)
Epidermal growth factor (EGF)
Fibroblast growth factor 1–7 (FGF1–7)
Insulin-like growth factor 1 and 2 (IGF1 and 2)
Transforming growth factor α and β (TGFα and β)

(EGF) and this is followed by the requirement of progression factors, such as insulin-like growth factor (IGF), to progress through the cycle. Normal cells require growth factors to remain viable but tumor cells can circumvent this process.

Two models of growth factor action have been recognized. In the autocrine model, binding of a growth factor to its receptor leads either to increased secretion of more growth factor from the same cell or up-regulation of the receptor promoting autostimulation of tumor growth. This model has been proposed, for example, for the way bombesin secretion acts in small-cell lung cancer. In the paracrine model, a growth factor released from one cell (e.g. stromal tissue cells), binds to a receptor on a neighboring cell (e.g. adjacent tumor cells). This in turn stimulates release of a paracrine factor to increase more growth factor production from the first cell. There is therefore reciprocal stimulation of growth factor production.

As indicated above growth factors function in a number of different ways. Following binding of EGF and PDGF to their receptors, cell division is stimulated, advancing the cell from stationary phase into the G_1 phase of the cycle [7]. Insulin-like growth factors (IGF1 and IGF2) are responsible for assisting in the passage of cells through G_1 of the cell cycle. They have endocrine functions, e.g. in mediating the role of growth factors, but also have both autocrine and paracrine functions during normal development and growth as well as in cancer development [8]. Transforming growth factor (TGF) β is considered both as a tumor suppressor and as a promoter of progression and invasion [9]. It is a potent inhibitor of normal stromal, hematopoietic and epithelial cell growth. Through its ability to inhibit cell proliferation it suppresses tumor development in its early stages. However, once tumors develop, they often become resistant to TGFβ growth inhibition and in the later stages of development of a cancer the tumor cells themselves secrete large amounts of the growth factor thereby enhancing invasion and metastases.

Growth factor receptors

A second group of proto-oncogenes encode either the growth factor receptors themselves or their functional homologs. The growth factor receptors (*Table 2.5*) link the information from the extracellular environment (the growth factors) to a number of different intracellular signaling pathways. These genes, when mutated, make up a large proportion of the oncogenes. The most important of the growth factor receptors with respect to malignant transformation are the transmembrane receptor tyrosine kinases although other receptors such as the hemopoiesis growth factor receptors and steroid receptors also play a role. At least 58 different transmembrane receptor tyrosine kinases have been recognized [10,11]. The protein tyrosine kinases show a high degree of sequence homology particularly in the catalytic domain and are found only in metazoans.

The receptor tyrosine kinases possess an extracellular ligand-binding domain, a transmembrane domain and one or two intracellular catalytic kinase domains responsible for transducing the mitogenic signal (*Figure 2.6*). Binding of the growth factor results, in most cases, in receptor dimerization. This in turn leads to activation of their intrinsic kinase activity with subsequent autophosphorylation of specific tyrosine residues. These phosphorylated residues mediate the specific binding of cytoplasmic signaling proteins containing src homology-2 (SH2) domains such as the GTPase activating protein for *RAS* and phosphatidylinositol 3-kinase (PI3-K) (see below) (*Table 2.6*). These create the assembly of a 'signal particle' on the inner surface of the membrane which transmits the signals that mediate the pleiotrophic responses to the growth factors via the intracellular signaling pathways (*Figure 2.7*). As these receptors are important in the development and progression of cancers they have become the focus of interest for the development of novel targets for anticancer therapy [12]. The majority of these receptors are well studied (see [10] for an

Table 2.5 Role of proto-oncogene products in the cell: growth factor receptors

EGFR	RON
HER2–4	MET
IGF-IR/IGF-IIR	EPHA2/EPHB2/EPHB4
PDGFR – α and β	FGFR1–4
CSF-1R	TRKA/TRKC
KIT	AXL
FLK2/FLT3	TEK
VEGFR1–3	RET
TIE	ROS
ALK	

extensive list of reviews of specific receptors). Several, such as EGFR, PDGFR and TGFR, were amongst the first to be studied and shown to have a major role in tumor development.

Over 20 years ago, *EGFR* was shown to be the proto-oncogene of the mutant constitutively active viral oncogene *v-erbB*. The epidermal growth factor receptor family, which consists of four receptors, EGFR itself (also known as HER1) and HER2–4, interacts with a wide range of downstream signaling pathways to regulate cell growth, differentiation and survival [13]. Overexpression of these genes has been implicated in a wide range of tumors and the mechanism of signal transduction by this family has revealed major therapeutic

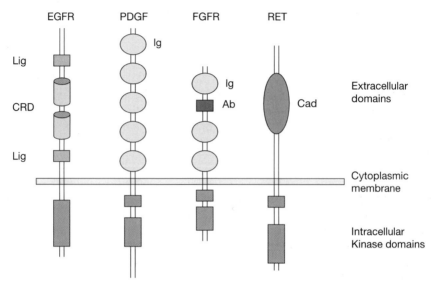

Figure 2.6

Structure of four representative transmembrane tyrosine kinase receptors showing the extracellular, transmembrane and cytoplasmic domains. Lig: Ligand binding domain, CRD: Cystine rich domain, Ig: Immunoglobulin like domain, Cad: Cadherin domain, Ab: Acid box.

Table 2.6 Role of proto-oncogene products in the cell: signal transducers

Membrane-associated/guanine nucleotide-binding proteins
RAS
GSP
GIP

Membrane-associated/cytoplasmic protein tyrosine kinases
ABL
SRC
FES
FGR
SYN
LCK

Cytoplasmic protein serine–threonine kinases
RAF
MOS
COT

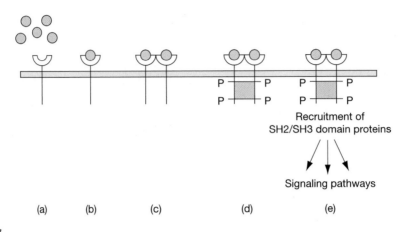

Figure 2.7

Activation of the transmembrane tyrosine kinase receptors leading to four downstream signaling pathways. (a) Growth factors and their receptor; (b) growth factor binding to its receptor; (c) receptor dimerization; (d) receptor phosphorylation on specific tyrosine residues; (e) receptor activation and downstream signaling.

targets. An antibody to HER2 (Herceptin) has been widely evaluated for the treatment of breast cancer [14].

Platelet-derived growth factor receptor (PDGF-R) was also one of the earliest growth factor/receptor pathways to be identified and shown to constitute a family of ligands and receptors, primarily PDGF-R α and β. Again these receptors have been shown to be novel therapeutic targets and the tyrosine kinase inhibitor Imatinib, originally developed for treatment of chronic myeloid leukemia by blocking activity of the BCR-ABL oncoprotein (see below, Sections 11.5 and 13.2.2), has also been shown to inhibit PDGF-R.

Transforming growth factor receptors are serine/threonine kinases rather than tyrosine kinases. There are two recognized receptors, types I and II, which operate via an unusual mechanism. TGFβ binds to TGFβ-RII and this complex then rapidly leads to phosphorylation of TGFβ-RI. TGFβ-RI in turn phosphorylates and activates members of the Smad proteins which transduce signals from cytoplasm to nucleus [15].

Signal transducers

The proteins recruited following receptor activation belong to the third group of proto-oncogenes, the signal transducers. Included in this group are the cytoplasmic nonreceptor tyrosine kinases, other proteins with enzyme activity such as phospholipase Cγ (PLCγ) and PI3-K and the group of adaptor proteins such as Grb2. Many of these proteins have SH2 and SH3 domains. The SH2 domain is highly conserved in proteins involved in signal transduction and recognizes phosphorylated tyrosine residues and the SH3 domain recognizes proline-rich motifs present in signaling proteins and cytoskeletal proteins.

Many of the receptors described above utilize common signaling pathways through which they can transmit their signals and although a variety of intracellular signal transduction pathways exist (see http://biocarta1.epangea.net/ for diagrams of all major signaling pathways), three major pathways are recognized. These are the PI3-kinase (PI3-K)/AKT pathway, the RAS/mitogen-activated protein kinase (MAPK) pathway and the JAK/signal transducers and activators of transcription (STATs) pathway (*Figure 2.8*).

Many growth factors such as IGF and PDGF as well as cytokines such as interleukin 3 transmit signals via the PI3-K pathway. PI3-K is composed of two subunits – the p110 catalytic

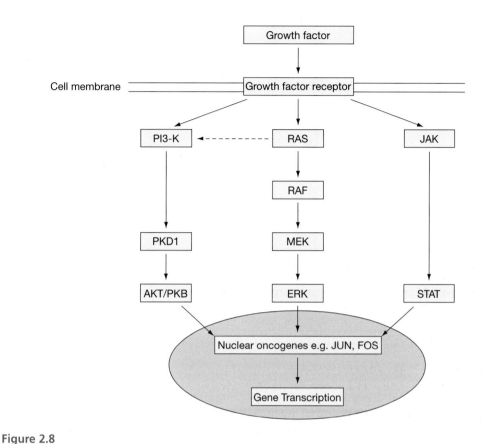

Figure 2.8

The three main signaling pathways activated by growth factor binding to receptors.

subunit and the p85 adaptor/regulatory subunit. Following receptor phosphorylation, binding of the p85 subunit to the growth factor receptor occurs via its SH2 domain. This in turn activates the p110 subunit leading to an increase in its enzymatic activity. The immediate result of this is the generation of phosphatidylinositol phosphates (PIPs). These lipid messengers are then binding sites for a further group of proteins which possess a pleckstrin homology domain which, in PI3-K signaling, primarily involves AKT, also known as protein kinase B, and 3'-phosphoinositide-dependent kinase-1 (PKD-1) both of which are central to PI3-K activity. Binding of AKT to the PIPs result in its translocation to the plasma membrane where it is itself phosphorylated to activate it. Phosphorylation is achieved through interaction with PKD-1. At least 13 substrates for AKT have been recognized [10] which fall into two main groups: (a) regulators of apoptosis such as BAD and the Forkhead transcription factors and (b) regulators of cell growth including protein synthesis and glycogen metabolism and cell cycle regulation [16]. The end result of this signaling pathway is therefore the promotion of cell survival and opposition of apoptosis.

Many of the growth factors that activate the PI3-K pathway also activate the RAS/MAPK pathway. The *RAS* gene is one of the best studied of all of the oncogenes and the family consists of three members *HRAS*, *KRAS* and *NRAS*. *RAS* encodes the G protein, p21, which cycles between a GDP-bound inactive protein and a GTP-bound active protein. Conversion between the active and inactive state is a tightly regulated event. The exchange of GDP for GTP to activate RAS is promoted by the nucleotide exchange factor, SOS [17]. Inactivation

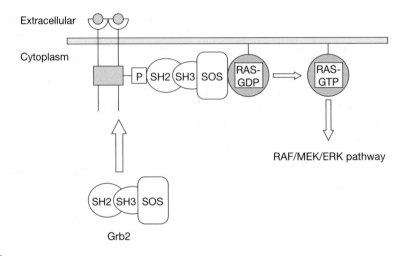

Figure 2.9

Activation of RAS and the downstream signaling pathway. Following receptor activation, the Grb2/SOS complex is recruited from the cytoplasm and binds to the receptor via the SH2 domain of Grb2. This converts inactive RAS bound to GDP to the active form bound to GTP. Once active RAS triggers the next stages of the pathway.

occurs by hydrolysis of GTP to GDP and is promoted by GTPase activating proteins called GAPs. Activation of RAS occurs following growth factor or cytokine interaction with cell surface receptors as described above. As SOS is a cytoplasmic protein, activation of RAS requires firstly that SOS is translocated to the cell membrane, a process mediated by the adaptor protein Grb2 with which SOS is complexed in the cytoplasm in its resting state. The signaling process therefore involves tyrosine phosphorylation of the growth factor receptor, binding of Grb2 to the receptor via its SH2 domain and translocation of SOS to the membrane. Interaction of inactive RAS with SOS leads to GDP/GTP exchange leading to its activation (*Figure 2.9*). An alternative scenario involves a third adaptor protein, Shc, which is phosphorylated by growth factors leading to its association with the Grb2/SOS complex. The functional consequences of this process are then the same as when Grb2/SOS interacts directly with the receptor.

The next stage of the pathway involves the MAP kinases (MAPKs). Activation of RAS results in a conformational change which allows it to interact with downstream effectors. It binds to the serine/threonine kinase, RAF, recruiting it from the cytoplasm to the plasma membrane. Activated RAF in turn then phosphorylates and activates its substrate MAPK/ERK kinase (MEK) on two serine residues in the kinase domain. MEK then binds to and activates a further kinase, ERK. Activated ERK can translocate to the nucleus thereby providing a link between extracellular signals, intracellular signaling pathways and the response in terms of transcriptional activation. The main targets of ERK are transcription factors such as MYC and FOS or components of the cell cycle machinery such as cyclin D1 [18]. This pathway has been implicated in contributing to many of the hallmarks of a cancer cell including immortalization, growth factor-independent growth, ability to invade and metastasize and avoidance of apoptosis. RAS is also active in other signaling pathways and in particular has been shown to activate the PI3-kinase pathway by binding to the catalytic subunit of PI3-K.

The Janus protein tyrosine kinases (JAKs) are cytoplasmic tyrosine kinases and are the main mediators of cytokine signaling pathways, such as the interleukin 3 pathway, and also of some receptor protein tyrosine kinases. The family is composed of four members, TYK2, JAK1, JAK2 and JAK3, and differs in structure from other tyrosine kinases particularly in the lack of an SH2 or SH3 domain. Their substrates are the STAT (signal transducers and activators of transcription) family of transcription factors of which seven are known and which are composed of a DNA-binding domain and a C-terminal SH2 domain [19]. Cytokine receptors are phosphorylated by JAKs following ligand binding. These phosphorylated residues bind STATs through their SH2 domains leading to STAT oligomerization. Dimeric STATs are released from the receptor and translocate to the nucleus where they activate transcription. As before, there is however cross talk between signaling pathways. PI3-K also interacts with STATs and growth factors such as EGF and PDGF will also activate the pathway following interaction with their own receptors. Aberrations in the JAK/STAT pathway are primarily associated with the development of some leukemias and B cell malignancies (see Section 11.5).

The cytoplasmic tyrosine kinases, such as SRC and ABL, are also important in receptor signaling pathways. Approximately half of the 32 known kinases have been implicated in human cancers [10]. The cytoplasmic tyrosine kinase domain is in the carboxy terminus of the protein and all the tyrosine kinase oncogene proteins show homology in this domain.

SRC is a nonreceptor protein kinase which transduces signals involved in processes such as proliferation, differentiation, motility and adhesion [20]. The activated form of the *SRC* cytoplasmic tyrosine kinase gene was one of the first oncogenes to be identified through studies on the Rous sarcoma virus, a chicken tumor virus which was the subject of intensive investigation in the 1960s and 70s. It was the discovery of reverse transcriptase, an enzyme, which led to major developments generally in molecular biology which was used to synthesize *v-src* and eventually led to identification of the cellular homolog *SRC*. The SRC subfamily has two homologous domains not found in the receptor tyrosine kinases family, the SH2 and SH3, domains as described above. As with receptor tyrosine kinases, there is tight control of SRC through intramolecular interactions. An important mechanism for its regulation is by control of its phosphorylation status particularly at two tyrosine residues. SRC is maintained in its inactive form by suppression of its kinase activity through phosphorylation of a specific tyrosine residue (Tyr530) in the carboxy terminus of the protein and interaction of this residue with the SH2 domain. Other intramolecular interactions between the SH3 domain and a linker region between the SH2 domain and the kinase domain also play a role in suppressing SRC activity. A second tyrosine (Tyr419) is also important in regulation of SRC and, when phosphorylated, is a positive regulator of SRC activity. Binding of ligands to receptor tyrosine kinases such as the PDGFR or dephosphorylation of Tyr530 by protein tyrosine phosphatases, results in SH2 displacement from the phosphorylated Tyr530 relieving the inhibitory constraints on the kinase. This results in autophosphorylation of the Tyr419 residue and stabilization of the active conformation. In turn this results in activation of downstream signaling pathways such as RAS/MAPK or PI-3 kinase.

The product of the *ABL* gene is also a cytoplasmic tyrosine kinase which is found in both nucleus and cytoplasm and the gene was one of the first to be identified through a consistent chromosomal translocation seen in chronic myeloid leukemia. The result of the chromosome translocation seen in leukemias is the production of the *BCR–ABL* fusion gene (see Sections 2.2.2 and 11.5). The product of the *ABL* proto-oncogene has a number of features in common with that of the *SRC* gene in that it has both SH2 and SH3 domains, a kinase domain and two C terminal domains – a DNA-binding domain and a domain through which it can bind to actin. The gene comprises two alternate first exons, 1a and 1b, with an exceptionally long intron between them. Two transcripts, 1a and 1b, are produced due to alternate splicing. The b form is myrostilated at an amino terminal glycine which directs the protein to the plasma membrane whereas the a form, containing the amino terminal part of the protein encoded by exon 1a, lacks the myrostilation signal and is predominantly present in the nucleus [21]. The ABL protein has a function in cell cycle regulation and is found complexed to the RB tumor suppressor protein during G_0 of the cell cycle, an association which suppresses the activity of ABL. During transition from G_1 to S, RB is phosphorylated and dissociates from ABL giving rise to the kinase activity of ABL. The activated protein can subsequently alter transcription, e.g. by phosphorylating RNA polymerase. Overexpression of ABL results in cell cycle arrest, suggesting a role for ABL as a negative regulator of cell growth. In addition, nuclear ABL has been shown to have a role in DNA damage-induced apoptosis.

Nuclear proto-oncogenes and transcription factors

The final group of proto-oncogenes are those involved in the control of gene expression by their action on DNA itself. This is the final site of action for messages sent from the growth factors and is the level at which control of growth and proliferation ultimately operates. Several of the proto-oncogene proteins have been shown to bind to DNA and presumably control the transcription of genes (e.g. MYC, FOS and JUN) (Table 2.7).

MYC was initially identified as the cellular homolog of the v-myc gene. The MYC family consists of a number of genes, three of which (MYC, MYCL and MYCN), have been associated with human cancers. MYC has a role in differentiation, in regulating cell proliferation and in apoptosis [22]. It can drive quiescent cells into continuous cycling and can prevent cells from exiting the cell cycle [23]. It appears to sensitize cells to a variety of apoptotic triggers rather than directly inducing apoptosis [24]. The MYC protein, p62, contains two domains, an N terminal transactivation domain localized to the first 143 amino acids and a C terminal DNA-binding and dimerization domain with a leucine zipper motif (HLH/LZ standing for basic helix-loop-helix leucine zipper). Immediately N terminal to the dimerization domain is a domain rich in basic amino acids which directly contacts specific DNA sequences within the DNA major groove [22]. MYC has been shown to act in conjunction with a second protein, MAX, and these heterodimers bind to target sites to transactivate genes via the MYC transactivation domain. Binding occurs at a specific DNA recognition sequence, the so-called E box element, that contains a central CAC(G/A)TG motif found in the promoter of a number of genes shown to be regulated by MYC. However the model is complicated by the identification of the MAD family of proteins. These proteins are capable of transcriptional silencing by histone deacetylation. Two of the family, MAD and MXI-1 (MAD 2), can also dimerize with MAX. These dimers bind to the same motif as MYC/MAX dimers. Changes in the level of expression of the four genes in response to growth stimuli lead to alterations in the composition of the two types of dimer which have opposite effects on regulation of expression. Stimulation of

Table 2.7 Role of proto-oncogene products in the cell – proteins acting in the nucleus and transcription factors

MYC
FOS
JUN
MYB
SKI
EVI1
REL

cells leads to the increase in expression of MYC and hence to the production of MYC-MAX dimers. Withdrawal of mitogens or the removal of differentiation stimuli leads to suppression of MYC expression and to an increase in the alternative dimers MAX-MAD or MAX-MXI-1. MYC/MAX dimers bind to DNA and MYC interacts with the TATA box binding protein TBP and transcriptional machinery leading to activation of target genes (*Figure 2.10*). In contrast, following DNA binding, MAX/MAD dimers recruit a number of proteins including those with histone deacetylation activity. This deacetylation process locks up the DNA in nucleosomes preventing transcription. Many genes have been implicated as the targets transactivated by MYC. Some, such as the cyclins and cyclin-dependent kinases, are involved in cell growth [25], others, such as LDH-A, are involved in growth and metabolism. The more recent finding is the association of MYC with apoptosis and it is assumed that MYC affects transcription of genes in the apoptotic pathway. In addition, MYC has been identified as an inducer of telomerase activity [22].

FOS and JUN are also transcription factors belonging to the AP-1 transcription family which dimerizes and regulates transcription from promoters containing AP-1 DNA-binding sites found in genes associated with proliferation and differentiation [26]. Dimerization occurs via a region known as the leucine zipper region and these proteins can then bind

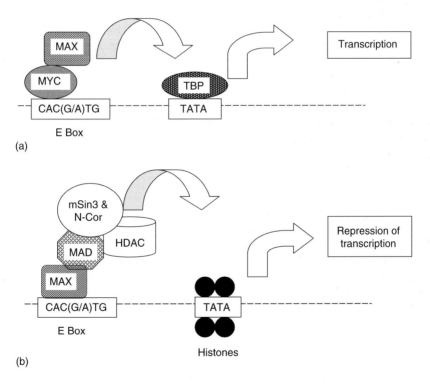

Figure 2.10

Mechanism for the role of MYC, MAX and MAD proteins in transcriptional activation. (a) The MYC-MAX heterodimer is bound to the E box CAC(G/A)TG motif. MYC can then interact with the TATA-binding protein (TBP) leading to transcriptional activation. (b) Withdrawal of mitogens or removal of differentiation stimuli leads to suppression of MYC expression and results in MAX-MAD dimers. These bind again to the central motif which results in recruitment of mSin3, the transcriptional co-repressor N-Cor and histone deacetylases (HDAC). HDAC then deacetylates histones locking them in the nucleosome and consequently results in repression of transcription.

to DNA through a region rich in basic amino acids in a sequence-specific manner. Both FOS and JUN also contain a number of activation domains. Activity of FOS and JUN is under tight control in the cell and they are regulated by phosphorylation of specific amino acids. It is now clear that Jun amino terminal kinase (JNK) is responsible for phosphorylating JUN at positions 63 and 73 enhancing JUN-dependent transcription.

As described above, the various groups of proto-oncogenes perform major functions in the cell. It might therefore be expected that any disruption in their expression would lead to disruption in the control of that cell. The various mechanisms by which the proto-oncogenes are activated to produce a gene with oncogenic potential are described in the next section.

2.2.2 Mechanisms of oncogene activation

There are three main ways, structural alteration, gene amplification and deregulated expression, by which proto-oncogenes are activated and examples are given in *Table 2.8*.

Structural alteration

The first mechanism is production of an abnormal product which can occur in a number of ways. Point mutations have been described in several oncogenes but have been studied most extensively in genes of the *RAS* family. Amino acid substitutions have been detected, particularly at positions 12, 13 and 61, in up to 30% of tumors including breast, lung and colon. Activation of the protein to an oncogenic form is also possible at positions 56, 63, 116 and 119, but these changes have not been seen in human tumors. These substitutions all alter the structure of the normal protein resulting in abnormal activity. Mutant RAS is locked in its activated state bound to GTP, circumventing the need for growth factor stimulation, resulting in constitutive signaling.

Point mutations which promote receptor dimerization in the absence of growth factor binding are a general mechanism of activation of some of the receptor tyrosine kinases. Classically this has been described in the *RET* gene in the inherited disorders, familial medullary thyroid cancers (FMTC) and multiple endocrine neoplasia types 2A and 2B (MEN2A and MEN2B, see Section 5.7.2) [10]. In FMTC and MEN2A, mutations in the extracellular domains at conserved cysteine residues cause formation of intermolecular disulfide bonds between RET molecules leading to constitutive dimerization and activation. In MEN2B, the common methionine to threonine mutation at codon 918 results in increased kinase activity of the protein without the need for dimerization. A point mutation in the transmembrane domain of the *NEU/HER2* proto-oncogene results in a valine to glutamic acid substitution causing receptor dimerization and kinase activation in the absence of ligands.

Larger structural alterations are found following chromosomal translocations as described in many lymphomas and leukemias [27]. One of the best studied of these is the

Table 2.8   Mechanisms of oncogene activation

Method of activation	Oncogene
Structural alteration	*ABL, HRAS, KRAS, SRC*
Amplification	*ERBB2, MYB, MYC*
Deregulated expression	
Insertional mutagenesis	*INT1, INT2*
Translocation	*MYC*

9;22 translocation found in CML which places the *ABL* oncogene on chromosome 9 next to the breakpoint cluster region (*BCR*) of the Philadelphia gene on chromosome 22. The fusion gene produced in this process produces a fusion protein with constitutive protein tyrosine kinase activity (*Figures 2.11* and *11.4*). The dimerization domain of the BCR region of the fusion protein moderates homo-oligomerization of BCR-ABL. This leads to cross-phosphorylation of tyrosine residues of the protein which subsequently form docking sites for several adaptor molecules thereby activating signaling pathways such as the RAS-MAPK, JAK-STAT and PI3-K pathways [21]. A further important aspect of BCR-ABL signal transduction is the cytoplasmic localization of the chimeric protein circumventing ABL's role in the nucleus to mediate DNA damage-induced apoptosis.

Subsequently, many other examples of translocations resulting in the creation of fusion genes have been identified and in many, but by no means all, cases these have been shown to involve transcription factors leading to disruption of transcriptional control. Alveolar rhabdomyosarcomas have been associated with 2;13 and 1;13 translocations (see Section 12.5). These translocations place the *PAX3* or *PAX7* genes, members of the paired box transcription factor family, next to the *FKHR* gene which also belongs to a transcription factor family known as the 'forkhead' family [28]. A fusion gene is produced which is made up of the DNA-binding domain of *PAX3* and *PAX7*, and the *FKHR* transactivational domain. The fusion gene products activate transcription from PAX binding sites at a level 10–100 fold that of the wild-type PAX proteins. This increased function results from the insensitivity of the FKHR activation domain to the inhibitory effects of the N terminal PAX domains. In addition there are increased levels of the fusion protein produced, in the case of the PAX7–FKHR fusion as a result of *in vivo* amplification mechanism and in the PAX3–FKHR protein due to a copy number-independent increase in transciptional rate. Finally the fusion proteins show exclusively nuclear localization.

Another cause of oncogenic mutation of proto-oncogenes can be production of a truncated form of the protein. A C to T transition mutation has been identified in *SRC* in a subset of colorectal tumors which results in the truncation of the SRC protein at tyrosine 530 [29]. This protein has increased tyrosine kinase activity as it lacks the C-terminal amino acids required for binding to the SH2 domain which establishes the inactive complex.

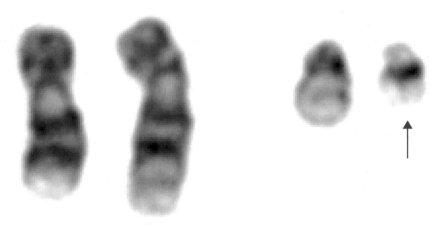

Figure 2.11

A partial karyotype showing the Philadelphia chromosome produced by the t(9:22) in CML. Photograph courtesy of Dom McMullan, West Midlands Regional Genetics Service, Birmingham, UK.

Amplification

The second mechanism of activation is overproduction of the normal protein by amplification of the proto-oncogene. For example, amplification of the *MYC* gene is found in some cancers (see *Figure 12.3*) seen microscopically as DMs or HSRs. Amplification of the *MYCN* gene is consistently associated with late-stage neuroblastomas (see Section 12.4). In breast cancer, amplification of the *HER2* gene is a common finding and has been shown in many studies to be an independent prognostic indicator predicting a more aggressive disease behavior in node-positive patients (see Section 9.2.2) [30]. Overexpression of EGFR has been shown to be a strong prognostic indicator in head and neck cancer, ovarian, cervical, bladder and esophageal cancers [31]. The *RAS* oncogene, which is frequently activated by point mutation, has also been found to be amplified in certain human tumors.

Deregulated expression

The third mechanism of activation occurs through loss of appropriate control mechanisms. This mechanism of activation is typified by activation of the *MYC* oncogene in Burkitt's lymphomas (see Section 11.3). Expression of *MYC* is normally under very fine control. In Burkitt's lymphoma, the *MYC* gene is translocated to one of the immunoglobulin loci in virtually every case of the disease. The translocation between chromosomes 8 and 14 involving *MYC* and the immunoglobulin heavy chain gene is seen in approximately 80% of cases of Burkitt's lymphoma (*Figure 2.12*) [32]. In the remaining cases the translocation involves either the κ light chain gene on chromosome 2 or the λ light chain gene on chromosome 22. Breakpoints can occur either in the first intron or exon 1 of *MYC*, immediately 5′ to the *MYC* gene or distant to the gene by up to 100 kb. In all cases the two coding exons, exons 2 and 3 of *MYC* are intact. The effect of the translocation appears therefore to involve the deregulation of *MYC* gene expression probably involving regulatory elements from the immunoglobulin loci. This results in overexpression of the *MYC* gene with continuous signaling and consequent cell proliferation. *MYC* is also involved in other hematological malignancies. In some cases of T cell ALL, the *MYC* gene is translocated into one of the T cell receptor loci. In multiple myeloma translocation of *MYC* also occurs into one of the immunoglobulin loci but here, unlike in Burkitt's lymphoma, the translocation is likely to be a secondary event.

There is no single consistent mechanism of activation of any one oncogene. Thus *RAS* is primarily activated by point mutation but amplification is also found; *MYC* is amplified in many tumors but abnormal expression is also associated with deregulation following

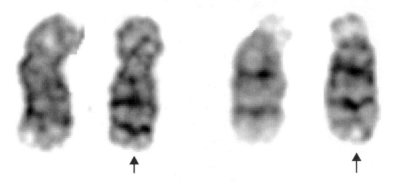

Figure 2.12

Karyotype from a Burkitt's lymphoma containing a t(8;14). Figure courtesy of Dom McMullan West Midlands Regional Genetics Service, Birmingham, UK.

chromosomal translocation. Whatever the mechanism for the activation of these genes, the end result is to produce a protein which can cause abnormal growth.

2.3 Oncogenes in human tumors

In most groups of tumors so far examined for the presence of an oncogene, such a gene has been found. A few of the most common tumors and associated oncogenes are shown in *Table 2.9* but the list is far from complete. The value of these oncogenes as diagnostic or prognostic markers is discussed fully from Chapter 5 onwards. Recently it has been shown that the *RET* oncogene is the causative gene in an inherited cancer syndrome, multiple endocrine neoplasia type 2A (MEN2A) [10]. Mutations in other proto-oncogenes are, however, very rarely the cause of familial cancers as they are likely to be lethal. In many tumors such as Burkitt's lymphoma or CML, the presence of the oncogene is likely to have an important, although not necessarily primary, role in the development of the tumor. In the majority of other tumors it is not clear exactly what role the presence of the oncogene has to play in the development of the cancer. Some oncogenes are likely to be detected not because they have been specifically activated, but because control of normal cell growth has been disrupted, leading to their subsequent activation.

Here we need to return to the concept of cancer as a phenotype which arises following a number of sequential alterations in the cell. It was stated earlier that a single mutational event was not enough for a tumor to develop. Examples of oncogenes have been described here however which act in a dominant manner and which in some cases have been detected only by their ability to transform the NIH-3T3 cell line. This experiment needs to be seen in perspective. The NIH-3T3 cell line is not normal but has undergone a number of mutational events during its establishment in culture. Insertion of a single oncogene into a totally normal cell line will not cause it to transform without considerable manipulation of its expression. Much evidence now suggests that there is collaboration between different oncogenes and that this is necessary to produce the fully transformed phenotype.

It is possible to divide oncogenes broadly into two groups based on their phenotypes in transformation assays. The first group, such as *MYC, FOS* or *JUN*, belongs to the immortalization group. The second group e.g. *RAS*, belongs to the transformation group. In model systems, it was initially demonstrated that whereas a single oncogene was insufficient for transformation, collaboration between genes, one from each group, could result in full transformation. Some of the observations have been demonstrated more directly in transgenic

Table 2.9   Oncogenes in human tumors

Tumor	Associated oncogene
Bladder	HRAS, KRAS
Brain	ERBB1, SIS
Breast	ERBB2, HRAS, MYC
Cervical	MYC
Colorectal	HRAS, KRAS, MYB, MYC
Gastric	ERBB1, HST, MYB, MYC, NRAS, YES
Lung	ERBB1, HRAS, KRAS, MYC, MYCL, MYCN
Melanoma	HRAS
Neuroblastoma	NMYC
Ovarian	ERBB2, KRAS
Pancreas	KRAS, MYC
Prostate	MYC
Testicular	MYC
Leukemia	ABL, MYC, BCR, BCL1, BCL2

mice. In this model, a fragment of DNA containing an oncogene linked to a suitable promoter can be injected into a fertilized mouse egg. Often this recombinant fragment of DNA is integrated into the mouse genome and may be expressed in many tissues or only a few depending on the tissue specificity of the promoter. When this was tried with either *MYC* or *RAS* alone, few of the progeny produced tumors, although the introduced gene was generally expressed. Only the occasional cell which had undergone further changes became transformed. Not all the progeny from a cross between two transgenic mice, one carrying *MYC* and the other *RAS*, developed tumors, although an increase in the number of tumors was seen compared with either parent [33].

This interaction between the oncogenes is only one of the processes involved in malignant development. We now know that a further group of genes, which act in a growth-regulatory, rather than a growth-promoting, fashion, are also involved with each other and with the oncogenes. These genes are the subject of the next chapter.

References

1. **Macdonald, F. and Ford, C.H.J.** (1997) *Molecular Biology of Cancer*, 1st Edition, BIOS Scientific Publishers, Oxford.
2. **Stehelin, D., Varmus, H.V., Bishop, J.M. and Vogt, P.K.** (1976) *Nature*, **260**, 170.
3. **Shih, C., Padhy, L.C., Murray, M. and Weinberg, R.A.** (1981) *Nature*, **290**, 261.
4. **Heim, S. and Mitelman, F.** (1987) *Hum. Genet.*, **75**, 70.
5. **Schwab, M. and Amler, L.C.** (1990) *Genes, Chrom. Cancer*, **1**, 181.
6. **Futrea, P.A., Kasprzyk, A., Birme, E. et al.** (2001) *Nature*, **409**, 850.
7. **Normano, N., Bianco, C., De Luca, A. and Salomon, D.S.** (2001) *Front. Biosci.*, **6**, D685.
8. **Sachdev, D. and Yee, D.** (2001) *Endocr. Relat. Cancer*, **8**, 197.
9. **Pasche, B.** (2001) *J. Cell. Physiol.*, **186**, 153.
10. **Blume-Jensen, P. and Hunter, T.** (2000) *Nature*, **411**, 355.
11. **Robinson, D., Wu, Y.-M. and Lin, S.-F.** (2000) *Oncogene*, **19**, 5548.
12. **Zwick, E., Bange, J. and Ullrich, A.** (2001) *Endocr. Relat. Cancer*, **8**, 161.
13. **Yardin, Y.** (2001) *Oncogene*, **61** (suppl. 2), 1.
14. **Rubin, I. and Yardin, Y.** (2001) *Ann. Oncol.*, **12**, S3.
15. **Kloos, D.U., Choi, C. and Wingender, E.** (2002) *T.I.G.*, **18**, 96.
16. **Talapatra, S. and Thompson, C.** (2001) *J. Pharmacol. Exp. Ther.*, **298**, 87.
17. **Jones, S.M. and Kazlauskas, A.** (2001) *Oncogene*, **19**, 5558.
18. **Kolch, W., Kotwaliwale, A., Vass, K. and Janosch, P.** (2002) *Exp. Rev. Mol. Med.* 25 April, http://www.expertreviews.org/02004386h.htm.
19. **Reddy, E.P., Korapati, A, Chaturvedi, P. and Rane, S.** (2000) *Oncogene*, **19**, 2532.
20. **Bjorge, J.D., Jakymiw, A. and Fujita, D.H.** (2000) *Oncogene*, **19**, 5620.
21. **Thijsen, S.F.T., Schuurhuis, G.J. and van Oostveen, J.W.** (1999). *Leukemia*, **13**, 1646.
22. **Dang, C.V.** (1999) *Mol. Cell. Biol.*, **19**, 1.
23. **Pelengaris, S., Rudolph, B. and Littlewood, T.** (2000) *Curr. Opin. Genet. Dev*, **10**, 100.
24. **Prendergas, G.C.** (1999) *Oncogene*, **18**, 2967.
25. **Hemeking, H., Rago, C., Schuhmacher, M. et al.** (2000) *P.N.A.S.*, **97**, 2229.
26. **Vogt, P.K.** (2002) *Nat. Rev. Cancer*, **2**, 465.
27. **Bohlander, S.K.** (2000) *Cytogenet. Cell. Genet.*, **91**, 52.
28. **Barr, F.G.** (2001) *Oncogene*, **20**, 5736.
29. **Irby, R.B. and Yeatman, T.J.** (2000) *Oncogene*, **19**, 5636.
30. **Slamon, D.J., Clark, G.M, Wong, S.G. et al.** (1987) *Science*, **235**, 177.
31. **Nicholson, R.I., Gee, J.M. and Harper, M.E.** (2001) *Eur. J. Cancer*, **37** (Suppl. 14), S9.
32. **Boxer, L.M. and Dang, C.V.** (2001).*Oncogene*, **20**, 5595.
33. **Sinn, E.** (1987) *Cell*, **49**, 465.

Tumor suppressor genes

3

3.1 Introduction

A second group of genes that plays an important role in tumorigenesis are the tumor suppressor genes. These are defined as genes involved in the control of abnormal cell proliferation and whose loss or inactivation is associated with the development of malignancy. They act therefore by effectively inhibiting or putting the brake on cell growth and cell cycling. These genes are recessive at the level of the cell although they show dominant inheritance when associated with a familiar cancer syndrome (see below). Both copies of the gene have to be mutated before tumors develop. Several dozen tumor suppressor genes have now been identified associated with human tumors (*Table 3.1*).

Table 3.1 Tumor suppressor genes associated with human cancers

Gene	Location	Function	Associated tumors	Familial cancer syndrome
APC	5q21	β-catenin binding, cytoskeleton stabilization, cell-cycle regulation	Colorectal	Familial adenomatous polyposis
BRCA1	17q21	DNA repair	Breast and ovarian cancer	Familial breast/ovarian cancer
BRCA2	13q12.3	DNA repair	Breast and some ovarian cancer	Familial breast/ovarian cancer
CDH1 (E-Cadherin)	17q22.1	Cell adhesion	Breast, colon lung, stomach	Familial gastric cancer
CDKN1C (p57)	11p15.5	Cyclin-dependent kinase inhibitor	Wilms' tumor, rhabdomyosarcoma	Beckwith–Wiedemann syndrome
CDKN2A (p16)	9p21	Cell cycle regulator at G_1/S	Melanoma, pancreatic	Familial melanoma
CYLD	16q12-13			Familial cylindromas
EP300	22q13.2	Transcription factor-binding protein	Colorectal, breast, pancreatic	
EXT1	8q24	Transmembrane glycoprotein – catalyzes polymerization of heparin sulfate	Exostoses, osteosarcomas	Multiple exostoses type 1
EXT2	11p12-11	Transmembrane glycoprotein – catalyzes polymerization of heparin sulfate	Exostoses, osteosarcomas	Multiple exostoses type 2
FHIT	3p14		Lung, stomach, kidney, cervix	Familial clear cell renal cancer

(continued)

Table 3.1 (*continued*)

Gene	Location	Function	Associated tumors	Familial cancer syndrome
MAP2K4	17p11	Protein kinase	Pancreatic, breast, colon	
MEN1	11q13	Transcriptional repressor	Parathyroid, pituitary, islet cell carcinoma,	MEN1
MLH1	3p21	Mismatch repair	Colorectal	HNPCC
MLH3	14q24	Mismatch repair	Colorectal	HNPCC
MSH2	2p22	Mismatch repair	Colorectal	HNPCC
MSH6	2p16	Mismatch repair	Colorectal	HNPCC
NF1	17q11	Regulator of RAS mediated proliferation	Neurofibromas, gliomas	Neurofibromatosis type 1
NF2	22q12	Cytoskeletal regulator and suppressor of cell adhesion	Meningiomas, schwannomas	Neurofibromatosis type 2
PMS1	2q31-3	Mismatch repair	Colorectal	HNPCC
PMS2	7p22	Mismatch repair	Colorectal	HNPCC
PRKAR1A	17q23-24	Role in cAMP pathway	Testicular, thyroid, breast ductal adenoma	Carney complex
PTCH	9q22	Regulator of sonic hedgehog	Basal cell carcinomas	Gorlin syndrome
PTEN	10q23	Protein tyrosine phosphatase	Hamartomas, prostatic, breast, endometrial	Cowdem syndrome, Bannayan–Zonana syndrome
RB1	13q14	Cell cycle inhibitor	Retinoblastomas, osteosarcomas	Familial retinoblastoma
RUNX3		Transcription factor	Gastric	
SDHD	11q23	Succinate dehydrogenase subunits	Pheochromocytoma, paraganglioma	Familial paraganglioma
SDHC	1q21			
SDHB	1p36-35			
SMAD4	18q21	Transducer of TGFβ signals	Colorectal, pancreatic	Juvenile polyposis
SMARCB1	22q11	Regulator of chromatin	Rhabdoid tumors	
STK11 (LKB1)	19q13	Serine/threonine kinase	Hamartomas, ovarian, prostate, breast	Peutz–Jegher syndrome
TP53	17p13	Transcriptional regulator	Breast, brain, colorectal, plus many more	Li–Fraumeni
TSC1	9q34	Maintainance of cytoskeleton	Brain, renal	Tuberous sclerosis
TSC2	16p13	Cell cycle regulator	Brain, renal	Tuberous sclerosis
VHL	3p26	Regulates proteolysis, responds to changes in oxygen	Renal, pheochromocytoma, hemangiomas	Von Hippel Lindau disease
WT1	11p13	Transcriptional regulator	Nephroblastoma	Wilms tumor

The literature abounds with different names for these genes which may cause some confusion including: ortho-genes, emerogenes, flatogenes and onco-suppressor genes. Most commonly, they are described as tumor suppressor genes or antioncogenes. As the genes do not always act directly on an oncogene, this latter term is a misnomer. The term tumor suppressor gene also has its limitations; as the control of growth is likely to involve a number of genetic mechanisms, it is unlikely that a single gene can abrogate all of them. However this term is the most commonly used and will be used here. As our understanding of how these genes function grows and as more are identified, defining exactly how a tumor suppressor functions is increasingly difficult. It is now complicated by the addition of extra terms such as caretaker, gatekeeper and landscaper genes as described later in the chapter.

This chapter will describe the method of detection of the tumor suppressor genes and the function of some of the better-understood protein products. The clinical aspects of these genes are considered in Chapter 5.

3.2 Evidence for the existence of tumor suppressor genes

Tumor suppressor genes are more difficult to identify than oncogenes. Introduction of an oncogene into an untransformed cell culture and identification of the resulting transformed colonies is relatively straightforward. It is not as easy to identify untransformed revertants on a background of transformed cells. Two different techniques have been used to establish the existence of tumor suppressor genes in man. These are: (1) the suppression of malignancy in somatic cell hybrids; and (2) a consistent loss of chromosomal regions, initially seen in hereditary cancers and subsequently also shown in sporadic cancers.

3.2.1 Suppression of malignancy by cell fusion

The earliest evidence for tumor suppressor genes predates the discovery of oncogenes by over 20 years. Harris and colleagues showed that when malignant cells were fused with normal diploid cells, the resulting hybrids were nontumorigenic as determined by their inability to grow in immunocompromised hosts [1,2] (*Figure 3.1*). This suppression of

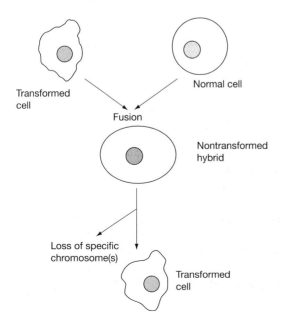

Transformed cell

Normal cell

Fusion

Nontransformed hybrid

Loss of specific chromosome(s)

Transformed cell

Figure 3.1

Production of somatic cell hybrids. Adapted from [2].

malignancy was dependent on retention of a specific chromosome. As the hybrids were unstable, there was random loss of chromosomes and when a particular chromosome was lost, malignant clones were once again capable of growth *in vivo*. The assumption was that the chromosome which was lost contained the tumor suppressor gene(s). Fusion of two malignant cells also gave rise to nontumorigenic hybrids suggesting complementation between lesions. Detailed cytogenetic analysis of hybrids has identified specific chromosomes involved in the suppression of the malignant phenotype. For example, the presence of chromosome 11 from the normal partner is necessary to maintain the suppression of the malignant phenotype when HeLa cervical carcinoma cells are fused with normal fibroblasts. This result was confirmed by introducing a single fibroblast chromosome 11 into HeLa cells by microcell fusion resulting in suppression of the malignant phenotype of the cervical cancer cells [3].

Although tumorigenesis is suppressed in hybrids between HeLa cells and normal fibroblasts, the hybrids still behave as transformed cells *in vitro*. This suggests that transformation and tumorigenicity are under separate genetic control. Other characteristics of tumor cells, such as immortalization and metastatic ability, can also be suppressed in hybrids, suggesting that a number of different genes are involved in the tumorigenic phenotype [3].

Reversion of the malignant phenotype can also be demonstrated in the presence of an activated oncogene. Nontumorigenic hybrids have been produced following fusion of cell lines carrying either activated *NRAS* or *HRAS* genes with normal fibroblasts. These phenotypically normal hybrids, and also tumorigenic revertants, still express the product of the respective oncogene [3].

3.2.2 Tumor suppressor genes in hereditary cancers: the retinoblastoma model

A second piece of evidence for the existence of tumor suppressor genes came from studies of hereditary cancers. These are cancers with a clear pattern of inheritance, usually autosomal dominant, a tendency for earlier age of onset of tumors than that seen for sporadic tumors and the frequent occurrence of multiple tumors, both synchronous and metachronous.

The prototype for studies on hereditary cancers is retinoblastoma, a childhood cancer of the retina which occurs in both sporadic and familial forms (*Figure 3.2*). Forty percent of cases are hereditary, transmitted as an autosomal dominant trait. Of these cases around 10–15% are transmitted from an affected parent, the remainder arising as *de novo* germ-line mutations, more usually on the paternal germ-line than the maternal [4]. Tumors frequently

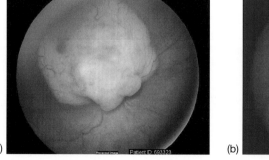
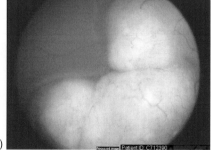

(a) (b)

Figure 3.2

Eyes from two patients with retinoblastoma tumors. Ilustration courtesy of the Retinoblastoma team, Birmingham Children's Hospital.

arise in both eyes, the average number of tumor foci being three to five. The remaining 60% of cases are sporadic and characteristically tumors are seen in only one eye. In 1971, from studies of age/incidence curves, Knudson [4] postulated that the disease arose from two sequential mutational events. In the hereditary form of the disease, one mutation is inherited in the germ-line and is phenotypically harmless, confirming the recessive nature of the mutation at the cellular level. A second 'hit', occurring in a retinal cell, causes the tumor to develop (*Figure 3.3*). As there are a large number of retinoblasts in the eye (over 10^7), which are all at risk because they already carry one mutation, a second 'hit' will occur frequently enough to cause a high proportion of tumors in at least one eye and often in both. In the sporadic form of the disease, both mutations occur in the somatic tissue (*Figure 3.3*). The probability of two mutations occurring in the same cell is low, therefore the disease is primarily unilateral [5].

This 'two-hit' hypothesis was subsequently confirmed by identification of mutations or deletions of the retinoblastoma gene itself. Before this could be done, the retinoblastoma locus had to be mapped to a specific chromosome. A number of patients with retinoblastoma were identified who had a cytogenetically visible deletion in the region of band q14 of chromosome 13 (*Figure 3.4*) and it was inferred that the retinoblastoma gene (*RB1*) lay in this region [6]. Linkage of the *RB1* gene to the polymorphic enzyme, esterase D, which had been mapped to 13q14, further supported this location [7]. Retinoblastoma patients with deletions of chromosome 13q14 also showed quantitative differences in the expression of esterase D [8]. These studies suggested that in tumors there was loss or deletion of part of chromosome 13 and it was assumed that there was a mutation in the *RB1* gene on the remaining copy of chromosome 13.

This pattern of mutation and allele loss was confirmed at the molecular level by Cavanee and colleagues [9] using the loss of heterozygosity test which has become widely used for

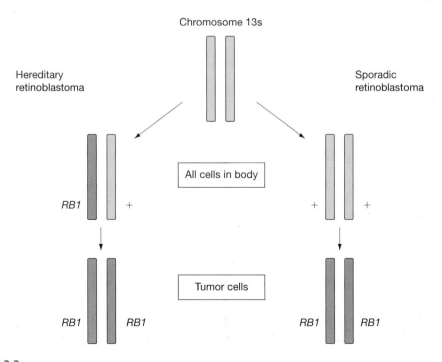

Figure 3.3

The origin of hereditary and sporadic retinoblastomas. *RB1* = mutant copy of retinoblastoma gene, + = wild-type copy of gene. Adapted from [2].

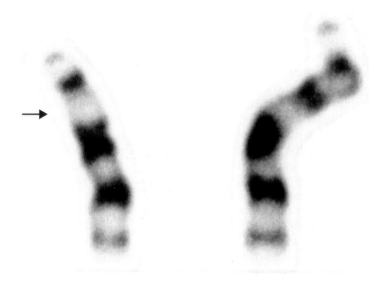

Figure 3.4

Deletion of 13q14 as seen in retinoblastomas. Photo courtesy of the West Midlands Regional Genetics Laboratory. Adapted from [2].

the detection of tumor suppressor genes. Initially, the loss of heterozygosity test depended on the use of restriction fragment length polymorphism (RFLP) analysis. These RFLPs are present throughout the genome and are seen as variable lengths of DNA fragments generated by restriction enzyme digestion of genomic DNA and are detected by DNA probes specific for the DNA region of interest (*Figure 3.5*). With the development of PCR, the technique more commonly makes use of microsatellite markers which are also distributed throughout the genome but which are more useful than RFLPs because of the greater number of alleles present at any one locus. At the time the retinoblastoma gene was studied, only RFLP analysis was available and the probes selected for their detection were located on chromosome 13q. Patients suitable for study were those who had different-sized fragments of DNA (alleles) on each of the two chromosome 13s in their somatic tissue, that is, they were heterozygous. When DNA from tumors from the same patients were analyzed, only one of the two alleles was present (*Figure 3.5*). This loss of heterozygosity (LOH) can occur by a number of possible mechanisms (*Figure 3.6*), including loss of the normal chromosome possibly followed by reduplication of the abnormal one, an interstitial deletion of the normal chromosome, or a recombination event resulting in two copies of the deficient allele. Most of the mechanisms shown in *Figure 3.6* result in loss of heterozygosity along the majority of the chromosome, the exception being a second point mutation [9]. Recently it has been shown that inactivation of the second copy of tumor suppressor gene can occur due to methylation and transcriptional silencing [10].

LOH studies helped to narrow the region in which the *RB1* gene lay. It was then isolated by a 'positional cloning' approach. A molecular clone was first identified which was shown to be deleted in retinoblastoma tumors [11]. Based on the assumption that this sequence must lie close to the *RB1* gene, a number of chromosome walks were carried out and fragments of DNA present in single copies were identified. These were used to probe DNA from humans and other species and one sequence was found to be conserved across species suggesting that it was perhaps a coding exon of a gene. Hybridization of this sequence to RNA from retinal cell lines showed that it recognized a 4.7 kb mRNA. Finally, the conserved sequence was used to isolate a cDNA clone containing the *RB1* gene [12]. There were several lines of evidence which verified that this was the correct sequence: (1) mutations were identified in the patients and their tumors [13]; (2) absence or abnormal expression of the

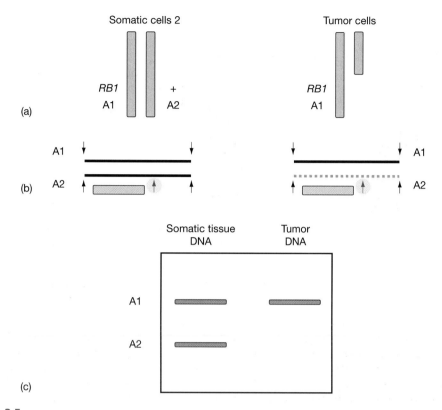

Figure 3.5

RFLP analysis as a means of detecting loss of heterozygosity. (a) On the left two chromosome 13s are shown, one carrying the normal copy of the retinoblastoma (+) and one carrying a mutated copy (*RB1*). The chromosomes on the right show a deletion of the wild-type allele. Near to the gene is an RFLP showing two alleles which are of different sizes when analyzed as shown in (b) and (c). (b) Black arrows indicate common restriction sites present on all chromosomes and the purple arrow indicates a polymorphic restriction site which creates the smaller allele A2. (c) Southern blot analysis with a probe from within the region of chromosome 13. Tumor DNA shows the presence of only one allele due to the deletion of the part of the chromosome – this results in only one band on the Southern blot, i.e. heterozygosity is lost. Adapted from [2].

RNA transcript was detected in tumors [14]; and (3) a structural change in the fibroblasts of a patient with bilateral disease and a second mutation in the tumor was detected, confirming Knudson's hypothesis [14]. The native RB protein was isolated using antisera raised against a synthetic fusion protein, which had been produced by inserting part of the *RB1* cDNA into an expression vector [15]. The tumor suppressor nature of RB was shown by the introduction of a single copy of *RB1* into tumor cell lines lacking the gene resulting in complete or partial suppression of the tumorigenic phenotype [16].

3.3   Retinoblastoma – *RB1*

The *RB1* gene spans approximately 180 kb of DNA and is made up of 27 exons which encode a 110 kDa nuclear phosphoprotein [17]. RB belongs to a three-member family of proteins, RB itself, p107 and p130 which together are termed the pocket proteins and which show

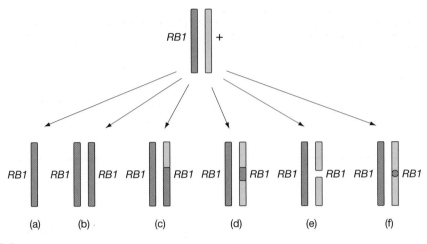

Figure 3.6

Mechanism of loss of heterozygosity. (a) Nondisjunction, (b) nondisjunction and reduplication, (c) mitotic recombination, (d) gene conversion, (e) deletion, (f) point mutation. Adapted from [2].

greatest homology in the pocket domain. All three show functional similarities but also show significant differences primarily with the proteins through which they interact. The RB protein contains 928 amino acids split into four domains, the N terminal domain, the R-motif, the A/B pocket and the C terminal domain [18]. The A/B domains together form a functional repressor motif through which the majority of RB's 100-plus associated proteins interact [19]. The C terminal is important for growth suppression and interacts with and regulates the A/B domain. This domain also contains most of the motifs through which RB is phosphorylated and contains a nuclear localization signal. The N terminal region is less well defined but appears to be important for the tumor suppressor function of RB, possibly by determining the conformation state of the protein. The protein has a gate-keeper role in cell proliferation, acting via transcription factors to prevent the transcriptional activation of a variety of genes, the products of which are required for the onset of DNA synthesis, the S phase, of the cell cycle [19,20]. In addition to cell cycle control, RB has a number of other roles regulating a variety of cellular processes including differentiation, DNA replication and apoptosis [19,20].

3.3.1 Cell cycle regulation of RB

The cell cycle consists of the DNA synthesis phase (S) and cell division at mitosis (M) separated by two gap intervals G_1 and G_2 (see Chapter 4). Upon growth factor stimulation, cells move from the resting phase, G_0, into G_1. Cycling will stop at this stage if mitogenic signals are removed prior to a specific point – the restriction point (R) – at the end of G_1 and will return to G_0. Once past this point, cycling will continue even if stimulants are removed. It is at R that many of the signaling pathways which assist passage through R converge on the RB signaling pathway [19]. RB function is regulated by phosphorylation in a cell cycle-specific manner and occurs at 16 serine and threonine sites recognized by the cyclin-dependent kinases (CDKs), complexed with the cyclins (*Figure 3.7*). It is hypophosphorylated in the G_0/G_1 phase of the cycle, hyperphosphorylated prior to G_1/S transition and throughout S, in at least a two-step process, and dephosphorylated as the cell leaves mitosis and goes back to G_0 or G_1 phase [21]. The phosphorylation state of the

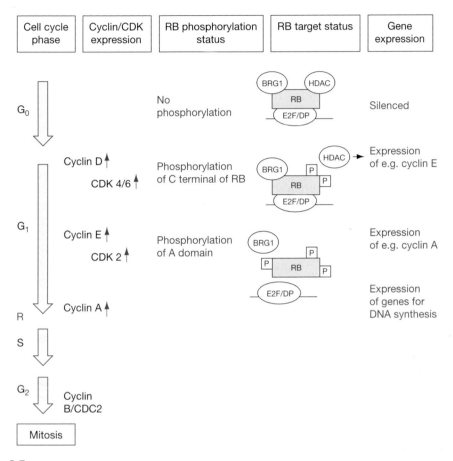

Figure 3.7

Mode of action of RB. During G_0 RB is underphosphorylated and is bound to the E2F-1 transcription factor complexed with DP-1 as well as other corepressors such as HDAC and BRG1. As cells exit G_0, cyclin D levels rise and CDK4 and 6 are activated. This leads to phosphorylation of RB on its C terminal domain. This dislodges proteins such as HDAC and releases repression of certain genes including cyclin E. The rise in cyclin E expression leads to CDK2 expression and further RB phosphorylation within the A domain of the pocket region. This in turn dislodges E2F and corepressors such as BRG1. All RB-induced repression is therefore removed and genes such as cyclin A and those needed for S phase entry DNA synthesis are expressed. The cell cycle checkpoint R is released in this process allowing cells to progress into S phase [adapted from 18,20].

RB protein therefore varies throughout the cell cycle and correlates with its capacity to interact with the transcription factors described below. Phosphorylation of RB occurs through the action of cyclin D1–3/CDK4–6 in early G_1, by cyclin E/CDK2 in late G_1 and early S and by cyclin A/CDK2 in S phase.

3.3.2 RB/E2F pathway

RB does not act directly to repress transcription but does so by interaction with transcription factors and corepressors, many of which have now been identified [19,21]. The first of these to be identified was E2F-1, a member of the E2F family (E2F-1–6). This is a family of sequence-specific DNA-binding transcription factors which had been shown to regulate expression of genes required for entry into S phase and for DNA synthesis (e.g. dihydrofolate reductase). Interaction of RB and E2F-1 is stabilized by binding DP-1, a further E2F-related protein. E2F activates transcription by activating general transcription factors such as TBP and TF11H and/or by tethering the histone acetylase CBP to promoters. E2F-1 interacts via its transactivation domain with underphosphorylated RB protein and can then repress transcription by a number of mechanisms. It can act directly by quenching the transcriptional activity of E2F so long as the two proteins remain bound. Second, it can quench two activators at the same time – E2F can tether RB to a promoter leaving it free to simultaneously bind to the activation domain of another activator. Third, it can repress activators indirectly by recruiting corepressors such as the histone deactylases (HDAC) which bind RB through its A/B domain. The end result of this is a less-accessible chromatin configuration thereby repressing transcription of a number of genes including cyclin E. A second type of chromatin-modifying enzyme which binds to RB is BRG1 and its related protein BRM. This process of transcriptional repression involves the alteration of higher-order chromatin structure regulated by nucleosome-regulated complexes. BRG1 and BRM are the mammalian homologs of two yeast proteins which comprise the best studied of these complexes. So long as RB remains in its hypophosphorylated state it will be associated with E2F and corepressors such as HDAC. This will maintain the silencing of the gene's need to progress through R. Once RB is phosphorylated in the cell cycle-dependent manner described above, these factors are released allowing initiation of transcription. A model for the control of RB function is shown in *Figure 3.7*.

3.3.3   RB and apoptosis

Evidence from studies on mouse models of RB indicated the role that this protein plays in apoptosis. Mice deficient in RB die during embryogenesis and display extensive apoptosis particularly in certain neurons, hematopoietic tissues, the lens and the liver [22]. The mechanism by which this occurs is still not fully understood but again E2F plays a major role. Ectopic expression of E2F has been shown to lead to the induction of apoptosis in tissue culture cells. E2F knockout mice also showed a decreased rate of thymic apoptosis and equally mice lacking both E2F and RB survive longer than embryos lacking RB alone and display a significant reduction in apoptosis [20,22,23]. Initial evidence has shown that RB-induced apoptosis involves the *p53* tumor suppressor gene (see below). Recent evidence has linked the RB- and p53-dependent apoptotic pathways by the identification of the *ARF* gene as shown in *Figure 3.8*.

3.3.4   RB in human cancers

Analysis of human tumors has shown that there is a wide spectrum of mutations leading to disruption of the RB/E2F pathway. The majority of *RB1* mutations result in truncated unstable proteins which leave E2F-1 free to continually initiate transcription. Interaction of viral oncoproteins with RB also inactivates it, similarly leaving E2F-1 free and promoting cell growth. *RB1* has been found to be mutated in a wide range of tumors and clinical aspects of mutations are discussed more fully in Chapter 5. Some of these mutations, such as those seen in osteosarcomas, were expected as these tumors frequently arise in retinoblastoma patients later in life. The unexpected observation is that mutations in the *RB1* gene can be detected in unrelated tumors including breast, bladder and lung and in fact the role of *RB1* in sporadic tumors is probably greater that the role in hereditary retinoblastomas.

   Much has been learnt about the function of RB over recent years and has shown that it is important in the development of many cancers. In addition, the 'two hit' model for *RB1* and studies on loss of heterozygosity have been central to the work leading to the identification of further tumor suppressor genes in other hereditary cancers.

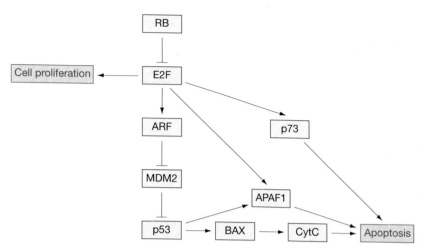

Figure 3.8

Pathway to RB/E2F-induced apoptosis. E2F released following RB phosphorylation results in the transcriptional activation of ARF. Accumulation of ARF blocks MDM2 and hence allows p53 to accumulate leading eventually to apoptosis through targets such as the proapoptotic protein BAX (see also Section 1.4).

3.4   TP53 (p53)

The *TP53* gene (also called *p53*) is located on the short arm of chromosome 17. Much is already known about the gene since its identification over 20 years ago primarily because of its major role in many cancers. Most cancer cells show a defect either in the *TP53* gene or in the pathway leading to *TP53* activation. Mutations in *TP53* are found in over half of sporadic cancers and also in inherited cancer syndromes, primarily Li–Fraumeni syndrome (see Chapter 5). In tumors without mutations, p53 is inactivated by viral proteins interacting with it or more commonly as a result of alterations of components of the p53 pathway [24].

p53 was originally thought to be a tumor antigen as it was found complexed to the SV40 large T antigen. Subsequently mutant *TP53* was shown to co-operate with *HRAS* to transform cells and was therefore believed to be an oncogene. However its ability to abolish the tumorigenic phenotype if transfected into tumor cell lines, the frequent finding that both alleles were inactivated in tumor cells and its association with hereditary cancers led to its reclassification as a tumor suppressor gene.

The p53 tumor suppressor belongs to a small family of proteins which includes two other members, p63 and p73. The proteins show overall structural and sequence homology [25,26] and there is some functional overlap but generally the physiological functions are distinct [27]. p63 and p73 code several isoforms, both have clear roles in normal development and one isoform of p73 at least functions as an oncogene. A number of different roles have been assigned to p53 including cell cycle arrest, differentiation, apoptosis, senescence and angiogenesis. In normal cells, *TP53* and the components of the p53 pathway are switched off and are only activated in response to stress to a cell causing the cell to break in order to allow it to repair damage to DNA or to send the cell into apoptosis. For this reason *TP53* has sometimes been referred to as 'The Guardian of the genome'.

3.4.1   The *TP53* gene and its protein product p53

TP53 is made up of 11 exons and encodes a 2.8 kb mRNA. The protein product of *TP53* is virtually undetectable in normal cells as it has a short half-life of around 20 minutes. It is

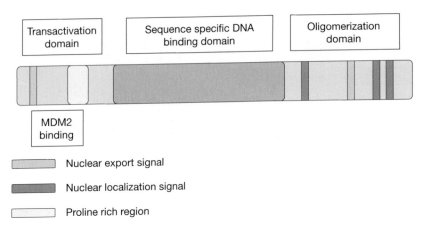

Figure 3.9

Structure of the p53 protein showing the five highly conserved regions. The regions at which protein partners bind to p53 are also indicated.

primarily located in the nucleus but can be detected in the cytoplasm in G_1 and also following DNA synthesis. The p53 protein contains three distinct regions with differing functions (*Figure 3.9*). The amino terminus contains many acidic amino acids and a large number of prolines. This domain is responsible for the transactivational properties of p53. The carboxyl terminus is hydrophilic and has many highly charged residues. This region also contains three nuclear localization signals and mutations in this region can result in a protein which is primarily located in the cytoplasm. The C domain is also responsible for the ability of p53 to form oligomers. NMR studies have described the three-dimensional structure of this region and have shown that two carboxyl-terminal peptides form a tight dimer and in turn these dimer pairs assemble into tetramers [28]. The central region of p53 is highly hydrophobic. It is highly conserved between species and the majority of mutations in human tumors are found here. It is this region which shows DNA binding activity and which interacts with the target sequences in the genes which undergo transcriptional transactivation. X-ray crystallography has shown that this region is folded into β sheets to form a sandwich with protruding residues which make contact with the major and minor grooves in the DNA to which it binds [28]. Mutations in this region either interfere with the protein–DNA interaction or affect the stability of the 3D structure.

3.4.2   Activation of *TP53*

Control of p53 during normal growth and development is essential and as mentioned above levels of p53 in normal cells are extremely low with the levels being maintained primarily through the rate of protein degradation. Integral to the process of maintaining p53 at low levels is the MDM2 protein. This protein targets p53 for degradation by ubiquitin-mediated proteolysis by functioning as a ubiquitin protein ligase, a process which is subject to a feedback loop [29]. p53 binds to the regulatory region of the *MDM2* gene to stimulate its transcription and translation. MDM2 protein in turn then binds to p53 and stimulates addition of ubiquitin to the carboxy terminus leading to p53 degradation. Consequently the level of p53 is lowered, transcription of MDM2 is reduced and in turn levels of p53 rise again. Clearly any abnormalities in either protein will have a dramatic effect on the control of the other. Efficient degradation of p53 also requires export from the nucleus which is dependent on the nuclear export signal regions of the protein.

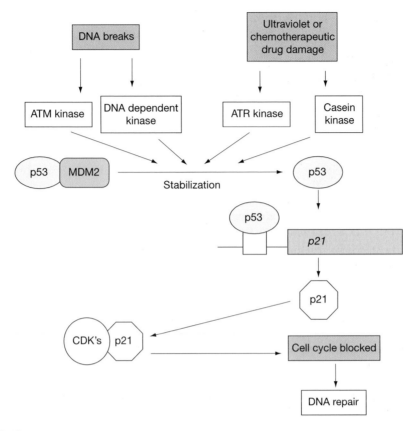

Figure 3.10

Mode of action of p53. Activation of the p53 network following stress results in activation of kinases which phosphorylate p53 resulting in release of MDM2 with concomitant stabilization of p53. p53 binds to p21 resulting in its transactivation. p21 then interacts with the cyclin-dependent kinases which prevent RB phosphorylation and thereby prevent cell cycle progression until damage is repaired.

Two main forms of cellular stress lead to activation of p53 for it to function as a transcriptional activator (*Figure 3.10*) [24]. The first of these, and probably the best studied, is the one initiated by DNA damage which may be caused either by ionizing radiation or by chemotherapeutic drugs, ultraviolet light or protein kinase inhibitors. There are slight differences in which pathway is activated dependent on the mechanism of DNA damage but, in general, the check point proteins which signal to p53 that damage has occurred and that cell cycling should be halted until DNA is repaired, are kinases. These include ATM (ataxia telangiectasia mutated), CHK1 and 2, DNA-PK (DNA-dependent protein kinase), JNK (JUN N-terminal kinase) and casein kinase 1 and 2 [29]. These kinases all phosphorylate p53 at amino terminal sites in the MDM2-binding region of the protein, the effect being to block the interaction of MDM2 and p53 allowing p53 stabilization. Mutations in *ATM* are associated with the inherited condition, ataxia telangiectasia, and mutations in *CHK2* have been found in a few families with Li–Fraumeni. Mutations in these genes therefore mean that they are no longer able to activate the p53 pathway to allow repair to DNA damage which consequently leads to the development of cancers.

The second pathway for activation of p53 occurs in the absence of DNA damage and results from deregulated oncogene expression [24,29]. This provides a fail-safe mechanism to eliminate cells with proliferative abnormalities. The oncogene proteins, e.g. RAS or MYC, stimulate expression of the *ARF* gene, the protein product of which interacts with MDM2 and inhibits its activity, again allowing levels of p53 to increase. The ARF protein also sequesters MDM2 into the nucleolus so that it cannot interact with p53 in the nucleus.

The activity of p53 can also be modulated in other ways apart from allowing an increase in protein levels. These include covalent modifications of the protein in a variety of ways including phosphorylation, dephosphorylation, sumoylation, acetylation and glycosylation. For example, inhibition of DNA binding of p53 is controlled by binding of the C terminal region to the central DNA-binding region. Acetylation of lysine residues or phosphorylation of serine residues in the carboxy terminal interferes with this folding, enhancing DNA binding. In addition, other proteins also bind to p53 and some of these are responsible for additional regulation of stability or activity. Examples include the MDM2 related protein, MDMX, which inhibits p53's transactivation function and, in contrast, there are also transcription cofactors such as p300 which enhance p53-mediated transcription.

Once activated, p53 can have a number of roles but the overwhelming function of this protein is DNA binding and transactivation of nearby genes. The region of p53 responsible for binding to specific sequences of these genes has been clearly defined and mutations in this area of p53 reduce its ability to activate transcription. Many genes which are directly controlled by p53 have now been identified [30] and searches of the human genome have identified over 4000 putative p53-binding sites [25]. Some of the genes proven to be activated by p53 are shown in *Table 3.2* and the few shown here indicate the complexity of p53 pathways. These genes, activated by p53, belong to several groups, the two main ones being cell cycle inhibition genes and genes involved in apoptosis as described below.

3.4.3 Activated p53 responses – cell cycle inhibition

A major effect of activated p53 is to block the cell cycle so that DNA damage can be repaired. A consensus-binding sequence in p53 target genes has been identified and shown to be a 10 bp repeated motif. In 1993, a gene was identified which is specifically induced by an

Table 3.2   p53 target genes and their function

Apoptosis	Cell cycle arrest	Angiogenesis	Autoregulation
APAF1	p21$^{WAF1/CIP2}$	TSP1	MDM2
BAX	CDKN1A	BAI1	TP73
FAS/APO1	14-3-3-σ	MMP2	CCNG1
FDXR	GADD45	MASPIN	
IGF-PB3	p53R2	KAI1	
KILLER/DR5	BTG2		
NOXA			
p53AIP1			
p53DINP1			
PERP			
PIDD			
PIG3			
PIG8/ei24			
PTEN			
PUMA			
WIP1			

Information taken from [25].

excess of wild-type p53 [31]. This gene, which contains the consensus-binding motif, was called *p21*. Transactivation of *p21* in turn leads to inhibition of cyclin-dependent kinases which block cell cycle progression at both G_1 and G_2 [24,29]. As described in Section 3.3.1 and Chapter 4, the cyclin–CDK complexes are responsible for the phosphorylation of RB at different stages of progression through the cell cycle. The next step of the pathway therefore suggests that inhibition of cyclin–CDK complexes by p21 prevents the phos-phor-ylation of RB, a step which is required for progression through the cell cycle (*Figure 3.10*). Degradation of p53 leads to a reduction in p21 levels, phosphorylation of RB and continuation through the S phase of the cell cycle.

A second target of p53 is the protein 14-3-3-σ which also contributes to G_2 arrest in epithelial cells. This protein belongs to a family of proteins involved in signal transduc-tion and cell cycle control which acts by sequestration of cyclinB1–CDC2 complexes. 14-3-3-σ binds to p53 and activates its sequence-specific transactivation again resulting in the blocking of cell cycle progression.

3.4.4 Activated p53 responses – apoptosis

Table 3.2 shows many of the target genes for p53 involved in apoptosis, many of which are likely to be specific for a particular cell system. None of the genes has however been shown to be a pivotal factor in the apoptotic pathway. Although transcription-independent mechanisms are involved in p53-mediated apoptosis, both transcriptional activation and repression also occur [29]. Several p53-activated apoptotic pathways are found in cells including death-activated pathways, mitochondrial pathways and inhibition of survival signals [25].

The first proapoptotic gene to be identified as a target for p53 was *BAX*, followed by others including *NOXA*, *PUMA* and *p53AIP1*. The protein products of these genes are localized in the mitochondria and promote loss of membrane potential and cytochrome C release thereby activating the APAF1/caspase9 cascade (see Section 1.4) [29]. Mitochondrial integrity may also be damaged by a further class of genes termed p53-inducible genes (PIGs), which code for redox potential-controlling enzymes. These reactive oxygen species pro-duced by PIGs cause damage to the mitochondria.

p53 is also implicated in the death receptor-induced pathway and expression of at least two receptors FAS/APO1 and DR5/KILLER are induced by p53.

Loss of survival proteins can also contribute to apoptosis. The antiapoptotic gene, *BCL2*, has been shown to be repressed by p53 and therefore contributes to apoptosis by blocking survival signals mediated by BCL2 [32]. Negative regulation of the IGF pathway by p53 and inhibition of integrin-associated survival signaling also sensitizes cells to apoptosis.

The choice as to whether a cell undergoes apoptosis or cell cycle arrest and DNA repair depends on a number of factors. Some may be independent of p53 such as extracellular survival factors, the existence of oncogenic alterations and the availability of additional transcription factors. However the extent of DNA damage may also contribute to the choice by affecting the level of activity of p53 induced. Activation of apoptosis has been associ-ated with higher levels of p53 than those required for cell cycle arrest which may reflect a lower affinity of cell cycle arrest target gene promoters for p53 [29]. In addition, the type of cell may affect the response to p53. Importantly it is vital to identify why transformed cells die in response to p53 whereas normal cells undergo cell cycle arrest and DNA repair as this may be of great potential for the development of cancer therapies.

3.4.5 Other functions of activated p53

p53 has a number of other functions in the cell. It has a role in several forms of DNA dam-age repair and cells lacking p53 are deficient in both nucleotide excision repair and base excision repair. Normal p53 also stimulates expression of genes which prevent the growth of new blood vessels needed to allow tumor growth [33]. Cells with p53 mutations are therefore more likely to be able to recruit new vessels and give them a growth advantage.

3.4.6 Mutations of p53

Loss of p53 function in tumors can occur by a number of different routes including mutation of the gene itself as well as mutation of genes which regulate p53, examples of which are shown in *Table 3.3*. Records of over 15 600 mutations in *p53* are now stored in the *p53* database (www.iarc.fr/p53/ and www.umd.necker.fr:2001/P53). Mutation of *p53* is found in around 50% of cancers. A total of 74% of these mutations are missense, a much higher proportion that seen in other tumor suppressor genes [34]. Around 95% of mutations in *p53* are found in the central domain of the protein. However much of the early mutation analysis concentrated on exons 5–8 and once more extensive analysis continues, the proportion of mutations outside this region may increase. Many of the *p53* mutations affect a CpG dinucleotide, indicating that the most common cause of mutations is the spontaneous deamination of 5-methylcytosine resulting in C to T transitions. Twenty-eight percent of mutations in *p53* are found at six codons, which in order of frequency are codons 248, 273, 175, 245, 249 and 282. Details of the breakdown of mutations into type, location and distribution between sporadic and inherited cancers, etc., can be found on the *p53* mutation database. The explanation as to how these mutant proteins act is generally believed to be by a dominant negative effect. Mutant p53 forms a tetramer with wild-type protein and it appears that mutant p53 has the ability to drive the wild-type protein into a mutant or inactive conformation [34]. Heteromerization decreases the ability of wild-type p53 to bind to its various targets and transactivate downstream genes (*Table 3.3*). Tumors with point mutations often show loss of heterozygosity of the wild-type allele suggesting that the efficiency of the dominant negative inhibition is not complete but may give an initial selective advantage. There is also evidence that p53 mutants show a gain of function. The involvement of p53 in sporadic and inherited cancers is discussed in subsequent chapters.

Table 3.3   Mutational spectrum of *p53* in tumors

Mechanism of p53 inactivation	Type of tumor	Effect of inactivation
Point mutation to substitute amino acids	Many including breast, colon, lung, esophagus, pancreas, liver	Prevents DNA binding and transactivation
Deletion of carboxy terminus	Occasional tumors	Prevents tetramer formation
Gain of function mutation	T-cell ALL	Interact and inhibit p73 and p63
Deletion of ARF	Colon, breast, lung	MDM2 inhibition lost so p53 not degraded
Amplification of MDM2	Sarcomas, brain	Increased degradation of p53
Viral infection Loss of components of apoptotic cascade, e.g.APAF1	Cervix, liver, lymphomas	Viruses bind and inactivate p53 Decreased apoptosis

3.5 *APC*

The initial identification of the *APC* gene (for adenomatous polyposis coli), came through studies of an inherited colorectal cancer condition, familial adenomatous polyposis (FAP, see Chapter 5). The observation of a cytogenetically detectable deletion in a patient with FAP and mental retardation led to the localization of the gene to chromosome 5q21 by family linkage studies [35]. Subsequently this gene has been shown to have a major role in sporadic colorectal cancer where it is mutated in the majority of cases at an early stage of adenoma development (see Chapter 7).

3.5.1 The *APC* gene and its protein product

A positional cloning strategy was used to identify the causative gene for FAP. The first gene to be identified in the region was the 'Mutated in Colorectal Cancer' gene (*MCC*). This was ruled out as the gene for FAP as no mutations were identified in FAP families, although it was mutated in sporadic cancers. The *APC* gene itself was identified in 1991 and was found to lie in close proximity to *MCC* [36,37]. A coding sequence of 8532 bp was identified, divided into 21 coding exons. Generally however, only 16 exons are referred to in mutation analysis, with the other exons being associated with alternate transcripts in specific tissues. The majority of exons are several hundred base pairs at most, whereas exon 15 is unusually long, encompassing 6.5 kb and comprising over 75% of the coding sequence. The most abundant transcript lacks the smallest exon, exon 10A [38]. Mutations are found distributed throughout the gene and the vast majority of these introduce premature stop codons resulting in the production of a truncated protein [39,40].

The commonest isoform of APC is a 2843 amino acid protein which shows little homology to other known proteins and is localized within the cytoplasm accumulating along the lateral margins or subapical regions of certain cells, particularly surface cells [39]. Since the identification of APC, an increasing number of protein partners have been identified and multiple functions assigned to it, including signal transduction through the WNT signaling pathway, intercellular adhesion, stabilization of the cytoskeleton, regulation of the cell cycle and apoptosis. Various domains can be mapped along its length, through which it interacts with these proteins [39,40] (*Figure 3.11*).

The first domain at the N terminal end of the protein contains a series of heptad repeat sequences, from amino acids 6–57, which are predicted to form α helical structures, capable of homodimerization [41]. Mutant, truncated proteins have been shown to form dimers with wild-type protein demonstrating the potential for dominant negative interference and thereby reducing the tumor suppressor function of the protein.

The region from amino acids 453–767 shows homology to the central repeat region of the *Drosophila* segment polarity protein, armadillo. This domain binds to the APC-stimulated guanine nucleotide exchange factor, Asef. This interaction enhances the binding to RAC, a member of the RHO family of small GTPases, which controls cell adhesion and motility of the actin cytoskeleton network [39,40]. The domain also binds to the regulatory subunit of protein phosphatase 2A (PP2A) an enzyme which also binds to axin via its catalytic subunit. This phosphatase is an antagonist of glycogen synthase kinase 3β (GSK3β)-dependent phosphorylation of β-catenin which marks the latter for degradation.

Two domains in APC have been shown to interact with β-catenin. Three 15 amino acid repeat regions are found between amino acids 1020 and 1169 and an imperfect repeat of 20 amino acid is found seven times between amino acids 1262 and 2033. Both the 15 and 20 amino acid repeat regions are involved directly in β-catenin binding whereas degradation is also associated with the 20 amino acid repeat region [42]. Binding of β-catenin only occurs once the 15 and 20 amino acid repeats have been phosphorylated by GSK3β.

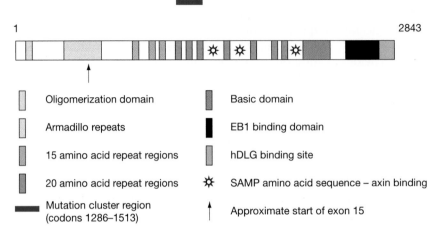

Figure 3.11

Structure of the APC protein showing the various domains and binding sites. The position of mutation cluster region is indicated (codons 1286–1513) along with the approximate position for the start of exon 15.

The 20 amino acid repeat region is also interspersed with SAMP (Ser-Ala-Met-Pro) repeats which form axin-binding sites.

APC has been shown to bind to the microtubular cytoskeleton through the basic domain in the C terminal region between amino acids 2200 and 2400. APC binding has been shown to stabilize microtubules both *in vitro* and *in vivo* [43]. Mutated APC very rarely contains the basic domain, so this interaction will be lost in tumors. A recent finding is the demonstration of APC localization at kinetochores, the microtubule attachment site of chromosomes in mitotic cells and evidence that APC mutants are defective in spindle formation suggests a further role for APC in cell division [44]. The C terminus also contains the binding site for the end binding protein (EB1) which is associated with centromeres and mitotic spindles. Loss of the EB domain does not prevent APC from binding to microtubules but it does so in a very indiscriminate manner. The role of the EB domain therefore appears to be to direct APC to microtubule tips. This feature of APC may be an explanation for the common occurrence of chromosomal aneuploidy seen in colorectal cancers.

The final domain in the C terminus has been shown to be the binding site for hDLG, the human homolog of the *Drosophila* disks large tumor suppressor protein. hDLG is normally found at the site of cell-to-cell contact in epithelial cells and presynaptic nerve termini of the CNS implicating a role for APC in neuronal behavior and function. Evidence also suggests that the APC/hDLG complex is responsible for the association of APC with cell cycle arrest from G_0/G_1 to S phase [45].

3.5.2 Function of the APC protein

Central to the function of APC is its binding to β-catenin, the first protein partner of APC to be identified which is involved in both the WNT signal transduction pathway and in cell adhesion [46]. APC functions as a negative regulator of β-catenin levels and acts in a complex called the destruction complex which is composed of APC, axin, conductin, together with the serine/threonine kinase, GSK3β. Phosphorylation of β-catenin at several serine and threonine residues by this complex, targets the protein for ubiquitin-mediated degradation [47].

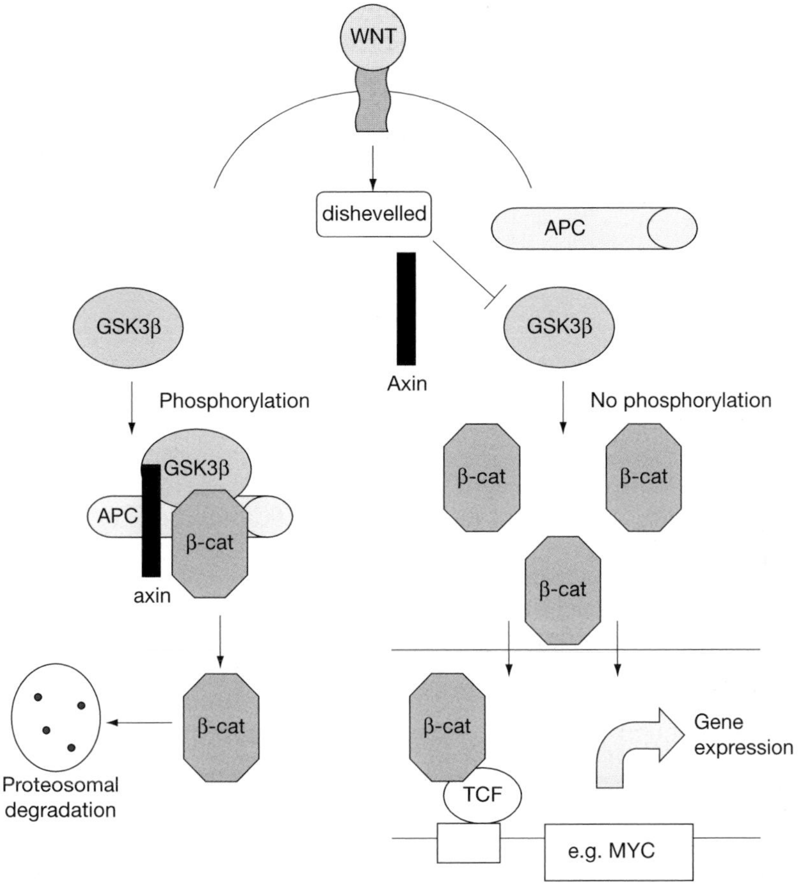

Figure 3.12

Mode of action of APC. Following WNT binding to 'frizzelled', 'dishevelled' prevents GSK3β phosphorylation of β-catenin preventing its degradation and allowing levels to rise. β-catenin then activates transcription of genes involved in cell cycle progression. In the absence of a signal from WNT, phosphorylation of β-catenin occurs leading to its degradation in the proteosome.

This process is controlled via the WNT signaling pathway. The pathway begins with the binding of secreted WNT molecules to its transmembrane receptor, 'frizzled'. This leads to phosphorylation and activation of the cytoplasmic protein 'dishevelled', a protein which forms a complex with axin leading to its dephosphorylation. This prevents axin from complexing with APC which in turn prevents the GSK3β phosphorylation of β-catenin. This represses β-catenin degradation causing its accumulation within the cell. Unphosphorylated β-catenin translocates to the nucleus where it interacts with the transcription factors, T cell factor (TCF) and lymphoid enhancer factor (LEF). These activate transcription of target genes such as *MYC* and *cyclin D1* which are involved in cell cycle entry and progression (*Figure 3.12*). In the absence of signaling from WNT, binding of β-catenin to axin and APC promotes its phosphorylation by GSKβ and targets its degredation, preventing any unwanted transcriptional activation. Mutations in APC therefore prevent β-catenin phosphorylation and result in constitutive signaling, thus leading to unregulated cell cycling. Recent evidence has also demonstrated that constitutive activation of the WNT signaling pathway leads to differentiation defects in embryonic stem cells [48].

The observation that APC binds β-catenin also suggests that it has a role in cell adhesion. This is further confirmed by the observation of APC at the sites of cell–cell adhesion. Three catenins α, β and γ all associate with cadherins which function as cell adhesion molecules. β-catenin interacts with E-cadherin to link the latter to α-catenin. APC and E-cadherin compete for binding sites on β-catenin. Binding of one protein excludes the other though it seems likely that overlapping regions of β-catenin mediate its roles in WNT signaling and cell adhesion. However, a disruption in balance of one pathway is likely to affect the other.

3.5.3   Mutations in *APC*

Germ-line and somatic mutations are found throughout the *APC* gene and several thousand have now been identified. Two mutations, both 5 basepair deletions at codons 1061 and 1309, are found in around 10% of FAP individuals but others may have only been identified on one or a few occasions. The majority, if not all, causative mutations result in a truncated APC protein. However several missense mutations have been identified which increase the risk of colorectal cancer [49]. The role of *APC* mutations in FAP is discussed further in Chapter 5.

Approximately 80% of somatic mutations are found clustered within a region of exon 15 from codons 1286–1513, termed the mutation cluster region (MCR, *Figure 3.11*). Inactivation of both copies of *APC*, either by mutation of each allele or mutation in one copy and deletion of the second is found in both adenomas and carcinomas, defining *APC* as a tumor suppressor gene [50]. Approximately 65% of both adenomas and carcinomas have mutations in *APC* and these have been found in adenomas as small as 3 mm. Loss of the wild-type allele has been demonstrated in 50–60% of colorectal tumors. Mutations in both copies of *APC* appear therefore to be an early step in colorectal carcinogenesis and *APC* has been classified as a 'gatekeeper' gene. In both inherited and sporadic tumors, a correlation has been established between the position of the first mutation and the nature of the second. Mutations around codon 1300 in the MCR are associated with loss of the wild-type allele, whereas first mutations either side of this region are associated with truncating mutations within the MCR in the second allele (see Section 7.4.2) [51].

3.6   *BRCA1* and *BRCA2*

Approximately 1 in 8–12 women will develop breast cancer and around 5% of all cases are due to an inherited predisposition. The autosomal dominant nature of breast cancer in some families has been recognized for over 100 years.

A gene for familial breast cancer was initially mapped to chromosome 17q21 in 1990 and was termed *BRCA1* [52]. An early study indicated the presence of heterogeneity by demonstrating that the majority of families with breast and ovarian cancer were likely to be caused by this gene whereas only a proportion of families with breast cancer alone were *BRCA1* associated [53]. The *BRCA1* gene was isolated in 1994 [54]. In the same year, *BRCA2* was mapped to chromosome 13 [55] and was isolated the following year [56]. Existing evidence suggests that at least one or more genes with similar effects to *BRCA1* and *2* remain to be detected.

3.6.1   The *BRCA1* and *2* genes and their protein products

BRCA1 is composed of 24 exons, 22 of which are coding. Exon 11 alone accounts for approximately 50% of the gene. *BRCA1* encodes a 1863 amino acid nuclear phosphoprotein

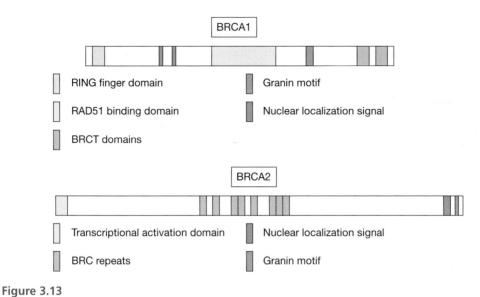

Figure 3.13

Structure of BRCA1 and 2 showing the functional domains.

in which a number of domains can be recognized (*Figure 3.13*). A RING finger motif has been identified at the N terminus which interacts with the BRCA1 associated RING domain protein, BARD1 [57,58]. Tumorigenic missense mutations in this domain abolish the interaction between BRCA1 and BARD1 indicating that BARD1 may be critical in mediating tumor suppression by BRCA1 [57]. The RING domain also interacts with the ubiquitin C-terminal hydrolase, BAP1 (BRCA1-associated protein 1). The C terminal region of the protein contains two BRCT (BRCA C terminal) domains. These repeats are commonly found in proteins involved in DNA repair and cell cycle regulation and have been shown to activate transcription [59]. The central region of the protein contains a RAD51 binding domain. RAD51 has a role in DNA repair and in particular is critical for DNA recombination and double-strand break repair [60]. A granin motif is also found in BRCA1 though the significance of this is controversial. Finally, a nuclear localization signal has been identified and this region has been shown to interact with the importin-α subunit of the nuclear transport signal receptor.

BRCA1 is localized in the nucleus and is widely expressed in many tissues, indicating that gene expression does not account for the tissue-restricted phenotype of breast and ovarian cancer. The transcription of *BRCA1* is induced late in the G_1 phase of the cell cycle and remains elevated during the S phase, which demonstrates a role for BRCA1 in DNA synthesis. In the mouse, loss of both copies of *BRCA1* is embryologically lethal and appears to be caused by a lack of cell proliferation indicating that this protein is essential for early embryonic development.

In the absence of BRCA1 protein, cells become sensitive to radiation damage that primarily causes DNA double-strand breaks. Cells lack both transcription-coupled repair and homology-directed DNA repair [58]. Further evidence for the role of BRCA1 in repair processes came from the observation of the interaction between BRCA1 and RAD51. In meiotic cells, these two proteins are found to colocalize at unsynapsed axial elements of the synaptonemal complex consistent with a role in recombination processes [60]. In mitotic cells the proteins are found together with BARD1 in foci within the nucleus at S phase. These foci disperse on DNA damage, a process that is accompanied by BRCA1 phosphorylation.

BRCA1, RAD51 and BARD1 are then found to localize on replication structures implying an interaction with damaged replicating DNA [61]. Three kinases so far, ATM, the ATM-related kinase, ATR, and CHK2 have all been shown to phosphorylate BRCA1. DNA repair by BRCA1 also involves BRCA2 as described below. Recent evidence suggests that BRCA1 is associated with a large complex named the BRCA1-associated genome surveillance complex (BASC) which is composed of several DNA repair proteins and tumor suppressors including MSH2, MSH6, MLH1, ATM and the RAD50-MRE11-NBS1 complex. It has been suggested that BASC may serve as a sensor of abnormal DNA structures and/or as a regulator of the postreplication repair process [62]. Therefore, the available evidence indicates that *BRCA1* is a 'caretaker' gene, like *p53*, which serves to maintain genomic integrity [63]. When this function is lost, it probably allows for the accumulation of other genetic defects that are themselves directly responsible for cancer formation.

BRCA1 is also involved in transcription regulation [57]. The BRCT domains of BRCA1 have been shown to have transcription-activating potential on their own when fused to reporter genes in yeast. Over the last few years a number of proteins have been shown to interact with BRCA1, all of which are involved in transcriptional activation including p53, RB, MYC, histone deacetylase (HDAC), RNA helicase A and CTIP, a protein which interacts with the transcriptional corepressor, CTBP, (C-terminal binding protein) which is used by a number of transcription factors.

BRCA2 is a 27-exon gene, 26 of which encode a 3418 amino acid protein [55]. Like *BRCA1*, *BRCA2* exon 11 is very large and accounts for almost 50% of the gene. Exon 10 is also large compared to other exons. BRCA2 is normally located in the nucleus and contains phosphorylated residues. The protein has several functional domains which include an N-terminal transcriptional activation domain, a C terminal nuclear localization signal and a central domain containing eight copies of 30–80 amino acid repeats called the BRC repeats encoded by exon 11 [64] (*Figure 3.13*).

There are many similarities between *BRCA1* and *2*. Like *BRCA1*, *BRCA2* is expressed in most tissues and cell types analyzed. *BRCA2* transcription is induced late in the G_1 phase of the cell cycle and remains elevated during the S phase again indicating some role in DNA synthesis. In order to study the function of *BRCA2*, knockout mice homozygous for mutations truncating *BRCA2* from the 5′ end of exon 11 have been created. This results in embryonic lethality characterized by a lack of cell proliferation similar to that seen in *BRCA1* knockouts.

BRCA2 appears to be involved in the DNA repair process. Like BRCA1, it interacts with the RAD51 protein, a key component in homologous recombination and double-strand break repair. This occurs through the first four highly conserved BRC repeats at the 5′ end of exon 11. The four repeats at the 3′ end of the exon are less conserved and do not, or only weakly, bind RAD51 [64]. As mentioned above, BRCA2 also interacts with BRCA1 as well as BARD1 and CTIP in DNA repair processes. *BRCA2*-deficient cells derived from mouse embryos are hypersensitive to radiation and radiomimetics and show poor DNA repair [65]. *BRCA2*-deficient cells are hypersensitive to methyl methanesulfonate treatment, which induces double-strand breaks [65]. *BRCA2* therefore also functions as a 'caretaker' of the genome.

Additional studies have attempted to attribute specific biochemical functions to the *BRCA2* gene product. These include the observation that the N terminal region of BRCA2 is capable of inducing transcription, potentially supporting its role as a modulator of cell growth. This region binds to P/CAF, a transcriptional coactivator with intrinsic histone deacetylase activity which enhances transcription by modifying chromatin as well as several transcription factors.

3.6.2 Mutations in *BRCA1* and *2*

Germ-line mutations are found throughout *BRCA1* and *2* (see Section 5.5.1). Both truncating and missense mutations have been found. Many mutations have only been identified in one or a few families though there are some exceptions with founder mutations having been identified in specific ethnic populations (see Section 5.5.1) [66,67]. Recently evidence of large genomic rearrangements, including both duplications and deletions, have also been identified, the most common of which is the duplication of exon 13 of BRCA1 [68,69]. Most of these are caused by recombination between Alu repeats which are commonly found in BRCA1. In addition a duplicated region at the 5′ end of the BRCA1 gene and containing the BRCA1 pseudogene has also been shown to be another source of illegitimate recombination [70].

Loss of the wild-type allele in familial breast and ovarian tumors has been found for BRCA1 [71] and introduction of BRCA1 into cancer cell lines has been shown to inhibit growth in vitro. Loss of wild-type BRCA2 has also been found in tumors of patients carrying BRCA2 germ-line mutations. These results indicate that both genes can be considered tumor suppressor genes.

In contrast to many other tumor suppressor genes there is no evidence for mutations of the BRCA1 and 2 genes in sporadic breast cancer. Instead, decreased expression of the BRCA1 gene has been shown to be common in sporadic tumors, which may be related to CpG island methylation.

3.7 WT1

The WT1 gene is mutated in Wilms' tumor and in conditions predisposing to this tumor. Wilms' tumor is a relatively common embryonal childhood renal tumor with both a hereditary and a nonhereditary form. Wilms' tumor is also associated with WAGR, an acronym for the complex including aniridia, genitourinary abnormalities and mental retardation, with Denys–Drash syndrome (DDS), Beckwith–Wiedemann syndrome (BWS), Perlman syndrome and finally with a dominantly inherited form of Wilms' tumor which occurs without other congenital abnormalities (see Section 12.3). Cytogenetic analysis of patients with WAGR syndrome showed deletions of chromosome 11 centered around 11p13, an observation which was helpful for the localization of the Wilms' tumor gene, WT1, which is the only cloned Wilms' tumor gene to date [72]. However Wilms' tumor is a genetically complex disorder and a number of other loci have been identified by LOH studies, by linkage and by the observation of chromosome abnormalities in BWS patients suggesting other susceptibility loci [73].

Comparison of constitutional deletions in patients with WAGR narrowed the critical region containing the gene to between the catalase gene and the gene for the follicle-stimulating hormone subunit at 11p13. Many DNA markers were generated in this region, one of which showed a homozygous deletion in a Wilms' tumor. The extent of the deletion was studied by pulsed field gel electrophoresis and candidate genes in the area which showed evolutionary conservation and were expressed in the appropriate tissues, were studied. Finally WT1 was isolated at the centromeric end of the deletion and the aniridia gene, PAX6, was subsequently identified at the more telomeric end.

The WT1 gene encodes multiple protein isoforms generated by alternate splicing [74]. The gene has ten exons and encodes a transcription factor with four zinc finger DNA-binding domains towards the carboxy terminus plus a transactivation domain in the proline/glutamine-rich amino terminal [75]. The WT1 protein shows a high degree of homology in the C terminus with the early growth response 1 gene (EGR1), a gene which is involved in the control of cell proliferation. WT1 has been shown to bind to the EGR1 consensus-binding sequences and a large number of genes with this sequence in their promotors have been suggested as targets for WT1 transcriptional regulation. However whether these genes are all affected by WT1 expression remains to be proven and only in a few examples such as amphiregulin – an epidermal growth factor family member – and BCL2 has this been confirmed [76].

WT1 is expressed in tissues of the developing kidney, the genital ridge, fetal gonad and mesothelium and is therefore involved in genitourinary development [77]. In knockout mice, loss of *WT1* leads to high levels of apoptosis in the metanephric blastema leading to failure of kidney development [78].

There is considerable variation in the phenotype associated with *WT1* mutations as might be expected from the number of conditions associated with it, as described in Chapter 12. In sporadic tumors carrying a mutation in *WT1*, characteristic loss or point mutation of the wild-type allele is seen confirming the tumor suppressor nature of the gene. *WT1* mutations appear to be an early event in tumorigenesis as mutations have been found in nephrogenic rests which are the precursors of Wilms' tumor [79]. However these lesions are not malignant suggesting that additional events are required for tumors to develop.

3.8   *NF1* and *NF2*

Neurofibromatosis1 (NF1) and neurofibromatosis2 (NF2) are genetically distinct conditions in which both benign and malignant tumors develop primarily affecting the nervous system. It has been suggested that as many of the features associated with both conditions are not related to tumor formation that the conditions should not be considered to be cancer predisposition syndromes but rather tumor suppressor gene disorders [80]. Two different genes have been identified as the cause of the two syndromes, *NF1* and *NF2*.

3.8.1   NF1

Neurofibromatosis type 1 is an autosomal dominant condition characterized by multiple neurofibromas, *café au lait* spots, hamartomas in the iris and an increased risk of cancers such as neurofibrosarcomas, astrocytomas and pheochromocytomas (see Section 5.11.1) [80].

The disease was localized to chromosome 17 by linkage analysis and also by the cytogenetic observation of translocations involving 17q11. These translocations were helpful in bringing about the rapid isolation of the *NF1* gene [81,82]. The large gene spans over 350 kb of DNA, produces a transcript of 11–13 kb and has 60 exons. It is widely expressed in a variety of human tissues. Four alternatively spliced NF1 transcripts have been identified.

The protein product of *NF1* is a 2818 amino acid protein called neurofibromin, which localizes to the cytoplasmic microtubules. It shares homology to the catalytic domain of the GTPase-activating protein GAP and as such it catalyzes the conversion of GTP-bound RAS to GDP-bound RAS (see Section 2.2.1) and therefore functions as a negative regulator of RAS-mediated proliferation. Mutations in *NF1* are likely to result in increased levels of active RAS, leading to tumor formation. In addition, neurofibromin has other functions, a feature indicated by the strong sequence conservation in regions outside of the area homologous to GAP. Studies in *Drosophila* have suggested a role for neurofibromin in regulation of protein kinase A signaling [83]. A recently identified domain located upstream of the GAP domain is a cysteine/serine-rich region with three cysteine pairs suggestive of an ATP-binding region, which also contains three potential cAMP-dependent protein kinase recognition sites [84]. The co-incidence of mutations in this region and the high degree of conservation between human and *Drosophila* suggest significant functional relevance of the region.

Over 80% of mutations so far identified are predicted to disrupt or inactivate neurofibromin [84–86]. Over 50% of mutations are *de novo* mutations. Reduced GAP activity is likely to lead to increased levels of active RAS p21 resulting in abnormal signaling through pathways described in Section 2.2.1.

NF1 is generally considered to be a tumor suppressor gene since its inactivation is associated with cell proliferation. LOH on chromosome 17 has been found in neurofibrosarcomas with mutations in the retained allele [87]. Other studies have been carried out on LOH in neurofibromas and have indicated that somatic deletions of *NF1* are present in only a proportion of benign dermal neurofibromas. However more recent work with mouse models has indicated that loss of both alleles is needed for development of plexiform neurofibromas [88].

3.8.2 NF2

NF2 is much less common than NF1 and characterized by bilateral vestibular schwannomas with predisposition to other tumors such as meningiomas, astrocytomas and spinal and cranial schwannomas (see Section 5.11.2) [89]. Cytogenetic analysis of meningiomas and subsequently schwannomas initially suggested that the *NF2* gene was likely to lie on chromosome 22 as this chromosome, or part of it, were frequently lost in these tumors. LOH studies confirmed these findings and eventually the gene was isolated by the discovery of deletions in tumor and constitutional DNA of a gene termed merlin or schwannomin [90].

Merlin, so called because of its similarity to three cell surface glycoproteins of the ERM family (moesin-, ezrin- and radixin-like protein), is a much smaller gene than *NF1* with 17 exons and encodes a protein of 595 amino acids. Merlin has been shown to contain an N-terminal domain which is thought to mediate binding to cell surface glycoproteins, an α helical domain and a C terminal domain which lacks the actin-binding domain found in other ERM proteins. Its function in the cell is to regulate both actin cytoskeleton-mediated processes as well as cell proliferation and operates in a signaling pathway linking the two processes [91].

Many mutations have been found in *NF2*, the majority of which are chain terminating [89]. The tumor suppressor nature of the gene is confirmed by the many studies which have shown LOH on 22q associated with an inactivating mutation in the other copy of *NF2* both in sporadic schwannomas and meningiomas as well as in familial cases [89]. Recent studies have shown a high rate of mutations in tumors other than those associated with NF2 such as malignant mesothelioma. *NF2* loss appears to be a relatively early event in tumorigenesis and again functions as a 'gatekeeper' in a similar manner to that described earlier for *APC*.

3.9 *VHL*

Von Hippel Lindau (VHL) is an autosomal dominant disorder associated with a wide variety of tumors including retinal angiomas, cerebellar, spinal cord and brain stem hemangioblastomas, renal cell carcinoma and pheochromocytoma [92].

The gene for VHL was localized to 3p25-26 [93] and isolated in 1993 [94]. *VHL* is a relatively small gene compared with many of the others described above, with only three exons encoding a 213 amino acid protein. The identification of at least five VHL-binding proteins has led to a clearer understanding of the function of *VHL* [95]. The VHL protein binds to elongin B and C as well as cullin-2 (CUL2), to form a complex which targets specific proteins for poly-ubiquitinylation and subsequent degradation by the proteosome. It also binds to the α subunit of hypoxia-inducible factor (HIF). In the presence of oxygen, HIF-α subunits are ubiquitinylated, which targets them for degradation, a process which is under the direction of VHL. In hypoxic conditions, HIF-α accumulates and dimerizes with the HIF-β subunit to activate transcription of hypoxia inducible genes such as vascular endothelial growth factor (VEGF) and erythropoietin. These proteins are both overproduced in the highly vascular tumors found as part of the VHL phenotype. Mutations in VHL prevent HIF-α degradation therefore leading to expression of the hypoxia-inducible genes without regard to oxygen availability. Finally VHL has been shown to be involved in fibronectin metabolism [95].

Over 150 mutations have been identified in *VHL* in all three exons. Approximately 60% of mutations are nonsense and missense mutations, whereas the remainder are whole gene or partial gene deletions [96–98]. The nature of mutations has been associated with specific phenotypes as discussed in Chapter 5. Founder mutations in *VHL* are uncommon but there are a small number of recurring mutations [96,97].

3.10 *TSC1* and *TSC2*

Tuberous sclerosis (TS) is an autosomal disorder which is characterized by the presence of hamartomas in multiple organs such as the brain, skin and kidneys (see Section 5.13) [99].

Linkage of TS was first reported in 1987 to chromosome 9q34 and the locus was termed *TSC1*. There was immediate evidence of heterogeneity and a second locus, *TSC2*, was identified in 1992 on 16p13.3. *TSC1* proved to be the more difficult gene to isolate as it lay in a very gene-rich region but mutation analysis of a number of candidate genes from within the region led to its eventual identification in 1997 [100]. The gene has an 8.5 kb transcript and has 23 exons, 21 of which are coding. The 1164 amino acid product is termed hamartin. This protein contains a transmembrane domain and a coiled coil region at the C terminus. Hamartin has been shown to bind ERM proteins via the coiled coil region and is involved in cell adhesion and RHO signaling [101]. It may also regulate the activity of the product of *TSC2*.

Identification of an unbalanced chromosome translocation which included a deletion of the short arm of chromosome 16 in a boy with mental handicap, seizures and autism helped in the more rapid isolation of *TSC2*. Positional cloning strategies eventually led to the isolation of the gene and a previously unknown protein, called tuberin, was predicted from the transcript [102]. *TSC2* is a large gene with 41 exons encoding a protein with 1807 amino acids. Tuberin shows a region homologous to the rap1 GTPase-activating protein, GAP3, which stimulates hydrolysis of GTP bound rap1a and b to their GDP-bound forms. Rap1a and b belong to the RAS superfamily of proteins involved in signal transduction as described in Chapter 2. Tuberin also contains a leucine zipper region likely to be involved in dimerization and a number of potential tyrosine kinase phosphorylation sites which may be involved in signaling via proteins with SH2 domains. A role in cell cycle regulation has also potentially been assigned to tuberin. Tuberin and hamartin have been shown to interact to form a complex and explain why deficiency of either protein in hamartomas deregulates cell growth [103]. Recent evidence has also shown that *TSC1* and *2* act in a parallel pathway which converges on the insulin pathway downsteam of *AKT* indicating that they are novel negative regulators of insulin signaling [104].

As is the case with the other tumor suppressor genes discussed in this chapter, mutations are found throughout *TSC1* and *2* which include single base substitutions, insertions and deletions [99]. LOH has been found in a wide variety of tumors including hamartomas, astrocytomas and renal cell cancers. It has however been documented far more frequently in *TSC2* than in *TSC1* which may reflect a lower frequency of *TSC1* disease [99].

3.11 *PTCH*

Gorlin syndrome or basal cell naevus syndrome (BCNS) is a relatively rare, autosomal dominant disorder associated with multiple basal cell carcinomas of the skin, and palmar and plantar pits [105,106] as well as a number of other associated features (see Section 5.12).

The gene was mapped to chromosome 9q in 1992 [107] and was isolated in 1996 [108]. The gene is *PTCH* which contains 23 exons spanning approximately 34 kb. The human PTCH protein is predicted to contain 12 hydrophobic membrane-spanning domains and two large hydrophilic extracellular loops. PTCH is involved in the sonic hedgehog (Shh) signaling pathway, a pathway which is important in vertebrate development but which if left uncontrolled leads to cancer. PTCH is the receptor for Shh and is an inhibitory regulator of the constitutively active transmembrane protein Smoothened (Smo). Binding of the ligand, Shh, to the receptor relieves PTCH repression of Smo. This activates the transcription of downstream genes such as the Gli1 transcription factor, PTCH itself and other target genes [109]. In the absence of Shh, the activity of Smo is negatively regulated by PTCH, preventing activation of Gli1. Mutations in PTCH, as seen in Gorlin syndrome, relieves repression of Smo without Shh binding leading to constitutive signaling.

Multiple mutations are found in the *PTCH* gene in the germ-line of patients with Gorlin syndrome. LOH has been found in basal cell nevi of Gorlin patients confirming *PTCH* as a classic tumor suppressor gene.

3.12 Interaction and differences between oncogenes and tumor suppressor genes

As mentioned in Chapter 2 and alluded to at various points in this chapter, no single genetic event causes tumors. Even retinoblastoma, which has been described as being caused by mutations in a single gene, does not contradict this because at least two hits are required in the gene for the tumor to develop. Other oncogenes, for example, *MYCN*, have also been shown to contribute to the development of retinoblastomas.

Colorectal cancer has been studied perhaps more than any other tumor when investigating the ways in which tumor suppressor genes and oncogenes might interact, and as such serves as a good model for other cancers. Colorectal cancer has several hereditary forms which have given some clues as to the genes involved in the development of sporadic forms of the cancer (see Chapter 7). In addition, it has a well-defined pattern of progression from adenoma to carcinoma. For these reasons, and also because it is one of the commonest forms of cancer, more is known about colorectal cancer than most other sporadic cancers. *Table 3.4* shows the major changes in oncogenes and tumor suppressor genes seen in colorectal cancer [110].

These alterations have been put together to suggest a model for tumorigenesis as described in detail in Section 7.2 [112]. This colorectal model is similar to that described

Table 3.4 Involvement of oncogenes and tumor suppressor genes in sporadic colorectal cancer

Gene	Location	Percentage of adenomas			Percentage of cancers
		Early	Intermediate	Late	
APC	5q21	60	60	60	60
*DCC**	18q	13	11	47	70
KRAS	12q	10	50	50	50
p53	17p	6	6	24	75

Data from [111] and personal communications.
*Note that although this was originally believed to involve the *DCC* gene this is now known not to be the case. See Section 7.3.4.

for other tumors such as small-cell lung cancer (SCLC), breast cancer and melanoma where, again, evidence for interactions between oncogenes and tumor suppressor genes has been observed. Such a model provides us with a framework on which to expand.

References

1. **Harris, H.** (1988) *Cancer Res.*, **48**, 3302.
2. **Macdonald, F. and Ford, C.H.J.** (1997) *Molecular Biology of Cancer*, 1st Edition, BIOS Scientific Publishers, Oxford.
3. **Stanbridge, E.J.** (1986) *BioEssays*, **3**, 252.
4. **Knudson, A.G. Jr.** (1971) *Proc. Natl Acad. Sci. USA*, **68**, 820.
5. **Knudson, A.G.** (1993) *Proc. Natl Acad. Sci. USA*, **90**, 10914.
6. **Yunis, J.J. and Ramsey, N.** (1978) *Am. J. Dis. Child.*, **132**, 161.
7. **Sparkes, R.S., Sparkes, M.C., Wilkson, M.G. et al.** (1980) *Science*, **208**, 1042.
8. **Connolly, M.J., Payne, R.H., Johnson, G. et al.** (1983) *Hum. Genet.*, **65**, 122.
9. **Cavenee, W.K., Dryja, T.P., Phillips, R.A. et al.** (1983) *Nature*, **305**, 779.
10. **Jones, P.A. and Laird, P.W.** (1999) *Nat. Genet.*, **21**, 163.
11. **Dryja, T.A., Rapaport, J.M., Joyce, J.M. and Petersen, R.A.** (1986) *Proc. Natl Acad. Sci. USA*, **83**, 7391.
12. **Friend, S.H., Bernards, R., Rogelj, S. et al.** (1986) *Nature*, **323**, 643.
13. **Dunn, J.M., Phillips, R.A., Becker, A.J. and Gallie, B.L.** (1988) *Science*, **241**, 1797.
14. **Fung, Y.K.T., Murphree, A.L., T'ang, A. et al.** (1987) *Science*, **236**, 1657.
15. **Lee, W.H., Bookstein, R., Hong, F. et al.** (1987) *Science*, **235**, 1394.
16. **Huang, H.-J.S., Yee, J.-K., Shew, J.-L. et al.** (1988) *Science*, **252**, 1563.
17. **Toguchida, J., McGee, T.L., Paterson, J.C. et al.** (1993) *Genomics*, **16**, 535.
18. **Hensey, C.E., Honf, F., Durfee, T. et al.** (1994) *J. Biol. Chem.* **269**, 1380.
19. **Zheng, L. and Lee, W.H.** (2001) *Exp. Cell Res.* **264**, 2.
20. **Classon, M. and Harlow, E.** (2002) *Nat. Rev. Cancer*, **2**, 910.
21. **DiCiommo, D., Gallie, B.L. and Bremner, R.** (2000) *Semin. Cancer Biol.*, **10**, 255.
22. **Hickman, E.S., Moroni, M.C. and Helin, K.** (2002) *Curr. Opin. Genet. Dev.*, **12**, 60.
23. **Nevins, J.R.** (2001) *Hum. Mol. Genet.*, **10**, 699.
24. **Vogelstein, B., Lane, D. and Levine, B.J.** (2000) *Nature*, **408**, 307.
25. **Vousden, K.H. and Lu, X.** (2002) *Nat. Rev. Cancer*, **2**, 594.
26. **Melino, G., De Laurenzi, V. and Vousden, K.H.** (2002) *Nat. Rev. Cancer,* **2**, 605.
27. **Yang, A. and McKeon, F.** (2000) *Nat. Rev. Mol. Cell Biol.*, **1**, 199.
28. **Haffner, R. and Oren, M.** (1995) *Curr. Opin. Genet. Dev.*, **5**, 84.
29. **Balint, E. and Vousden, K.** (2001) *Br. J. Cancer*, **85**, 1813.
30. **Kannan, K., Kaminski, N., Rechavi, G. et al.** (2001) *Oncogene*, **20**, 2225.
31. **El-Deiry, W.S., Tokino, T., Velculescu, V.E. et al.** (1993) *Cell*, **75**, 817.
32. **Hickman, E.S., Moroni, M.C. and Helin, K.** (2002) *Curr. Opin. Genet. Dev.*, **12**, 60.
33. **Hendrix, M.J.** (2000) *Nat. Med.*, **6**, 374.
34. **Sigal, A. and Rotter, V.** (2000) *Cancer Research*, **60**, 6788.
35. **Bodmer, W.F., Bailey, C.J., Bodmer, J. et al.** (1987) *Nature*, **328**, 614.
36. **Groden, J., Thliveris, A., Samowitz, W. et al.** (1991) *Cell*, **66**, 589.
37. **Kinzler, K.W., Nilbert, M.C., Su, L.-K. et al.** (1991) *Science*, **253**, 661.
38. **Xia, L., St. Denis, K.A. and Bapat, B.** (1995) *Genomics*, **28**, 589.
39. **Sieber, O.M., Tomlinson, I.P. and Lamlum, H.** (2000) *Mol. Med. Today*, **6**, 462.
40. **Fearnhead, N.S., Britton, M.J. and Bodmer, W.F.** (2001) *Hum. Mol. Genet.*, **10**, 721.
41. **Su, L.K. Johnson, K.A., Smith, K.J. et al.** (1993) *Cancer Res.*, **53**, 2728.
42. **Rubinfield, B., Souza, I., Porfiri, E. et al.** (1996) *Science*, **272**, 1023.
43. **Mimori-Kiyosue, Y. and Tsukita, S.** (2001) *J. Cell. Biol.*, **154**, 1105.
44. **Fodde, R.J., Kuipers, C., Rosenberg, R. et al.** (2001) *Nat. Cell Biol.*, **3**, 433.
45. **Baeg, G.H., Matsumine, A., Kurodu, T. et al.** (1995) *EMBO J*, **14**, 5618.
46. **Rubinfeld, B., Souza, B., Albert, I. et al.** (1993) *Science*, **262**, 1731.
47. **Polakis, P.** (2000) *Genes and Dev.*, **14**, 1837.

48. **Kielman, M.F., Rindapaa, M., Gaspar, C.** *et al.* (2002) *Nat. Genet.*, **32**, 594.
49. **Frayling, I.M., Bech, N.E., Ilyas, M.** (1998) *Proc. Natl Acad. Sci. USA*, **95**, 10722.
50. **Nagase, H. and Nakamura, Y.** (1993) *Hum. Mutat.*, **2**, 425.
51. **Lamlum, H., Ilyas, M., Rowan, A.** *et al.* (1999) *Nat. Med.*, **5**, 1071.
52. **Hall, J.M., Lee, M.K., Morrow, J.** *et al.* (1990) *Science*, **250**, 1684.
53. **Easton, D.F., Bishop, D.T., Ford, D., Crockford, G.P., Breast Cancer Linkage Consortium** (1993). *Am. J. Hum. Genet.* **52**, 678.
54. **Miki, Y., Swensen, J., Shattuck-Eidens, D.** *et al.* (1994) *Science*, **266**, 66.
55. **Wooster, R., Neuhausen, S.L., Mangion, J.** *et al.* (1994) *Science*, **265**, 2088.
56. **Wooster, R., Bignell, G., Lancaster, J.** *et al.* (1995) *Nature*, **378**, 789.
57. **Wu, L.C., Wang, Z.W., Tsan, J.T.** *et al.* (1996) *Nat. Genet.* **14**, 430.
58. **Lee, J-S. and Chung, J.H.** (2001) *Exp. Rev. Mol. Med.* www.ermm.cbcu.cam.ac.uk/01003131h.htm.
59. **Koonin, E.V., Altschul, S.F. and Bork, P.** (1996) *Nat. Genet.* **13**, 266.
60. **Scully, R., Chen, J., Plug, A.** *et al.* (1997) *Cell*, **88**, 65.
61. **Scully, R., Chen, J., Ochs, R.L.** *et al.* (1997) *Cell*, **90**, 425.
62. **Wang, Y., Cortez, D., Yazdi, P.** *et al.* (2000) *Genes Dev.* **14**, 927.
63. **Zhang, H., Tombline, G. and Weber, B.L.** (1998) *Cell*, **92**, 433.
64. **Chen, P.L., Chen, C.F., Chen, Y.** *et al.* (1998) *Proc. Natl. Acad. Sci. USA*, **95**, 5287.
65. **Arnold, M. and Goggins, M.** (2001) *Exp. Rev. Mol. Med.* www.ermm.cbcu.cam.ac.uk/0100309Xh.htm.
66. **Abeliovich, D., Kaduri, L., Lerer, I.** *et al.* (1997). *Am. J. Hum. Genet.*, **60**, 505.
67 **Thorlacius, S., Olafsdottir, G., Tryggvadottir, L.** *et al.* (1996) *Nat. Genet.*, **13**, 117.
68. **The BRCA1 exon 13 duplication Screening Group** (2000) *Am. J. Hum. Genet.*, **67**, 207.
69. **Montagna, M., Dalla Palma, M., Menin, C.** *et al.* (2003) *Hum. Mol. Genet.*, **12**, 1055.
70. **Puget, N., Gad, S., Perrin-Vidoz, L.** *et al.* (2002) *Am. J. Hum. Genet.*, **70**, 858.
71. **Smith, S., Easton, D.G., Evans, D.G. and Ponder, B.A.** (1992) *Nat. Genet.*, **2**, 128.
72. **Riccardi, V.M., Sujansky, E., Smith, A.C. and Franke, U.** (1978) *Paediatrics*, **61**, 604.
73. **Brown, K.W. and Malik, K.T.A.** (2001) *Exp. Rev. Mol. Med.* www.ermm.cbcu.cam.ac.uk/01003027h.htm.
74. **Haber, D.A., Sohn, R.L., Buckler, A.J.** *et al.* (1991) *PNAS*, **88**, 9618.
75. **Gessler, M., Poustka, A., Cavenee, W.** *et al.* (1990) *Nature*, **343**, 774.
76. **Mayo, M.W., Wang, C.Y., Drouin, S.S.** *et al.* (1999) *EMBO J.*, **18**, 3990.
77. **Van Heyningen, V. and Hastie, N.D.** (1992) *Trends Genet.* **8**, 16.
78. **Kreidberg, J.A., Samola, H., Lorning, J.M.** *et al.* (1993) *Cell*, **74**, 679.
79. **Charles, A.K., Brown, K.W. and Berry, P.J.** (1998) *Am. J. Pathol.*, **153**, 991.
80. **Gutmann, D.H.** (2001) *Hum. Mol. Genet.*, **10**, 747.
81. **Visklochil, D., Buchberg, A.M., Xu, G.** *et al.* (1990) *Cell*, **62**, 187.
82. **Wallace, M.R., Marchuk, D.A., Andersen, L.B.** *et al.* (1990) *Science*, **249**, 181.
83. **Guo, H.F., The, I., Hannan, F., Bernards, A. and Zhong, Y.** (1997) *Science*, **276**, 795.
84. **Fahsold, R., Hoffmeyer, S., Mischung, C.** *et al.* (2000) *Am. J. Hum. Genet.*, **66**, 790.
85. **Shen, M.H., Harper, P.S. and Upadhyaya, M.** (1996) *J. Med. Genet.*, **33**, 2.
86. **Ars, E., Serra, E., Garcia, J.** *et al.* (2000) *Hum. Mol. Genet.*, **9**, 237.
87. **John, A.M., Ruggieri, M., Ferner, R. and Upadhyaya, M.** (2000) *J. Med. Genet.*, **37**, 44.
88. **Zhu, Y. and Parada, L.F.** (2002) *Nat. Rev. Cancer*, **2**, 616.
89. **Evans, D.G.R., Sainio, M. and Baser, M.E.** (2000) *J. Med. Genet.*, **37**, 897.
90. **Trofatter, J.A., MacCollin, M.M., Rutter, J.L.** *et al.* (1993) *Cell*, **72**, 791.
91. **Gutmann, D.H., Haipek, C.A., Burke, S.P.** *et al.* (2001) *Hum. Mol. Genet.*, **10**, 825.
92. **Maher, E.R., Iselius, L., Yates, J.R.W.** *et al.* (1991) *J. Med. Genet.*, **28**, 443.
93. **Maher, E.R., Bentley, E., Yates, J.R.W.** *et al.* (1991) *Genomics*, **10**, 957.
94. **Latif, F., Tory, K., Gnarra, J.** *et al.* (1993). *Science*, **260**, 1317.
95. **Kaelin, W.G.** (2002) *Nat. Rev. Cancer*, **2**, 673.
96. **Chen, F., Kishida, T., Yao, M.** *et al.* (1995) *Hum. Mut.*, **5**, 66.
97. **Zbar, B., Kishida, T., Chen, F.** *et al.* (1996) *Hum. Mut.*, **8**, 348.
98. **Stolle, C., Glenn, G., Zbar, B.** *et al.* (1998) *Hum. Mut.*, **12**, 417.
99. **Cheadle, J., Reeve, M.P., Sampson, J.R. and Kwiatkowski, D.J.** (2000) *Hum. Genet.*, **107**, 97.

100. **Van Slegtenhorst, M., deHoogt, R., Hermans, C. *et al.*** (1997) *Science*, **77**, 805.
101. **Lamb, R.F., Roy, C., Diefenbach, T.J. *et al.*** (2000) *Nat. Cell. Biol.*, **2**, 281.
102. **European Chromosome 16 Tuberous Sclerosis Consortium** (1993) *Cell*, **75**, 1305.
103. **Van Slegtenhorst, M., Nellist, M., Nagelkerken, B. et al.** (1998) *Hum. Mol. Genet.*, **7**, 1053.
104. **Gao, X. and Pan, D.** (2001) *Genes Dev.*, **15**, 1383.
105. **Gorlin, R.J.** (1987) *Medicine*, **66**, 98.
106. **Evans, D.G.R., Ladusans, E.J., Rimmer, S. *et al.*** (1993) *J. Med. Genet.*, **30**, 460.
107. **Farndon, P.A., Del Mastro, R.G., Evans, D.G.R. and Kilpatrick, M.W.** (1992) *Lancet*, **339**, 581.
108. **Johnson, R.L., Rothman, A.L., Xie, J. *et al.*** (1996) *Science*, **272**, 1668.
109. **Bale, A.E. and Yu, K.P.** (2001) *Hum. Mol. Genet.*, **10**, 757.
110. **Oliner, J.D., Kinzler, K., Melitzer, P.S. *et al.*** (1992) *Nature*, **358**, 80.
111. **Cho, K.R. and Fearon, E.R.** (1995) *Curr. Opin. Genet. Dev.*, **5**, 72.
112. **Fearon, E.R. and Vogelstein, B.** (1990) *Cell*, **61**, 759.

Cell cycle control genes and mismatch repair genes

4

4.1 Introduction

A crucial decision in every proliferating cell is the decision to continue with a further round of cell division or to exit the cell cycle and return to the stationary phase. Similarly quiescent cells must make the decision whether to remain in the stationary phase (G_0) or to enter into the cell cycle (see Chapter 1). Entry into the cycle occurs in response to mitogenic signals and exit in response to withdrawal of these signals. To ensure that DNA replication is complete and that any damaged DNA is repaired, cells must pass through specific checkpoints. Tumor cells undergo uncontrolled proliferation either due to mutations in the signal transduction pathways as described in Chapters 2 and 3 or because of mutations in the regulatory mechanism of the cell cycle.

The first half of this chapter deals with further groups of genes intimately involved in the positive and negative regulation of the cell cycle which interact with the oncogenes and tumor suppressor genes and which may in some cases be considered to be such in their own right. The second half deals with genes associated with DNA repair mechanisms.

4.2 Cyclins and cyclin-dependent kinases

Each of the stages of the cell cycle is carefully controlled by the activation or inactivation of the cyclin-dependent kinases (CDKs), a group of serine/threonine kinases which form heterodimeric complexes composed of the CDK which is the catalytic subunit, and their regulatory subunits, the cyclins [1]. The cyclins can influence when their partners, the CDKs, are active by a combination of regulation of cyclin abundance and cyclin localization [2]. However other levels of control exist which include the phosphorylation status of the CDK and the activity of cyclin-CDK inhibitors (CKIs).

Nineteen mammalian cyclins have now been identified (*Table 4.1*), each of which is required at different stages of the cell cycle [1]. The cyclins are broadly classified into G_1

Table 4.1 Mammalian cyclins and cyclin-dependent kinases

Cyclin	CDK	Stage
A1–2	CDC2 (CDK1), CDK2,3	S, G_1, M
B1–3	CDC2 (CDK1)	M
C	CDK8	G_1
D1–4	CDK4,6	G_1
E1–2	CDK2,4,5,6	G_1/S
F	?	G2/M
G1–2	GAK	G_1, S
H	CDK7	throughout
K	?	?
T1–2	CDK9, CDH10	?

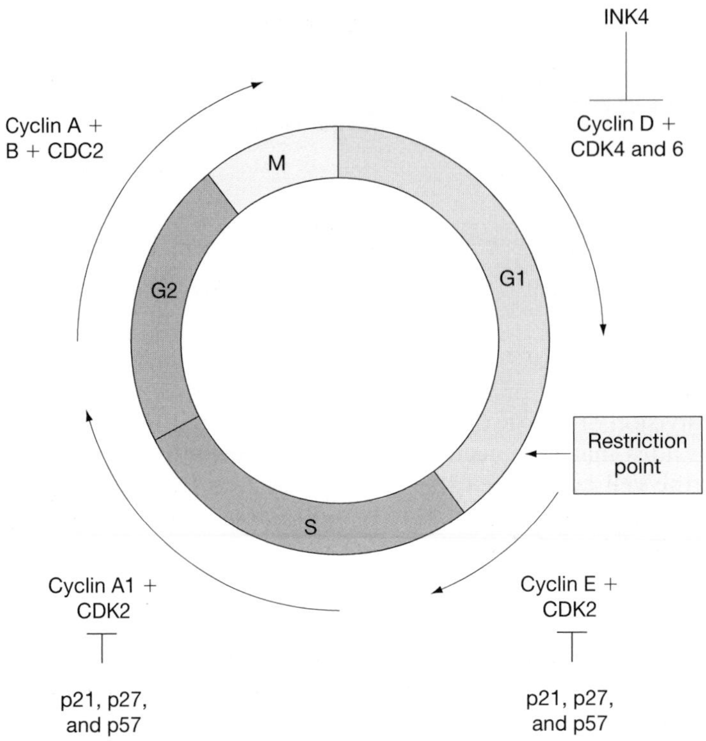

Figure 4.1

The stages of the cell cycle and expression of the cyclins CDKs and their inhibitors, the CKIs.

and mitotic cyclins, according to the stage of the cycle in which they are expressed. They all contain a homologous region of about 100 amino acids called the cyclin box and it is this region of the protein which binds to the cyclin's partner, the CDK [3]. Each cyclin undergoes a characteristic pattern of synthesis and degradation dependent on the stage of the cycle in which it acts (*Figure 4.1*).

The G_1 cyclins are relatively short lived (approximately 30 minutes) and their induction, synthesis and assembly with the CDKs is dependent on continued mitogenic signaling [1]. The mitotic cyclins are degraded during or towards the end of mitosis by the proteosome after ubiquitination by the anaphase-promoting complex (APC). This pattern of degradation of the cyclins is important as it necessarily inactivates the CDKs.

Co-ordination of nuclear import and export is also important in cell cycle control and can be controlled by the protein's nuclear localization sequence. Cyclin D1 is localized to the nucleus during G_1 but translocated to the cytoplasm once DNA replication begins. In the cytoplasm it is targeted for proteolysis by SCF ubiquitin ligase – a multiprotein complex which ligates ubiquitin to proteins due for degradation by the proteosome. This process of nuclear transport is mediated by glycogen synthase kinase-3-β (GSK-3-β), in a process akin to that seen for β-catenin in the WNT signaling pathway (Section 3.5.1) [2,4]. Cyclin A is also localized to the nucleus upon synthesis and remains there until it is degraded in mitosis. Cyclin E is found together with CDK2 in Cajal bodies, subcellular organelles associated with histone gene clusters, at the beginning of the G_1/S boundary. Cyclin B1 is localized to the cytoplasm during S phase and G_2 phase but is translocated to the nucleus at the beginning of mitosis.

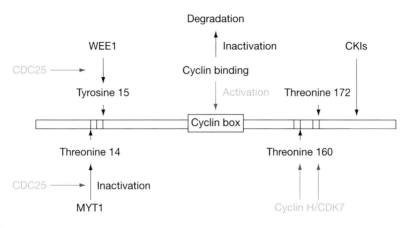

Figure 4.2

Activation and inhibition of the cyclin-dependent kinases showing the amino acids which are phosphorylated and dephosphorylated. Cyclin binds to the cyclin box; the exact position of binding of the CKIs remains unknown. Steps involving activation are shown in purple and those which are inhibitory are shown in black.

CDKs are serine/threonine kinases which can phosphorylate multiple substrates such as RB and p53. Eleven mammalian CDKs have so far been identified (*Table 4.1*). Control of the CDKs comes through binding of their protein partners, the cyclins, which gives them partial activity and by phosphorylation at specific residues on the kinases to fully activate them [2,4,5]. This phosphorylation occurs at a conserved threonine residue within the T loop, a short peptide sequence within the kinase, to fully open its catalytic cleft (*Figure 4.2*). This is carried out by the cyclin H-CDK7 complex, also known as the CDK activating complex (CAK), a serine/threonine kinase involved in transcription and DNA repair [6]. In addition, the action of cyclin/CDKs can also be inhibited by phosphorylation which occurs on threonine 14 or tyrosine 15 and is controlled and maintained by the protein kinases, WEE1 and MYT1, thereby keeping the CDK in its inactive state [2,4]. This inhibition is relieved when a member of the CDC25 family of phosphatases dephosphorylates these residues to trigger entry into mitosis (see below and *Figure 4.2*).

4.3   Cyclin-dependent kinase inhibitors

Additional control of cyclins and CDKs occurs by their association with a group of inhibitor proteins known as the cyclin-dependent kinase inhibitors (CKIs). There are seven different CKIs in mammalian cells, which are grouped into two different classes according to structural similarities, the CIP/KIP family and the INK4 family [7] (*Table 4.2*). They regulate CDK activity either by physically blocking activation or by blocking substrate/ATP access [5].

The CIP/KIP family is comprised of p21 (WAF1), p27 (KIP1) and p57 (KIP2). These inhibitors form heterotrimeric complexes with all members of the G_1/S class of CDKs. However in stochiometric amounts *in vivo*, they only inhibit the cyclin E/CDK2 complexes. Induction of these CKIs occurs in response to a number of cellular processes. The first CKI to be identified was p21, the effector induced by p53 to allow DNA damage induced G_1 arrest (see Section 3.4.3 and below). Tumor cells lacking a functional p53 are unable to induce p21, leading to elevated genetic instability. This inhibitor also associates with PCNA, a subunit of DNA polymerase delta. Both mechanisms inhibit DNA synthesis

Table 4.2 Cyclin-dependent kinase inhibitors

Inhibitor	Target	Chromosomal location	Regulator
p15	CDK4,6	9p21	TGFβ
p16	CDK4,6	9p21	?
p18	CDK4,6	1p32	?
p19	CDK4,6	19p13	?
p21	CDK2,3,4,6	6p21	p53, TGFβ
p27	CDK2,4,6	12p12–13	Rapamycin cAMP
p57	CDK2,3,4	11p15	?

and point to a role for p21 as a major regulator of replication [8]. p27 is induced by a number of antimitogenic signals including cell-to-cell contact, TGF-β stimulation, serum withdrawal and cAMP accumulation, all of which suggest that it plays a contributing role in the decision to commit to the cell cycle or to withdraw [8]. It specifically interacts with cyclin E/CDK2 in the nucleus thereby inhibiting a complex involved in G_1 progression. p27 can be regarded as a potential tumor suppressor gene as it functions as a growth inhibitor and abnormal levels have been detected in many tumors. There is evidence that it is haplo-insufficient for tumor suppression, with loss of one copy of the gene contributing to cancer development [1]. p57 is an imprinted gene which has been associated with Wilms' tumor and also with Beckwith–Wiedemann syndrome (see Sections 3.7 and 12.3), a condition associated with overgrowth and a predisposition to tumors. It has a role in the control of the cell's commitment to cell cycling or withdrawal to the stationary phase.

The second class of inhibitor is the INK4 (Inhibitor of CDK4) family comprised of the ankyrin repeat proteins. The four members include p15, p16, p18 and p19. These inhibitors bind to CDK4 or CDK6 preventing their association with, and activation of, cyclin D [8]. *p16* is a well-recognized tumor suppressor gene associated with many cancers. Interestingly the *p16* gene also encodes a second protein, the tumor suppressor, ARF, which acts in the p53 pathway by interacting with MDM2 to block degradation of p53 [9]. The gene encoding p16 has three exons, 1α, 2 and 3. However, there is an alternative first exon, 1β, lying upstream of 1α, which is also spliced to exons 2 and 3. The initiation codons of exons 1α and 1β are in different reading frames and hence, when spliced to exons 2 and 3, encode completely different proteins, either p16 or ARF. Overexpression of p18 leads to cell cycle arrest and B cell differentiation. It is closely related to p19, overexpression of which results in G_1 arrest by cyclin D1/CDK4 inactivation. The last member of the INK4 family is p15, which is closely related to p16 and like the CIP family member, p27, is induced by TGF-β and also causes G_1 arrest.

4.4 Progression and control of the cell cycle

The cell cycle is controlled by a complex pattern of synthesis and degradation of regulators together with careful control of their spatial organization in specific subcellular compartments. In addition, checkpoint controls can modulate the progression of the cycle in response to adverse conditions such as DNA damage. Cells either enter G_1 from G_0 in response to mitogenic stimulation or follow on from cytokinesis if actively proliferating (i.e. from M to G_1). Removal of mitogens allows them to return to G_0. The critical point between mitogen dependence and independence is the restriction point or R which occurs during G_1. It is here that cells reach the 'point of no return' and are committed to

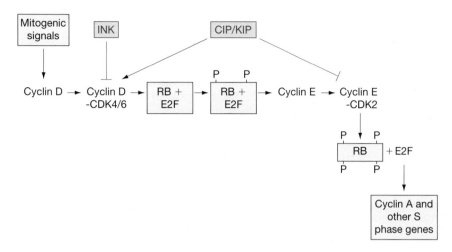

Figure 4.3

Regulation of the G_1 to S transition.

a round of replication. Synthesis of the D-type cyclins begins at the G_0/G_1 transition and continues so long as growth factor stimulation persists. This mitogen stimulation of cyclin D is in part dependent on *RAS* activation, a role which is highlighted by the ability of anti-RAS antibodies to block the progression of the cell cycle if added to cells prior to mitogen stimulation [7]. The availability of cyclin D activates CDK4 and 6 and these complexes then drive the cell from early G_1 through R to late G_1, largely by regulation of RB. As described in Section 3.3, RB exists in a phosphorylated state at the start of G_1 complexed to a large number of proteins [10]. Cyclin D-CDK4/6 activation begins phosphorylation of RB during early G_1. This initial phosphorylation leads to release of histone deacetylase activity from the complex alleviating transcriptional repression. The E2F transcription factor remains bound to RB at this stage but can still transcribe some genes including *cyclin E*. Levels of cyclin E therefore increase and lead to activation of CDK2 which can then complete phosphorylation of RB [11]. Complete phosphorylation of RB then results in the release of E2F to activate genes required to drive cells through the G_1/S transition (*Figure 4.3*).

The CKIs also play a role in control of cell cycle progression at this stage and in response to antimitogenic signals, oppose the activity of the CDKs and cause cell cycle arrest. INK4 inhibitors bind to CDK4/6 to prevent cyclin D binding and CIP/KIP inhibitors similarly inhibit the kinase activity of cyclin E-CDK2. CIP/KIP inhibitors also interact with cyclin D-CDK4/6 complexes during G_1 but, rather than blocking cell cycle progression, this interaction is required for the complete function of the complex and allows G_1 progression. This interaction sequesters CIP/KIP, preventing its inhibition of cyclin E-CDK2 and thereby facilitating its full activation to contribute to G_1 progression. In the presence of an antimitogenic signal, levels of cyclin D-CDK4/6 are reduced, CIP/KIP is released which can then interact with and inhibit CDK2 to cause cell cycle arrest [1,9].

Cells which have suffered DNA damage are prevented from entering S phase and are blocked at G_1. This process is dependent on the tumor suppressor gene *p53* (see Section 3.4.3) and p21. Activation of *p53* by DNA damage results in increased p21 levels which can then inactivate cyclin E-CDK2 to prevent phosphorylation of RB and inhibit release of E2F to promote transcription of genes involved in DNA synthesis. This causes the cell cycle to arrest in G_1. Clearly, loss or mutation of *p53* will lead to loss of this checkpoint control and cells will be able to enter S phase with damaged DNA.

After cells have entered S phase, cyclin E is rapidly degraded and CDK2 is released. In S phase, a further set of cyclins and CDKs, cyclin A-CDK2, are required for continued DNA replication [12]. Two A-type cyclins have been identified to date, cyclin A1 is expressed during meiosis and in early cleavage embryos whereas cyclin A2 is present in all proliferating cells. Cyclin A2 is also induced by E2F and is expressed from S phase through G_2 and M until prometaphase when it is degraded by ubiquitin-dependent proteolysis. Cyclin A2 binds to two different CDKs. Initially during S phase, it is found complexed to CDK2 following its release from cyclin E and subsequently in G_2 and M it is found complexed to CDC2 (also known as CDK1). Cyclin A2 has a role in both transcriptional regulation and DNA replication and its nuclear localization is crucial to its function. Cyclin A regulates the E2F transcription factor and in S phase, when E2F directed transcription is no longer required, cyclin A directs its phosphorylation by CDK2 leading to its degradation. This down-regulation by cyclin A2 is required for orderly S phase progression and in its absence apoptosis occurs. Recently cyclin A as well as cyclin E have been shown to be regulators of centrosome replication and are able to do so because of their ability to shuttle between nucleus and cytoplasm [13].

The final phase of the cycle is M phase, which comprises mitosis and cytokinesis [14] (*Figure 4.4*). The purpose of mitosis is to segregate sister chromatids into two daughter

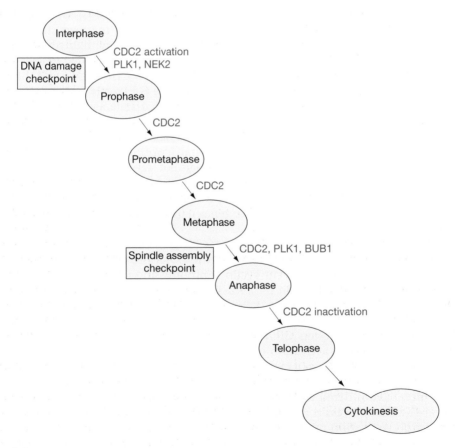

Figure 4.4

The stages of M phase in humans showing the stages of mitosis, the stages at which kinases are involved and the various checkpoints which control mitotic progression.

cells so that each cell receives a complete set of chromosomes, a process that requires the assembly of the mitotic spindle. Mitosis is split into a number of stages which includes prophase, prometaphase, metaphase, anaphase and telophase. Cytokinesis, the process of cytoplasmic cleavage, follows the end of mitosis and its regulation is closely linked to mitotic progression. Mitosis involves the last of cyclin/CDKs, cyclin B1 and CDC2 as well as additional mitotic kinases. These include members of the Polo family (PLK1), the aurora family (aurora A, B and C) and the NIMA family (NEK2) plus kinases implicated at the mitotic checkpoints (BUB1), mitotic exit and cytokinesis [14].

Entry into the final phase of the cell cycle, mitosis, is signaled by the activation of the cyclin B1-CDC2 complex also known as the M phase promoting factor or MPF. This complex accumulates during S and G_2 but is kept in the inactive state by phosphorylation of tyrosine 15 and threonine 14 residues on CDC2 by two kinases, WEE1 and MYT1. WEE1 is nuclear and phosphorylates tyrosine 15 whereas MYT1 is cytoplasmic and phosphorylates threonine 14 (see Figure 4.2). At the end of G_2, the CDC25 phosphatase is stimulated to dephosphorylate these residues thereby activating CDC2 [2]. These enzymes are all controlled by DNA structure checkpoints which delay the onset of mitosis if DNA is damaged. Regulation of cyclin B1-CDC2 is also regulated by localization of specific subcellular compartments. It is initially localized to the cytoplasm during G_2 but is translocated to the nucleus at the beginning of mitosis. A second cyclin B, cyclin B2, also exists in mammalian cells and is localized to the Golgi and endoplasmic reticulum where it may play a role in disassembly of the Golgi apparatus at mitosis.

A further checkpoint exists at the end of G2 which checks that DNA is not damaged before entry into M. Once more p21 activation by p53 can arrest the cell cycle as at the end of G_1. In addition, the CHK1 kinase can phosphorylate CDC25 to create a binding site for the 14-3-3 protein, a process which inactivates CDC25, thereby preventing dephosphorylation of CDC2 and halting the cell cycle [15]. Tumor cells can enter mitosis with damaged DNA, suggesting a defect in the G_2/M checkpoint. Tumor cell lines have been shown to activate the cyclin B-CDC2 complex irrespective of the state of the DNA.

Activation of cyclin B1-CDC2 leads to phosphorylation of numerous substrates including the nuclear lamins, microtubule-binding proteins, condensins and Golgi matrix components which are all needed for nuclear envelope breakdown, centrosome separation, spindle assembly, chromosome condensation and Golgi fragmentation respectively [14]. During prophase, the centrosomes – structures which organize the microtubules and which were duplicated during G_2 – separate to define the poles of the future spindle apparatus, a process regulated by several kinases including the NIMA family member NEK2, as well as aurora A. At the same time centrosomes begin nucleating the microtubules which make up the mitotic spindle. Chromatin condensation also occurs accompanied by extensive histone phosphorylation to produce well-defined chromosomes. Nuclear envelope breakdown occurs shortly after centrosome separation. The nuclear envelope is normally stabilized by a structure known as the nuclear lamin which is composed of lamin intermediate filament proteins. This envelope is broken down as a result of hyperphosphorylation of lamins by cyclin B-CDC2.

During prometaphase, the microtubules are captured by kinetochores, the structure which binds to the centromere of the chromosome. Paired sister chromatids interact with the microtubules emanating from opposite poles resulting in a stable bipolar attachment. Chromosomes then sit on the metaphase plate where they oscillate during metaphase.

Once all bipolar attachment is complete anaphase is triggered. This is characterized by simultaneous separation of all sister chromatids. Each chromosome must be aligned in the center of the bipolar spindle such that its two sister chromatids are attached to opposite poles. If this is correct, the anaphase-promoting complex (APC) together with CDC20 is activated to control degradation of proteins such as securin. This in turn activates the separin protease which cleaves the cohesin molecules between the sister chromatids allowing them to separate [14].

At this stage there is one final checkpoint, the spindle assembly checkpoint, at the metaphase to anaphase transition, which checks the correct assembly of the mitotic apparatus and the alignment of chromosomes on the metaphase plate [14]. The gatekeeper at this checkpoint is the APC complex. Unaligned kinetochores are recognized and associate with the MAD2 and BUB proteins which can prevent activation of APC and cell arrest at metaphase preventing exit from mitosis. In tumor cells abnormalities of spindle formation are found, suggesting that checkpoint control is lost.

Mitotic exit requires that sister chromatids have separated to opposite poles. During telophase, nuclear envelopes can begin to form around the daughter chromosomes and chromatin decondensation occurs. The spindle is also disassembled and cytokinesis is completed. The control of these processes requires destruction of both the cyclins and other kinases such as NIMA and aurora family members by ubiquitin-dependent proteolysis mediated by APC. Daughter cells can now re-enter the cell cycle.

4.5 Cyclins, CDKs and CKIs in cancer

Mutations in cell cycle regulators are a common feature of human tumors. Alterations include overexpression of cyclins such as cyclin D and E and the CDKs such as CDK4 and 6 as well as loss of CKI such as p15 and 16 [4]. Tumor-associated changes in the cell cycle regulators include chromosome alterations such as amplification of *cyclin D1* or *CDK4*, translocation of *CDK6* and deletion of *INK* or *RB*. In addition point mutations are found in *CDK4* and *6*, in *RB* and in *p16*. More recently epigenetic inactivation of *INK4* or *RB* by promotor hypermethylation has been identified. Such alterations are discussed in subsequent chapters. Rapid advances in the last few years have begun to show how disruption of cell cycle processes are involved in tumorigenesis. Components of the cell cycle have consequently begun to be investigated as targets for cancer therapy and a variety of inhibitors of the CDKs have been investigated [4].

4.6 Mismatch repair genes

The importance of mismatch repair (MMR) genes has been highlighted by the discovery that both sporadic and hereditary colorectal cancers show defects in these genes [16]. Hereditary nonpolyposis colon cancer (HNPCC) accounts for around 10–15% of all colorectal cancer (see Chapter 5). The first of six causative genes for this disease was mapped to chromosome 2p21 by linkage studies. A clue to the function of the gene which subsequently helped in its isolation came from studies of the tumors in these patients. When DNA from tumors was compared with DNA from normal tissues, the tumor DNA showed widespread alterations in short repeated sequences – microsatellite repeats (see Section 14.2.5) – distributed throughout the genome [17]. These were seen as additional alleles over and above the usual one or two alleles identified in the normal tissue DNA (*Figure 4.5*). This finding suggested that replication errors, caused by slippage of DNA polymerase, had occurred during tumor development and had not been repaired. This phenotype was termed replication error positive (RER positive) or microsatellite instability. Similarly, RERs were seen in sporadic colorectal cancers as well as other tumors such as endometrial, breast, prostate, lung and stomach [18–20]. The mechanism underlying this observation was suggested by previous studies of bacteria and yeast in which it had been shown that defects in mismatch repair genes resulted in instability of microsatellite repeats.

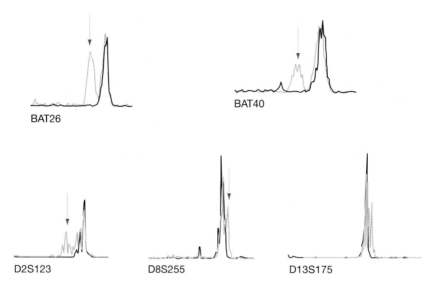

Figure 4.5

Microsatellite instability (MSI) in colorectal cancer. Five microsatellite markers have been used to look for MSI in tumor DNA. Each marker has been analyzed using PCR with fluorescently labeled primers and the products separated by electrophoresis on a DNA sequencer. Alleles detected in normal tissue DNA are shown in black and tumor DNA alleles are shown in purple. Figure courtesy of Yvonne Wallis, West Midlands Regional Genetics Laboratory.

Table 4.3   Mismatch repair genes in bacteria, yeast and humans

MutS homologs		MutL homologs	
E. coli	MutS	E. coli	MutL
S. cerevisiae	MSH1	S. cerevisiae	MLH1
	MSH2		MLH2
	MSH3		MLH3
	MSH4		PMS1
	MSH5		PMS2
	MSH6		
Human	MSH2	Human	MLH1
	MSH3		MLH3
	MSH5		PMS1
	MSH6		PMS2

The best-studied mismatch repair system is the DNA adenosine methylase (DAM)-instructed Mut HLS pathway in *Escherichia coli* (*Table 4.3*) [21]. DNA polymerase mis-incorporation errors occur during DNA replication and will result in the introduction of mismatches at approximately one in every million nucleotides. These are normally repaired post-replication. Mismatch repair in *E. coli* occurs via excision of the newly syn-thesized strand containing the mismatch (*Figure 4.6*) The newly synthesized strand is identified by the lack of methylation (methylation of this strand by the DAM methylase normally occurs post-replication). The mismatch is initially recognized by the MutS protein and is stabilized by the subsequent binding of MutL. MutH is then activated by this binding and the protein, which has single-stranded nuclease activity, nicks the

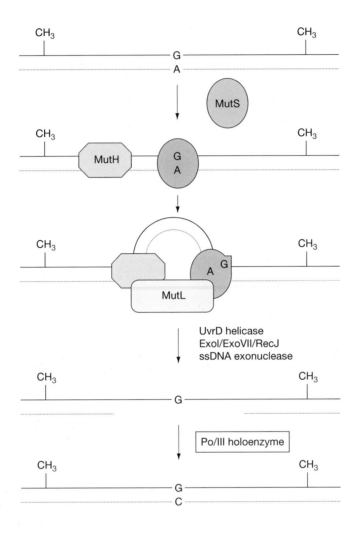

Figure 4.6

Mismatch repair in *E. coli*. The mismatch on the newly replicated strand is recognized initially by the MutS protein. MutL subsequently binds to stabilize the structure and MutH cuts the unmethylated strand. The UvrD protein together with one of the single-stranded exonucleases removes 1–2 kb. The new strand is resynthesized by the *pol*III holoenzyme.

unmethylated strand of DNA. The nicking is followed by the excision step involving the ATP-dependent helicase UvrD and one of the single-stranded exonucleases (*exo*I, *exo*VII or *rec*J) to remove several thousand nucleotides. The DNA is then resynthesized by the polymerase III holoenzyme and finally the DNA nick is sealed by DNA ligase.

Saccharomyces cerevisiae has a similar mismatch repair system to that found in *E. coli* (*Table 4.3*) Six *MutS* homologs have been identified which function in the mitochondria and the nucleus. Mutations in *MSH1* and *2* in particular lead to a high spontaneous mutation rate. *MSH4* and *5* do not function in MMR but play a role in meiosis. There are several *MutL* homologs in yeast, notably *MLH1-3*, *PMS1* and *PMS2*. *MLH1* acts in conjunction with *MSH2* and *PMS1* in the major mismatch repair pathway. *MLH2* does not appear to be involved in MMR but *PMS1* and *2* both appear to play a role. In *S. cerevisiae*, mutations in

Table 4.4 Human mismatch repair genes involved in HNPCC

Gene	Location	Protein size (amino acids)
hMSH2	2p21	934
hMLH1	3p21	756
hPMS1	2q31–33	932
hPMS2	7p22	862
hMLH3	14q24	1453
hMSH6	2p15–16	1360

the mismatch repair genes lead to destabilization of repetitive DNA by up to 700-fold [22]. This finding was a preliminary clue to suggest that the RER phenotype see in colorectal cancers might be caused by defects in MMR and eventually led to the identification of the genes involved in HNPCC.

Six mismatch repair genes have been identified in humans (*Table 4.4*). *hMSH2* lies at 2p21, the location initially identified in HNPCC families by linkage analysis [23]. Linkage analysis suggested the position of a second HNPCC gene at 3p21 and this was cloned by a degenerate PCR approach [24]. Two further human homologs of *MutL*, *hPMS1* and *hPMS2*, were then identified by the same method and mapped to chromosomes 2q and 7p respectively [25]. Although *PMS1* was originally thought to play a role in HNPCC, it has recently been shown that the family in which the *PMS1* mutation was identified also has a *MSH2* mutation and the *PMS1* mutation is likely to be a chance finding. There is currently no role for *PMS1* in HNPCC predisposition [16]. A fifth human mismatch repair gene, *MSH6*, also known as *GTBP* (for G/T binding protein), was identified which forms dimers with *hMSH2* [26]. This gene is estimated to lie within one Mb of *hMSH2* and may have been produced by duplication of the primordial *MutS* gene. Lower levels of MSI have been found in tumors of patients with *hMSH6* mutations compared with the high levels seen in tumors caused by germ-line mutations of the other mismatch repair genes [27]. *hMLH3* was isolated by homology to *hMLH1* with which it was subsequently shown to interact [28].

In humans, the mismatch repair proteins combine as heterodimers with MSH2 being central to the role of mismatch recognition and MSH3 and MSH6 directing the specificity [29]. Single base mispairs are bound by MSH2/MSH6 dimers called hMutSα whereas large insertion–deletion loops are recognized by MSH2–MSH3 complexes called MutSβ. Single nucleotide loops can be recognized by either complex. The recognition of the mismatches leads to the ATP-dependent formation of a repair complex containing MLH1 and PMS2, or occasionally MLH1/MLH3 or MLH1/PMS1, together with proliferating cell nuclear antigen (PCNA), replication protein A (RPA), replication factor C (RFC), one or more exonucleases, a DNA helicase and polymerase δ and ε to repair the DNA. The mismatch repair genes also have a number of other functions which include repair of branched DNA structures, the direction of non-MMR proteins in nucleotide excision and other forms of DNA repair and involvement in meiotic crossing over.

Several studies have indicated how defects in these genes might lead to the development of tumors by showing that loss of these genes leads to a mutator phenotype. In other words, mutations in mismatch repair genes result in a higher than normal mutation rate, allowing the accumulation of mutations in other genes such as *p53* or the adenomatous polyposis coli gene, *APC* [30,31]. Other target genes include those involved in apoptosis and signal transduction [32,33]. In addition, increased mutational inactivation of genes involved in double-strand break repair may contribute to an increase in chromosomal aberrations in MMR-deficient cells [34]. This function of the MMR genes has led to the concept of the genes as 'caretakers' of the genome as opposed to genes such as *APC*, which are considered to be 'gatekeepers'. Caretakers act to suppress growth by ensuring the fidelity of DNA by effective repair of DNA damage and prevention of genomic instability. Loss of the caretaker function leads to cancer by increasing the mutation rate thereby increasing the chances that a gatekeeper gene function will be lost.

As already described in Chapter 3, the tumor suppressor genes are associated with loss of heterozygosity leading to defects in both copies of the gene. Initially no evidence of LOH was found at either 2p21 or 3p21, suggesting that the mismatch repair genes operated by a different mechanism. However it is now clear that these genes also follow a version of the two-hit hypothesis leading to cells which are deficient in mismatch repair. Several pieces of evidence support this. First, analysis of a number of tumors from individuals with germ-line mutations in either *hMSH2* or *hMLH1* demonstrated somatic mutations in the second copy of the gene [25,35,36]. Second, cell lines with a mutation in only one copy of *hMSH2* were found to be proficient in mismatch repair whereas those which were hemizygous for a mutation were repair deficient [37,38]. Finally, loss of the wild-type allele has been shown to occur in 25% of tumors from patients with germ-line mutations in *hMLH1* [39]. LOH has however not been described at the *hMSH2* locus [40].

Hypermethylation of the promotor region of *hMLH1* has also been shown as an alternative mechanism of inactivation of the MMR genes and tumor cells in these cases showed the RER phenotype and lack of MLH1 protein [41,42]. Mutations of *hMLH1* plus promoter hypermethylation may be the major cause of the RER phenotype in sporadic RER-positive colorectal cancers. Again evidence of inactivation by promotor methylation was not seen in *hMHS2*.

The use of the oncogenes, tumor suppressor genes, cell cycle regulators and mismatch repair genes in the diagnosis and in determining the prognosis of some of the commonest cancers is discussed in the following chapters.

References

1. **Sherr, C.J.** (2000) *Cancer Res.*, **60**, 3689.
2. **Pines, J.** (1999) *Nat. Cell. Biol.*, **1**, E73.
3. **Morgan, D.O.** (1995) *Nature*, **374**, 131.
4. **Malumbres, M. and Barbacid, B.** (2001) *Nat. Rev. Cancer*, **1**, 222.
5. **Pavletich, N.P.** (1999) *J. Mol. Biol.*, **287**, 821.
6. **Nigg, E.A.** (1996) *Curr. Opin. Cell. Biol.*, **8**, 312.
7. **Sherr, C.J. and Roberts, J.M.** (1999) *Genes Dev.*, **13**, 1501.
8. **Vidal, A. and Koff, A.** (2000) *Gene*, **247**, 1.
9. **Ortega, S., Malumbres, M. and Barbacid, M.** (2002) *Biochem. Biophys. Acta*, **1602**, 73.
10. **Morris, E.J. and Dyson, N.J.** (2001) *Adv. Cancer Res.*, **82**, 1.
11. **Payton, M. and Coates, S.** (2002) *Int. J. Biochem. Cell. Biol.*, **34**, 315.
12. **Yam, C.H., Fung, T.K. and Poon, R.Y.** (2002) *Cell. Mol. Life Sci.*, **59**, 1317.
13. **Jackman, M., Kubota, Y., den Elzen, N. et al.** (2002) *Mol Biol. Cell*, **13**, 1030.
14. **Nigg, E.A.** (2001) *Nat. Rev. Mol. Cell Biol.*, **2**, 21.
15. **Smits, V.A. and Medema, R.H.** (2001) *Biochem. Biophys. Acta,* **1519**, 1.
16. **Peltomaki, P.** (2003) *J. Clin. Oncol.*, **21**, 1174.
17. **Aaltonen, L.A., Peltomaki, P., Leach, F.S. et al.** (1993) *Science*, **260**, 812.
18. **Ionov, Y., Peinado, M.A., Malkbosyan, S. et al.** (1993) *Nature*, **363**, 558.
19. **Thibodeau, S.N., Bren, G. and Schaid, D.** (1993) *Science*, **260**, 816.
20. **Wooster, R., Clenton-Jansen, A.M., Collins, N. et al.** (1994) *Nat. Genet.*, **6**, 152.

21. **Fishel, R. and Kolodner, R.D.** (1995) *Curr. Opin. Genet Dev.*, **5**, 382.
22. **Strand, M., Prolla, T.A., Liskay, R.M. and Petes, T.D.** (1993) *Nature*, **365**, 274.
23. **Fishel, R., Lescoe, M.K., Rao, M.R.S.** *et al.* (1993) *Cell*, **75**, 1027.
24. **Papadopoulos, N., Nicolaides, N.C., Wei, Y.-F.** *et al.* (1994) *Science*, **263**, 1625.
25. **Nicolaides, N.C., Papadopoulos, N., Liu, B.** *et al.* (1994) *Nature*, **371**, 75.
26. **Papadopoulos, N., Nicolaides, N.C., Liu, B.** *et al.* (1995) *Science*, **268**, 1915.
27. **Wu, Y., Berends, M.J.W., Mensink, R.G.J.** *et al.* (1999) *Am. J. Hum. Genet.*, **65**, 1291.
28. **Lipkin, S.M., Wang, V., Jacoby, R.** *et al.* (2000) *Nat. Genet.* **24**, 2.
29. **Kolodner, R. and Marsischky, G.T.** (1999) *Curr. Opin. Genet. Dev.*, **9**, 89.
30. **Bhattacharya, N.P., Skandalis, A., Ganesh, A.** *et al.* (1994) *Proc. Natl Acad. Sci. USA*, **91**, 6319.
31. **Lazar, V., Grandjouan, S., Bognel, C.** *et al.* (1994) *Hum. Mol. Genet.*, **3**, 2257.
32. **Rampino, N., Yamamoto, H., Ivanov, Y.** *et al.* (1997) *Science*, **275**, 967.
33. **Liu, W., Dong, X., Mai, M.** *et al.* (2000) *Nat. Genet.*, **26**, 146.
34. **Giannini, G., Ristori, E., Cerignoli, F.** *et al.* (2002) *EMBO Rep.*, **3**, 248.
35. **Leach, F.S., Nicolaides, N.C., Papadopoulos, N.** *et al.* (1993) *Cell*, **75**, 1215.
36. **Liu, B., Parsons, R.E., Hamilton, S.R.** *et al.* (1994) *Cancer Res.*, **54**, 4590.
37. **Parsons, R., Li, G.-M., Longley, M.J.** *et al.* (1993) *Cell*, **75**, 1227.
38. **Umar, A., Boyer, J.C., Thomas, D.C.** *et al.* (1994) *J. Biol. Chem.*, **269**, 14367.
39. **Hemminki, A., Peltomaki, P., Mecklin, J.-P.** *et al.* (1994) *Nat. Genet.*, **8**, 404.
40. **Tomlinson, I.P.M., Ilyas, M. and Bodmer, W.F.** (1996) *Br. J. Cancer*, **74**, 1514.
41. **Wheeler, J.M.D., Beck, N.E., Kim, H.C.** *et al.* (1999) *Proc. Natl Acad. Sci.*, **96**, 10296.
42. **Wheeler, J.M.D., Loukola, A. Aaltonen, L.A.** *et al.* (2000) *J. Med. Genet.*, **37**, 588.

Hereditary cancers

5

5.1 Introduction

Genetic factors are likely to be the primary determinants of the cancer seen in around 5–10% of cancer patients. The identification of some of these cancer susceptibility genes has revolutionized our understanding of cancer in general as described in other chapters. However it has also allowed us to improve our management of patients in those families with a genetic predisposition by offering presymptomatic testing and hence early intervention and treatment for the disease. This chapter will describe the clinical usefulness of the tumor suppressor genes whose biological function was described in Chapter 3 as well as give an example of one oncogene which has been associated with familial cancers.

5.2 Inherited cancer syndromes

Inherited cancer syndromes can be recognized by a number of features such as an earlier age of onset than is normal for the cancer, the presence of multiple tumors or bilateral disease, a familial clustering of cancers and segregation of the disease in a mendelian manner. Some of the hereditary conditions and their associated genes were listed in Chapter 3 in *Table 3.1*.

Recognition of a genetic predisposition in a family means that they can be offered screening clinically. However in the majority of these syndromes the disease segregates in an autosomal dominant manner which means that 50% of the offspring will develop the disease and 50% will not. Screening by clinical methods necessarily means that all individuals will have to have regular invasive procedures which can be unpleasant and stressful for the patient as well as costly for the provision of health services. The identification of the genes causing many of the inherited cancers means that we can now adopt a more targeted approach to helping each family. Once a family has been identified as having an inherited predisposition to cancer, the causative gene or genes can be screened and the mutation responsible identified. At-risk family members can then be offered a test to determine if they have inherited the defective gene and if they have, they can then continue to be screened clinically and treated at an early stage before the cancer develops. Those who have not inherited the defective gene need no longer be included in the screening program.

5.3 Procedures for genetic testing

Genetic testing for inherited cancers requires appropriate indications and a testing strategy plus provision of both pre- and post-test counseling. For this reason, it is generally recommended that testing should be carried out within the structure of a genetics center where multidisciplinary teams are available to provide both clinical and scientific expertise [1,2]. This enables a full pedigree of the family to be drawn up, confirmation of diagnoses to be made, information on individual risks to be available to families and support to be on hand when results are given. There are a number of ethical issues which also should be considered before predictive testing is carried out. One of these is the age at which testing should be done. In general, testing should not be carried out on children until they are at an age to understand the process and make their own decision as to whether they want the test [3]. Incorporation of a positive result in a child's medical record could for example also have problems in later life such as with employment and insurance issues. However some conditions, such as familial adenomatous polyposis, develop during adolescence and it may therefore be appropriate to carry out the testing in childhood [1,4,5]. In other conditions such as hereditary nonpolyposis colon cancer, the majority of individuals do not develop the disease until they are in their 20s or 30s so predictive testing can be delayed until individuals reach adult life. Screening options in some inherited cancers may be straightforward, as in hereditary colon cancers, in which a premalignant lesion can be identified at an early stage and appropriate surgery can be carried out. A more difficult issue to address is how to deal with situations for which clinical screening is not proven, e.g. in Li–Fraumeni syndrome. This condition is associated with a wide spectrum of tumors (see below) for which effective screening is not readily available. In this situation, it may be that individuals may prefer to be aware of their risks so that they can plan their life and make informed decisions about issues such as childbearing even though the available clinical screening may not be able to help them directly. Similarly in familial breast cancer, mammography in premenopausal women is not entirely efficient. In addition it is now clear that the penetrance of some genes such as the *BRCA1* and *2* genes associated with familial breast and ovarian cancer is not 100%. Women therefore have to be made fully aware of their options before deciding on a treatment such as prophylactic mastectomy [6]. Finally the option of prenatal diagnosis needs careful consideration for a condition with adult onset and for which treatment such as prophylactic surgery can potentially offer at least a partial cure.

In each family, DNA from an affected individual should always be tested in the first instance to identify the causative mutation. If there is no living affected individual in a family from whom DNA can be obtained, it is still feasible to carry out mutation detection if formalin-fixed paraffin blocks of tissue removed at the time of surgery are available from which DNA can be extracted [7]. Unaffected at-risk individuals should not be tested until a mutation is identified since no method of mutation detection is 100% sensitive. The absence of an identifiable change in an at-risk individual does not therefore necessarily mean that the individual is no longer at risk if the familial mutation has not been identified prior to testing.

5.4 Familial colon cancer

In the USA around 150 000 cases of colorectal cancer are diagnosed each year and it is the cause of death in around 60 000. In the UK, it accounts for around 15 000 deaths per annum and as such the disease remains a major cause of morbidity and mortality in the Western world. A hereditary component will be present in 10–15% of cases of colorectal cancer. The underlying mechanism in many of these remains unknown but in recent years it has been well established for two conditions. These are familial adenomatous polyposis (FAP) and hereditary nonpolyposis colon cancer (HNPCC) which together account for up to 5% of cases of colorectal cancer. Mutations in the causative genes for these two diseases place mutation carriers at a lifetime risk of developing colorectal cancer of 80–100% [8]. Both conditions are inherited in an autosomal dominant manner and thus 50% of the offspring of an affected parent are at risk of developing the disorder. The identification of the genes responsible for these syndromes means that it is now possible to offer patients presymptomatic diagnosis and assist in the early removal of potentially malignant lesions preventing the development of cancer and the early death of the individual.

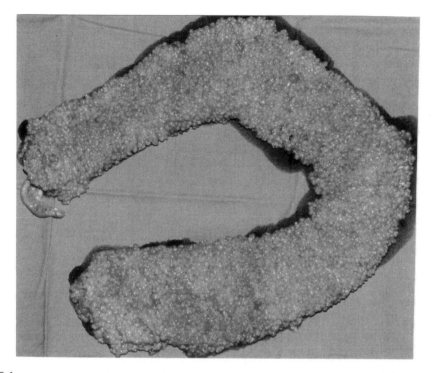

Figure 5.1

Colon of a 12-year-old FAP patient showing multiple polyps throughout the colon.

5.4.1   Familial adenomatous polyposis (FAP)

FAP accounts for around 1–2% of all colorectal cancer and has an incidence of approximately 1 in 10 000. It is characterized by the presence of hundreds to thousands of polyps throughout the colon and rectum and the presence of these features can be used to confirm the diagnosis (*Figure 5.1*). Without surgical intervention it is close to 100% certain that at least one of these polyps will become malignant usually by the third or fourth decade of life. Before the introduction of intensive screening and molecular testing was introduced, cancer was often already present by the time the patient was investigated and frequently led to early death [9]. In addition to the colorectal symptoms, a number of extracolonic features are associated with the disease including osteomas and epidermoid cysts. Gardner syndrome was initially taken to refer to the association of colorectal polyps together with epidermoid cysts and osteomas but it is now accepted to be a variant of FAP. Congenital hypertrophy of the retinal pigment epithelium (CHRPE) is found in approximately 80% of families and, if present, may be a useful indicator of a gene carrier [4]. Desmoid tumors are a significant cause of morbidity and mortality in FAP patients and are particularly found following surgery. Hereditary desmoid disease has also been described which shows autosomal dominant inheritance of desmoids without the occurrence of colorectal polyposis [10]. Upper GI adenomas also occur in some families as well as hepatoblastomas, periampullary cancers, medulloblastomas and thyroid cancers [11]. Attenuated FAP is characterized by the presence of less than 100 polyps, a later onset of carcinoma and usually with a lack of extracolonic features [12]. Approximately 20% of all cases of FAP are due to new mutations.

Table 5.1   Distribution of mutations in *APC*

Exon or segment (of exon 15)	Distribution of mutations on a worldwide database*
Exon 3	8
Exon 4	20
Exon 5	19
Exon 6	27
Exon 7	9
Exon 8	34
Exon 9	19
Exon 10	7
Exon 11	19
Exon 12	10
Exon 13	29
Exon 14	40
Segment 2	399
Segment 3	910
Segment 4	28
Segment 5	9

* = data taken from APC database at
http://perso.curie.fr/Thierry.Soussi/APC.html.

Prevention from early death from colorectal cancer requires reliable screening procedures. The recommended treatment for FAP is generally a total colectomy with ileorectal anastomosis. Subsequent monitoring should then include lifelong surveillance of the rectal stump and regular upper gastrointestinal surveillance. 'At-risk' individuals should be monitored by regular, usually annual, sigmoidoscopy from the early teens until at least 40 years of age [9].

The gene for FAP, the *APC* gene, was isolated in 1991 as detailed in Chapter 3. The isolation of the gene enabled direct mutation analysis of the families to be carried out. Many mutations have so far been identified [13,14] and are spread throughout the coding region of *APC*. The vast majority, >99%, of mutations are truncating mutations caused either by point mutations or small insertions or deletions. The majority of mutations have only been seen in a single family. However two mutations have been found worldwide and account for up to 10% of cases of FAP. These are both 5 bp deletions occurring at codon 1061 and codon 1309. A database of *APC* mutations can be found at http://perso.curie.fr/Thierry.Soussi/APC.html and these are summarized in *Table 5.1*.

No missense mutation has as yet been shown to be the primary determinant of the disease although some have been implicated as predisposing to colon cancer [15,16]. One particular variant, I1307K is found in around 6% of the Ashkenazi Jewish population and carriers are at a several-fold increased risk of developing colorectal cancer compared to the general population. This mutation creates a poly A tract which is hypermutable and undergoes slippage to produce frameshift mutations *in vitro* and may act because of dominant negative influences of mutations in this critical region of *APC in vivo* [15]. However it remains possible that other susceptibility factors outside of the hypermutable tract may contribute to the predisposition to colorectal cancer [17]. A second missense mutation E1317Q has been identified which is also associated with increased risk of colorectal adenomas. This variant appears also to have a dominant negative effect on the APC/β-catenin pathway [16].

Mutation analysis of DNA from an affected individual normally depends on the use of a technique such as SSCP, dHPLC or DGGE (see Chapter 14.2.3) to study the first 14 small exons and PTT is most commonly used to study exon 15 in four overlapping fragments. The two common mutations can be screened for by direct PCR [18]. Even if affected individuals

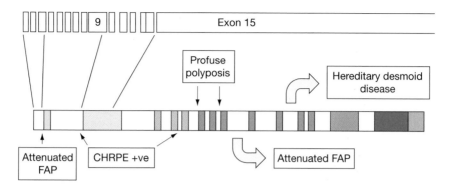

Figure 5.2

Genotype–phenotype correlations in *APC* showing regions in which mutations cause attenuated APC (AAPC), profuse polyposis, CHRPEs or desmoid disease.

are deceased, it is still possible to obtain mutation results as DNA extracted from tumor blocks has been shown to be suitable for PCR analysis [7]. A germ-line mutation has been identified in 20–90% of FAP patients in published studies [13,14]. The variation in pick up rates reflects the extent to which the gene has been analyzed and the technique used, although will also reflect the accuracy of the clinical diagnosis of the patient referred. The majority of analyses have concentrated on searching for point mutations or small insertions or deletions of *APC*. Recent evidence also suggests that large deletions may also contribute to the mutation spectrum and these may increase further the mutation detection rate [19].

Several studies have examined genotype–phenotype correlations in *APC* (*Figure 5.2*) [13,18,20]. An early association was made between the position of mutation and the presence or absence of CHRPE with lesions occurring in those families with mutations located between codons 457 and 1444 [18,21]. Desmoid tumors are primarily, although not solely, associated with mutations at the 3′ end of the gene [22]. Mutations from codons 1250 to 1464 have been associated with a profuse phenotype with more than 5000 polyps present [23] and the commonly seen mutation at codon 1309 is associated with a severe phenotype [24]. Attenuated FAP is associated with mutations near the extreme 5′ or 3′ ends of the gene or in the alternatively spliced region of exon 9 [12]. Mutations in the two ends of the gene result in an unstable protein which is rapidly degraded leaving cells with half of the level of the wild-type protein. Tumors will eventually develop however, due to the decrease in the level of APC but do so at a later age than in classical FAP as there is some functional protein present. However, no genotype–phenotype correlation is absolute and cannot be used clinically to predict the phenotype. For example, a particularly severe phenotype has been seen in two families with mutations in exon 11. In these families, affected individuals were diagnosed as young as 5 years of age and cancer was detected in one individual age 9 years [5]. The correlations can be used however for mutation screening purposes to indicate the region in which to begin scanning [13].

The complex relationship between genotype and phenotype may be even more complex in FAP and additional genes, which could modify the phenotype, have been identified [25]. The first modifier locus for FAP was identified by studies of the MIN mouse – a mouse that carries a truncating mutation in *APC* and consequently develops multiple intestinal polyps [26]. The number of polyps was shown to depend on the genetic background and different crosses showed a significant decrease in polyp number. These differences suggested the existence of a modifier gene which was identified and termed *MOM1* (for <u>M</u>odifier <u>o</u>f <u>Min</u>1). One candidate gene for *MOM1* in humans, *PLA2G2A*, encodes the secreted form of phospholipase A2 [24] although mutation analysis has so far failed to detect any abnormalities, casting doubt on its role. A second modifier locus has recently been identified in further crosses of the mice, termed *MOM2* [27]. Linkage has placed this locus on mouse chromosome 18 which shows synteny with human chromosome 18q, a region of the genome which shows frequent LOH in colorectal tumors. To date no candidate gene has been isolated from this region although a number of genes commonly implicated in colorectal cancer have been excluded [28].

A number of treatment strategies including the use of nonsteroidal anti-inflammatory drugs such as sulindac, starch and aspirin are now under investigation in individuals proven to be *APC* mutation carriers with a view to try to prevent or slow down the development of polyps although results to date show significant differences in efficacy [29,30].

5.4.2 Hereditary nonpolyposis colon cancer (HNPCC)

HNPCC is more common than FAP, accounting for 3–5% of colorectal cancers. It is subdivided into three clinical forms called Lynch type 1, Lynch type 2 and Muir Torre syndromes. As with FAP, it is associated with colorectal cancer but the mean age of onset is later than in FAP at around 45 years of age. Other features include: (1) predominantly right-sided tumors; (2) the presence of synchronous and metachronous tumors; (3) tumors with poor differentiation; and (4) tumors which behave rather more indolently than other colorectal cancers [31]. In Lynch type 2 syndrome, as well as colorectal cancer, there is an increased risk of other cancers such as endometrial, ovarian, stomach, small bowel and hepatobiliary cancers as well as transitional cell tumors of the renal pelvis and ureter (*Figure 5.3*). In

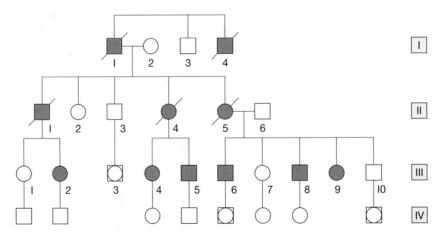

Figure 5.3

Pedigree of a family with HNPCC showing multiple tumors at several sites as well as the colon. This family have Lynch type 2 syndrome. I-1 died from colorectal cancer age 55, II-1 died from colorectal cancer age 60. II-4 developed endometrial cancer at 40 years of age and colorectal cancer at 50, II-5 developed endometrial cancer at 46 years, bladder cancer at 52 years and small bowel cancer at 55 years. In the third generation, III-2 developed endometrial cancer at 36 years and colorectal cancer at 44 years, III-4 had two synchronous colorectal cancers at 45 years which were resected and two further cancers in the remaining colon at 54 years, III-5 had colorectal cancer at 50 years, III-6 also developed colon cancer at 46 years, III-8 was first diagnosed with cancer of the ureter at 49 years and subsequently had two synchronous colorectal tumors at 52, III-9 developed endometrial cancer at 41 years. In the fourth generation there are six individuals at 50% risk of developing HNPCC and four whose parents are so far unaffected who are still at 25% risk.

Muir Torre syndrome all of these cancers are found along with sebaceous adenomas, carcinomas and keratoacanthomas. Turcot syndrome, which is associated with polyps in the colon plus glioblastomas, has been considered to be a separate entity but there is an overlap as some cases have been shown to have mismatch repair mutations and others have *APC* mutations. Treatment for HNPCC-affected individuals is generally subtotal colectomy with ileorectal anastomosis or hemicolectomy with surveillance of the remaining bowel. Because of the high risk of endometrial cancer, prophylactic hysterectomy has also been recommended.

Unlike FAP, HNPCC cannot be distinguished purely on clinical features. It was originally defined by a series of criteria termed the Amsterdam criteria (AC), which were described to facilitate research before the isolation of the causative gene (*Table 5.2*) [32]. However the Amsterdam criteria do not take into account other features of the disease, such as extracolonic cancers, and are now generally considered too strict for assessing families suitable for molecular testing now that the genes have been identified. Less-strict criteria have been identified termed the Bethesda criteria (*Table 5.2*) [33].

As described in Chapter 4, six mismatch repair genes have now been identified which are mutated in families with HNPCC. Mutations in *MSH2* and *MLH1* account for approximately 50% and 40% of mutations, respectively [34], with the majority of the remainder being found in *MSH6* [35] and *MLH3* [36]. *PMS1* and *2* were originally also believed to cause HNPCC but the significance of *PMS1* is now under debate and *PMS2* mutations are found primarily in patients with a Turcot-like phenotype [37]. Distribution and nature of mutations can be found on the HNPCC mutation database (www.nfdht.nl). Truncating mutations, caused by point mutations or small insertions and deletions, are common but, in addition, a large number of missense mutations have also been identified. These are technically the more difficult to deal with as it is not usually clear whether they are the cause of disease or merely polymorphisms. Functional analysis is now beginning to

Table 5.2 Amsterdam and Bethesda criteria for diagnosis of hereditary nonpolyposis colon cancer

Amsterdam criteria	Bethesda criteria
At least three affected relatives with colorectal cancer, one of whom must be a first-degree relative of the other two	Individual with cancer in families meeting Amsterdam criteria
	Individuals with two HNPCC-related cancers, including synchronous and metachronous cancers or associated extracolonic cancers
At least two successive generations must be affected	Individuals with colorectal cancer and a first-degree relative with colorectal and/or HNPCC-related cancers and/or a colorectal adenoma; one of the cancers diagnosed <45 years and the adenoma diagnosed <40 years
Cancer should be diagnosed in one individual under the age of 50 years	
FAP should be excluded	Individuals with colorectal cancer or endometrial cancer diagnosed at age <45 years
	Individuals with right-sided colorectal cancer with an undifferentiated pattern on histopathology diagnosed at age <45 years
	Individuals with signet-ring-cell-type colorectal cancer diagnosed <45 years
	Individuals with adenomas diagnosed at age <40 years

clarify those which can be regarded as pathogenic. Otherwise presymptomatic diagnosis for missense mutations is not recommended [38–40]. Mutations have primarily been found within the coding region of the genes but recently promoter mutations have also been identified [41]. Large genomic deletions also occur in 10–25% of HNPCC kindreds and it is therefore important to screen for these when carrying out mutational analyses [42,43]. There are no common mutations as such in HNPCC although a number of founder mutations have been recognized, particularly in Finland [44], in Newfoundland [45] and in the Ashkenazi Jewish population [46].

Mutation analysis of the mismatch repair genes is usually by a scanning technique such as SSCP or dHPLC [47] (see Chapter 14.2.3) or by direct sequencing of the genes. The protein truncation test (PTT) has also been used but is complicated by the existence of a number of splice variants within the gene [48]. Deletions can be detected by multiplex PCR and dosage techniques [49] or by multiplex ligation-dependent probe amplification (MLPA) [50]. Mutation detection rates of around 50–60% have been reported in AC families but are only found in 10–15% of those which do not fulfil the criteria [51].

Unlike FAP, there are no very clear genotype–phenotype correlations in HNPCC. Some loose associations have however been made. There is an increased risk of extracolonic cancers in individuals with *MSH2* mutations compared to *MLH1* mutations [52]. *MSH6* mutations have been associated with an atypical phenotype occurring in smaller pedigrees and associated with a later age of onset, a predominance of endometrial and ovarian cancer and a lower incidence of microsatellite instability [53]. As might be expected the phenotypes of mutation-positive families differ from those without mutations. Those without mutations show a lower frequency of multiple cancers as well as a lower frequency of extracolonic cancers and a higher age of onset of cancers [54].

As mutation analysis can be very time consuming, several studies have tried to develop a rational, efficient and reasonably cost-effective strategy to screen for mutations in the genes associated with HNPCC. Algorithms such as the Amsterdam criteria can be used to select patients for mutation analysis but these may be too narrow and small families or those with a significant number of extracolonic cancers, not considered by the AC, may be excluded from analysis. Mutations in these families will therefore never be detected if screening is limited to this group [55]. As described in Chapter 4, the majority of HNPCC tumors exhibit the feature of microsatellite instability (MSI). It has therefore been suggested that a consolidation of these two strategies, i.e. combination of family information plus the presence of MSI should be considered to select those individuals to be put into mutation screens [56]. The Bethesda criteria described in *Table 5.2* were developed to select those individuals to be tested for MSI. Those who showed instability could then proceed to mutation screening [33]. Immunohistochemical analysis in combination with MSI analysis can also be used to identify individuals suitable for mutation analysis. Antibodies are now available for mismatch repair proteins and can be used to look for lack of protein indicating a mutation in a specific gene (*Figure 5.4*). This approach also narrows down the gene to be screened for a mutation [57,58]. A development of this approach is to use tissue microarrays to simultaneously screen a number of tumors [58].

Preliminary therapeutic studies for HNPCC families have recently been initiated. One study has so far shown that MSI in colorectal cells carrying a MMR defect is considerably reduced following exposure to sulindac or asprin [59].

5.4.3 Other syndromes predisposing to colorectal cancer

A number of other conditions, predisposing to colorectal cancer, for which a gene has now been isolated include juvenile polyposis syndrome (JPS), Cowden syndrome (CS) and Peutz–Jegher syndrome (PJS) all of which are included in the group of hamartomatous polyposis syndromes.

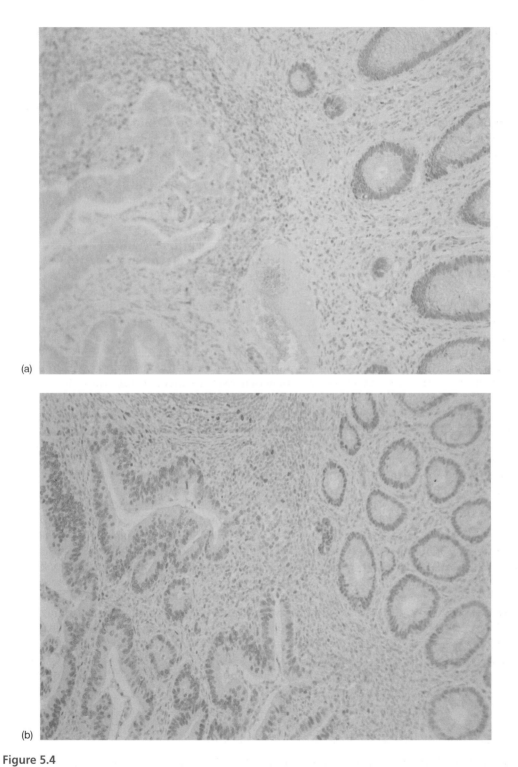

(a)

(b)

Figure 5.4

Immunostaining for MLH1 protein. (a) shows loss of MLH1 staining whereas there is no loss of MSH2 staining in (b). Tumors with loss of antibody staining are those which also exhibit MSI. Figure courtesy of Dion Morton, Department of Surgery, University of Birmingham.

A small proportion of children who develop polyps in their colons have juvenile polyposis which is caused in some cases by mutations in the *SMAD4* gene [60]. This is an 11-exon gene encoding a protein with a highly conserved C terminus which is a cytoplasmic mediator involved in the TGF-β signaling pathway. Loss of the C terminus of SMAD 4 results in loss of TGF-β signaling [61]. A common 4 bp deletion in exon 9 of *SMAD4* has been found in up to 25% of patients with JPS [62]. However JPS is genetically heterogeneous and individuals without *SMAD4* mutations have been found to have mutations in a further gene in the TGF-β pathway, *BMPR1A*, which encodes the bone morphogenetic protein receptor 1A [63]. This is a serine–threonine kinase receptor involved in phosphorylation of the SMAD family proteins. Mutations in this gene have been found in around 40% of individuals with JPS. At least one more gene remains to be identified in JPS.

Cowden syndrome (CS) is a rare syndrome associated with hamartomas in multiple organs including breast, thyroid, skin and gastrointestinal tract. Gastrointestinal polyps are found in 35–40% of patients, breast cancer develops in 25–50% of women with the condition and thyroid cancer occurs in a small proportion. A tumor suppressor gene termed *PTEN* (Phosphatase and Tensin homolog deleted on chromosome 10) was shown to be the gene for CS after mutations were identified in several cases [64]. *PTEN* acts in the PI-3 kinase pathway dephosphorylating phosphatidylinositol 3,4,5 tri-phosphate to phosphatidyl-inositol -di 4,5 phosphate. Mutations in *PTEN* lead to the accumulation of the triphosphate which activates the antiapoptotic protein kinase B, AKT [65]. Germ-line mutations have now been identified in over 80% of cases of CS spread throughout the 9 exon gene although some clustering is seen within exon 5 [66].

Peutz–Jegher syndrome (PJS) is again a rare hamartomatous polyposis syndrome with hamartomas in the intestine and an increased risk of cancer of the intestine, pancreas, breast, ovary, testis, uterus and cervix. The causative gene was mapped to chromosome 19q and was subsequently shown to be the *STK11* (for Serine Threonine Kinase 11) gene also known as *LKB1* [67,68]. As its name suggests, the STK11 protein is a kinase, which functions as a suppressor at the G_0 to G_1 checkpoint of the cell cycle. Mutations have been detected in the majority of cases of PJS with some hot spots in exons 1 and 7 of the 9 exon gene. Mutations result in the loss of the kinase function of the protein. This together with the evidence of LOH in PJS polyps indicates that this kinase functions as a tumor suppressor gene in contrast to other kinases involved in familial cancer syndromes such as *MET* and *RET* which have gain-of-function mutations reflecting their role as oncogenes.

A novel gene predisposing to colorectal cancer has recently been identified through studies of a family with multiple colorectal adenomas and carcinoma in whom germ-line mutations of *APC* or MMR genes had been excluded [69]. Analysis of tumor DNA in this family showed a highly significant increase in G to T transversions in the *APC* gene. Such transversions are caused by oxidative damage to DNA and it was known in *E. coli* that this type of damage is repaired by three enzymes, mutM, mutY and mutT, belonging to the base excision repair pathway. One of the human homologs of the bacterial genes is *MYH*. Germ-line DNA of affected individuals was therefore analyzed for mutations in this gene and patients were shown to be compound heterozygotes for the nonconservative missense mutations, Tyrosine to Cysteine at codon 165, and Glycine to Aspartic acid at codon 382. Follow-up studies in further families with multiple colorectal adenomas confirmed these findings [70,71]. Screening of *MYH* is now recommended in families with multiple adenomas without an *APC* mutation.

5.5 Familial breast and breast/ovarian cancer

Major susceptibility genes account for approximately 5–10% of all breast cancer. In women under 35 years of age, inherited breast cancer accounts for around 35% of cases

Table 5.3 Clinical syndromes associated
with breast cancer

Site-specific breast cancer
Breast/ovarian cancer
Cowden syndrome
Lynch type 2 syndrome
Li–Fraumeni syndrome
Peutz Jegher syndrome
Klinefelter syndrome
Bloom syndrome

although it accounts for <1% of cases diagnosed over 80 years of age. The syndromes associated with inherited breast cancer are shown in *Table 5.3*. In addition, common genetic variants are likely to account for a significant proportion of cases. These include polymorphisms in the *CYP17* gene which encodes a cytochrome 450 enzyme and in the glutathione transferase (GST) family [72].

5.5.1 *BRCA1* and *BRCA2*

As described in Chapter 3, these two genes, accounting for a significant proportion of familial breast cancer, were identified in 1994 and 1995 respectively. *BRCA1* mutations predispose to both breast and ovarian cancer. *BRCA2* is primarily associated with breast cancer and less frequently with ovarian cancer. However *BRCA2* gene carriers are at risk of other cancers including prostate cancer, pancreatic cancer, laryngeal cancer and melanomas [73]. In addition, *BRCA2* is associated with male breast cancer [74].

Screening of individuals with *BRCA1* or *2* mutations includes monthly breast examination and annual mammography from age 25–35. Screening for ovarian cancer is limited as the sensitivity of tumor detection is poor but generally includes ultrasound and measurement of the tumor marker CA125 from age 25–35 [75,76]. Patients with *BRCA1* and *2* mutations can also be offered prophylactic mastectomy and oophorectomy as a method of reducing cancer risk but cancers have still occurred following both procedures [75].

Counseling for those at risk of familial breast cancer is particularly difficult because of the reduced penetrance of *BRCA1* and *2* mutations and the reduced efficacy of clinical screening compared for example to familial colorectal cancer. Risks of developing breast cancer for carriers of a *BRCA1* mutation have been calculated to be 51% by age 50 rising to 85% by 70 years of age. The risk of ovarian cancer for *BRCA1* mutation carriers is approximately 30% by age 60 [77]. The risk of breast cancer for carriers of a *BRCA2* mutation is 60% by age 50 and 71% by age 60 [73,74]. The risk of male breast cancer for *BRCA2* mutation carriers is 6.3% by age 70. Female carriers of *BRCA* mutations have been identified who remain healthy with no evidence of cancer over 80 years of age. This information needs to be made clearly available to patients undergoing predictive testing and who may be considering prophylactic mastectomy and oophrectomy [78].

Several approaches have been used to detect mutations in *BRCA1* and *2*. DNA sequencing is possible and is used in the commercial setting. Most diagnostic laboratories carry out a cascade system of mutation screening. Regions of the genes in which mutations are clustered are screened first. A multiplex heteroduplex analysis has been developed covering four regions of *BRCA1* in which 24% of mutations can be found [79]. Following this, exon 11 of *BRCA1* and exons 10 and 11 of *BRCA2* can be analyzed using the protein truncation test (PTT) [80]. Together these tests would be expected to detect around 50% of all mutations in *BRCA1* and *2*. Further screening of both genes is then usually carried out by SSCP, DGGE or dHPLC.

Over 600 *BRCA1* mutations and 450 *BRCA2* mutations have been identified and are listed on the breast cancer information core (BIC) database (www.nhgri.nih.gov/Intramural_research/Lab_transfer/Bic/). Although some *BRCA1* and *2* mutations listed in the database have been seen on more than one occasion (including founder mutations described below), most are private. In both genes, nonsense and missense mutations have been identified, the latter being problematic as it can be difficult to prove that they are truly causative as was described above for HNPCC. A 6 kb duplication of exon 13 of *BRCA1* has been identified in many different populations and should be screened for routinely [81]. This mutation is found in geographically diverse populations and a large study from 19 countries has shown that the mutation is likely to have been derived from a common ancestor. Other large-scale genomic rearrangements have been reported in recent years and may be found in up to 30% of families making dosage analysis of *BRCA1* and *2* an important part of any mutation testing strategy.

In genetically isolated populations, a number of founder mutations have been found. In the Ashkenazi Jewish population, one mutation, 185delAG, is found in 20% of women with early-onset breast cancer and has a carrier frequency of 0.9% in the Ashkenazi population [82]. This is in contrast to the general population where the prevalence of all *BRCA1* mutations is around 0.2%. Further studies have identified two other common mutations in the Ashkenazi population. 5382insC in *BRCA1* is found at a carrier frequency of 0.13% and 6174delT in *BRCA2* is found at a carrier frequency of 1.52% [83]. In total, 60% of Ashkenazi Jewish women with early-onset ovarian cancer and 30% with early-onset breast cancer carry one of these three mutations [84]. Given the frequency of these mutations in the Ashkenazi population it is not surprising that double heterozygotes have also been found. 995del5 in *BRCA2* has been shown to be a founder mutation in the Icelandic population where, in addition to female breast cancer, it also accounts for 40% of male breast cancer in Iceland detected over the last 40 years [85]. Two *BRCA1* genomic deletions comprising 36% of all families with a *BRCA1* mutation have been shown to be founder mutations in the Dutch population [86]. Two mutations, C4446T in *BRCA1* and 8765delAG in *BRCA2* have been identified as founder mutations in the French Canadian population [87].

Some genotype–phenotype correlations have been described in *BRCA1* and *2*. As already indicated above, the presence of ovarian cancer is more likely to indicate the presence of a *BRCA1* mutation. The segregation of male breast cancer, pancreatic, prostate and laryngeal cancer as well as melanomas within a family is more likely to indicate a *BRCA2* mutation [88]. Within *BRCA1*, mutations in the 3' end of the gene are associated with a lower risk of ovarian cancer (*Figure 5.5*) [89]. If ovarian cancer is seen in families with a *BRCA2* mutation, the causative mutations are clustered in a 3.3 kb region of exon 11 bordered by nucleotides 3035 and 6629 [90]. As in FAP, the correlations are not absolute and can only be used as a guide to target mutation screening.

5.5.2 Other syndromes predisposing to breast cancer

Since the identification of *BRCA1* and *2* two other loci on chromosomes 8p and 13q have been identified as possible locations for *BRCA3* [91,92]. However the causative genes have not yet been isolated and one study has indicated that the locus on 13q is unlikely to be a major breast cancer gene [93].

Li–Fraumeni syndrome (LFS) is a rare autosomal dominant condition characterized by the presence of a range of tumors, including breast cancer (see Section 5.6) [94,95]. It is caused by mutations in the tumor suppressor gene *TP53*. The risk of developing breast cancer before the age of 45 in gene carriers is 18-fold higher than for the general population [96]. Overall, however, *TP53* mutations account for less than 1% of all breast cancers [97].

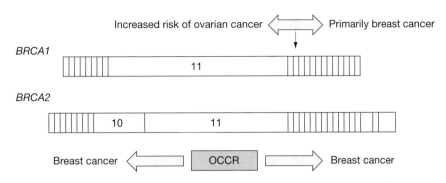

Figure 5.5

Genotype–phenotype correlations in *BRCA1* and *2*. In *BRCA1*, mutations within the first two thirds of the gene predispose to ovarian cancer as well as breast cancer whereas breast cancer alone is mainly found with mutations in the last third of the gene. Mutations within a 3.3 kb region of exon 11 of *BRCA2*, the ovarian cancer cluster region (OCCR), account for ovarian cancer in *BRCA2* families.

CHK2 is a cell cycle checkpoint kinase which phosphorylates p53 and BRCA1 in response to DNA damage [98]. Recent evidence has implicated the 1100delC mutation, a truncating mutation which abrogates the kinase activity of the protein, to be a low-penetrance susceptibility allele for breast cancer [99]. This variant has been found in 1.1% of healthy individuals but in 5.5% of individuals with breast cancer in non *BRCA1* or *2* mutation families. Cancers from patients with *CHK2* mutations have been shown to have reduced immunostaining for the protein in their tumors [100]. Several mutations have been found elsewhere in the gene but compared to 1100delC do not appear to be major contributors to familial breast cancer [101].

Contradictory data on the role of the ataxia telangectasia gene (*ATM*) gene in familial breast cancer exist. Patients with ataxia telangiectasia rarely survive long enough to develop cancers. Carriers of mutations in *ATM* have been described as having an increased risk of breast cancer with an odds ratio of 3.3–8 [72]. However others have found no such increased risk. The explanation for the opposite findings has been ascribed to variations in the function of different classes of *ATM* mutations [102]. Truncating mutations are unstable with low levels of abnormal protein present in cells but carriers of these mutations will still have 50% of ATM protein present from the normal allele and consequently have a normal phenotype with no increased breast cancer risk. Missense mutations result in a stable but functionally abnormal protein which complexes with the wild-type protein in a dominant negative manner and consequently has a more severe effect on the phenotype than the truncating mutations. A recent study of familial breast cancer patients without *BRCA1* or *2* mutations found mutations in *ATM* in over 8% of cases suggesting that there is a role for this gene in hereditary breast cancer [103].

As described in Section 5.4.3, the hamartomatous polyposis syndromes, Cowden disease and Peutz–Jeghers syndrome are both associated with an increased risk of breast cancer as well as the GI symptoms. However mutations in *PTEN* and *STK11* are rare in breast cancers other than in these conditions.

5.6 Li–Fraumeni syndrome (LFS)

This condition predisposes to a wide range of cancers in addition to breast cancer as discussed above including brain tumors, soft tissue sarcomas, osteosarcomas, leukemias and adrenocortical tumors. [94,95]. The condition is defined by very strict clinical criteria [95] but a group of Li–Fraumeni-like families has also been described [104]. More than 90% of gene carriers are expected to develop tumors by age 70 [104]. As already indicated it is caused by mutations in the tumor suppressor gene, *TP53*. Mutations are located primarily in exons 5–8 although they have increasingly been found in other exons, as mutation scanning has been extended [105]. Mutations in *TP53* have been found in 70% of all LFS families and 20% of Li–Fraumeni-like families when the entire gene including noncoding regions is sequenced [106]. More recently mutations in the *CHK2* gene have also been identified in Li–Fraumeni families [107,108].

Although identification of mutations in *TP53* or *CHK2* allows presymptomatic diagnosis to be offered to at-risk relatives, the lack of effective screening for the majority of associated tumors make testing for this condition very complex and ethically difficult.

5.7 Multiple endocrine neoplasia

The endocrine neoplasia syndromes are categorized by the presence of germ-line mutations which predispose hormone-secreting cells to become malignant. Mutations may be either loss of function mutations in a tumor suppressor gene such as seen in multiple endocrine neoplasia type 1 or, unusually for inherited cancers, gain-of-function mutations in an oncogene as seen in multiple endocrine neoplasia type 2.

5.7.1 Multiple endocrine neoplasia type 1 (MEN1)

MEN1 is an autosomal dominant condition associated with malignancies of the endocrine tissues which includes the parathyroids in over 90% of cases, the endocrine pancreas in 30–80% of cases and more rarely, the anterior pituitary [109]. The quoted incidence is quite variable from 1 in 10 000 to 1 in 100 000. There is considerable inter- and intra-familial variation in the phenotype. Biochemical screening can detect disease approximately 10 years before patients are symptomatic. Screening of 'at-risk' individuals is therefore important as many of the complications of the condition can be treated. This should begin around 10 years of age and is continued at 5-yearly intervals to include measurement of serum calcium levels, parathyroid hormone levels, pituitary hormone levels, and pancreatic hormones as well as routine clinical investigations. Treatment for tumors is normally surgical.

The gene was initially mapped to chromosome 11q13 [110] but it took a further 9 years before the gene was eventually isolated [111]. It has 10 exons, the first of which is noncoding and encodes a 610 amino acid protein called menin which is located primarily in the nucleus. Menin has been shown to interact with the AP1 transcription factor, JunD, which is an inhibitor of cell growth [112]. Mutations have been found in exons 2–10 and are primarily nonsense mutations or insertions and deletions predicted to produce a truncated protein resulting in the loss of function of menin [113,114]. Germ-line mutations have been found in up to 90% of cases of familial MEN1 as well as in the majority of sporadic cases. At the present time there are no genotype–phenotype correlations. Once individuals have been identified as gene carriers, clinical and biochemical evaluation can be commenced after the age of 5 years as described. This will enable early detection of tumors and thereby allow appropriate management of the patient.

5.7.2 Multiple endocrine neoplasia type 2 (MEN2)

MEN2 is characterized by the presence of medullary thyroid cancer (MTC) in association with pheochromocytoma. There are three clinical variants, MEN2A, MEN2B and familial medullary thyroid cancer (FMTC). MEN2A is the most common variant and is associated with pheochromocytoma and parathyroid tumors as well as MTC. In MEN2B parathyroid involvement is rare. FMTC is characterized by the presence of MTC alone [115]. All forms may be inherited as an autosomal dominant disorder but can also occur sporadically. The frequency of the inherited form of the disease is 1 in 25 000 to 1 in 30 000. Tumors associated with the disorders can be detected by biochemical screening and this can result in early diagnosis and hence reduce morbidity and mortality. Clarification of the status of 'at-risk' individuals can be difficult so the use of a DNA test to identify gene carriers is valuable.

The gene for MEN2 was mapped to the pericentromeric region of chromosome 10 in 1987 [116] and initially it was assumed that the gene responsible would be a tumor suppressor gene as is the case for the majority of other familial cancer syndromes. However the candidate gene based on genetic and physical mapping studies was the *RET* gene, which is a transforming oncogene. This gene is made up of at least 21 exons spanning 55 kb of genomic DNA and as described in Section 2.2.1 encodes a transmembrane receptor kinase. Mutations are identified in *RET* in cases of MEN2A, 2B and FMTC [117,118]. In MEN2A, gain-of-function mutations in five cysteine residues of the extracellular domain of *RET* are found in over 98% of cases, with mutations at codon 634 being present in 83% [119]. A strong genotype–phenotype correlation was found between the mutation at codon 634 and the presence of pheochromocytoma [119]. Mutations in non cysteine residues at codons 768, 790, 791, 804 and 891 located in the intracellular domain have been found at lower frequencies and are predominantly seen in FMTC. Two mutations are exclusively associated with the MEN2B phenotype. A single missense mutation at codon 918 in which methionine is replaced by threonine is found in over 95% of cases and alanine is replaced with phenylalanine at codon 883 in a further 3% of cases. The mutations encode the catalytic site of RET and may therefore alter substrate specificity.

As mutations in MEN 2 are clustered, it is relatively easy to identify them. It has therefore been easy to establish screening programs into which patients can be entered so that treatment can be initiated at an early stage.

5.8   Von Hippel Lindau syndrome (VHL)

This is an autosomal dominant disorder with an incidence of around 1 in 35 000 [120]. It is a multisystem disorder with affected individuals developing a wide variety of tumors including retinal angiomas, cerebellar, spinal cord and brain stem hemangioblastomas, clear cell renal cell carcinomas (RCC), pheochromocytomas and occasionally extra-adrenal paragangliomas [120]. The condition has been subdivided into three; type 1 is associated with renal cell cancer and CNS hemangioblastomas but no pheochromocytoma, type 2 has all of type 1 tumors plus pheochromocytoma. Type 2 can be further subdivided into type 2A which has a low risk of RCC, type 2B which has a high risk of RCC and type 2C which has just pheochromocytoma. The penetrance of the disease is around 90% by 60 years of age. Systematic screening and early detection of tumors have been shown to reduce morbidity and mortality [121]. Screening of both affected and at-risk individuals is fairly intensive involving annual examinations, direct and indirect ophthalmoscopy, MRI scans at 3-yearly intervals, annual renal ultrasound scans and annual 24 h urine collection for VMAs [121].

Over 150 mutations have been identified in VHL in all three exons (www.bham.ac.uk/ICH/vhl.htm and www.umd.necker.fr.2005). Approximately 60% of mutations are nonsense and missense mutations whereas the remainder are whole gene or partial gene deletions [122,123]. A combination of direct sequencing for detection of nonsense and missense mutations plus either Southern blotting, quantitative fluorescent PCR analysis or MLPA for detection of large rearrangements is capable of detecting close to 100% of mutations (123). Many mutations identified are different in different families though a few common mutations are seen such as that at codon 238.

Some genotype–phenotype correlations do exist. A major phenotypic difference between VHL families is the presence or absence of pheochromocytomas. Nearly all families without pheochromocytomas (type 1) have deletions or nonsense mutations. The presence of pheochromocytomas is associated in >90% of cases with missense mutations, particularly the mutation at codon 167 [124]. This finding suggests that a mutant protein must be full length to produce pheochromocytomas. The tyrosine to histidine mutation at codon 169 has been associated with a highly penetrant form of VHL without a significant risk of increased mortality compared to the general population [125]. Up to 50% of families with pheochromocytoma only have been shown to have mutations in VHL and hence mutation screening in this group is indicated [126]. Germ-line mutations have also been found in up to 5% of isolated cases of hemangioblastoma presenting at less than 50 years of age indicating that VHL mutation screening should also be offered to this group of patients [127].

5.9 Familial renal cell cancer

Hereditary renal cell carcinoma is a dominantly inherited form of renal cancer characterized by the predisposition to multiple papillary renal tumors. It is genetically distinct from VHL and from renal cell cancers caused by translocations between chromosomes 3 and 8. The causative gene was mapped to the long arm of chromosome 7 and subsequently germ-line mutations were identified in the *MET* oncogene [128]. These missense mutations were found within the tyrosine kinase domain of *MET* in codons homologous to those in other receptor tyrosine kinases such as *RET* which have been shown to be disease-causing. These mutations were shown to lead to constitutive activation of the MET protein. Papillary renal cell carcinoma (RCCP) has also been mapped to several additional loci including *PRCC* at 1q21, *TFE* at Xp11, *RCCP3* at 17q21, and *RCC17* at 17q25.

5.10 Familial paraganglioma

Hereditary paraganglioma is a dominantly inherited disorder characterized by highly vascularized tumors which are generally slow growing and comprised of neural crest-derived chief cells which produce and secrete catecholamines. Tumors are most frequently found in the carotid body, a small oxygen-sensing organ located at the bifurcation of the carotid artery in the head and neck and in the adrenal medulla in the abdomen [129]. Paragangliomas (PGL) may be found alone or in conjunction with pheochromocytomas. Skipping of generations and paternal transmission of the tumors provided evidence that at least a subset of these tumors might be imprinted [130].

PGLs are genetically heterogeneous with three genes, *SDHB*, *SDHC* and *SDHD*, having been shown to harbor disease-causing mutations [131]. Mutations in *SDHA* are associated with juvenile encephalopathy and play no role in PGL. These genes encode subunits of the mitochondrial complex II, the smallest complex of the respiratory chain. The products of *SDHC* and *SDHD* anchor the gene products of the other two components, SDHA and SDHB, which form the catalytic core, to the inner mitochondrial membrane. Mutations in the *SDHD* gene were the first to be associated with PGL with the identification of missense and nonsense mutations in four families with familial PGL [132]. As loss of the wild-type allele was found in the tumors, *SDHD* was suggested to be a tumor suppressor gene. Mutations have subsequently been demonstrated in all four exons of *SDHD*

[131,133]. Two mutations, R92Y and L139P have been shown to account for 94% of PGL in Dutch families indicating a founder effect in this population [133].

A single family with nonchromaffin head and neck PGLs was shown to have a mutation in *SDHC* which affects the initiating methionine [134]. Again LOH was shown in the tumor indicating the tumor suppressor nature of this gene. Finally, inactivating mutations in *SDHB* were identified in two out of three families with both PGLs and pheochromocytoma as well as in families with pheochromocytoma alone and in a case of sporadic PGL [135]. Mutations in *SDHB*, *SDHC* and *SDHD* have now all been shown to contribute to a significant fraction of non familial tumors and consequently any patient with a PGL should be investigated for SDH mutations regardless of family history [131].

5.11   Neurofibromatoses

5.11.1   Neurofibromatosis type 1 (NF1)

NF1 is an autosomal dominant condition with an incidence of 1 in 3000 characterized by multiple neurofibromas, café-au-lait spots and Lisch nodules of the iris. Young children can develop glial tumors of the optic pathway. Most adults develop benign tumors of Schwann cells and fibroblasts – dermal and plexiform neurofibromas – and are at increased risk of cancers such as neurofibrosarcomas, astrocytomas and pheochromocytomas. The mutation rate in *NF1* is extremely high with 30–50% of cases shown to have *de novo* mutations [136]. The phenotypic expression is highly variable even within a single family with the same mutation and mild cases often can go undiagnosed.

The *NF1* gene is extremely large with 60 exons making mutation analysis very difficult (see Section 3.8). In general mutation scanning techniques such as SSCP, dHPLC, PTT and DGGE have been used to search for mutations as well as direct sequencing [137]. Mutations are found scattered throughout the gene although a small percentage of recurrent mutations have been found (www.nf.org/nf1gene/nf1gene.home.html). Over 80% of mutations are expected to cause truncations of the protein product, neurofibromin, though large deletions have also been found [136,137]. Nonsense and frameshift mutations lead to a decrease or complete loss of mutant transcript [136]. In contrast to truncating mutations, missense mutations cluster in two distinct regions, the GAP related domain and a region upstream comprising exons 11–17.

Genotype–phenotype correlations are difficult given the large size of the gene and spread of mutations together with the variable expression of the disease and consequently no specific associations have been found. Large deletions removing all of *NF1* plus flanking sequences have been identified in several NF1 patients with a variable number of physical anomalies which could not be correlated with the extent of the deletion but all had a large number of neurofibromas for their age. The variability of expression may be caused either by mutations in the three genes embedded within *NF1*, *EVI2A*, *EVI2B* and *OMPG*, or by other genes near the *NF1* locus, deleted in these patients [136].

The demand for molecular testing is relatively low for two reasons. Firstly, the majority of cases of NF1 can be established on clinical criteria even in early childhood so presymptomatic diagnosis on molecular grounds is not a very high priority. As the phenotype is highly variable and mutation analysis cannot be used to predict the severity of the disease, prenatal diagnosis is rarely requested [136].

5.11.2   Neurofibromatosis type 2 (NF2)

NF2 is an autosomal dominant disorder with an incidence of approximately 1 in 40 000 and is characterized by the presence of bilateral acoustic neuromas (vestibular schwannomas) as well as other tumors such as schwannomas of the other nerves, meningiomas and low-grade CNS tumors. The majority of morbidity from the tumors results from their treatment [138]. The penetrance of NF2 is very high with over 95% of gene carriers becoming symptomatic by age 60. Approximately half of the cases represent new mutations.

The *NF2* gene, called merlin or schwannomin, was isolated in 1993. Mutation detection techniques such as SSCP, DGGE and direct sequencing have found mutations in 35–66% of cases. Mutations are spread throughout the large gene with no mutational hotspots (http://neuro-trials1.mgh.harvard.edu/nf2). The majority are truncating mutations and include large deletions [139] though missense mutations are also found. Genotype–phenotype correlations exist. Mutations that cause premature truncation of the NF2 protein present with severe phenotypes [140]. In contrast, missense mutations have mild NF2. Patients with 5′ mutations have more severe disease than those with mutations at the 3′ end of the gene [141]. Up to 20% of cases of NF2 are mosaics and generally will have a milder disease [142]. Mosaic cases may also explain why the mutation detection rate is relatively low as these will be less easy to detect by the majority of scanning techniques.

Management of NF2 requires a multidisciplinary team, of which presymptomatic testing is an integral part since early diagnosis and detection of tumors is crucial. Mutation analysis is the optimal method for presymptomatic testing but given the difficulties in mutation detection, linkage with flanking DNA markers remains a useful additional test. Prenatal testing has been carried out for NF2 but the demand for this remains low [138].

5.12 Gorlin syndrome

Gorlin syndrome or basal cell nevus syndrome (BCNS) is relatively rare with a prevalence of 1 in 56 000 but is the commonest cause of inherited skin cancer. It is characterized by the presence of a few to over 100 basal cell naevi on the face, upper trunk and neck (*Figure 5.6*) [143]. Other cancers include medulloblastoma and meningioma. In addition there are

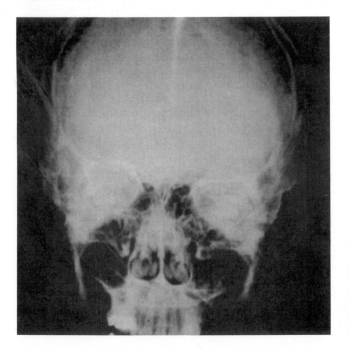

Figure 5.6

Calcification of the falx as seen in Gorlin syndrome. Figure courtesy of Professor Peter Farndon, West Midlands Regional Genetics Service, Birmingham.

many other associated features including frontal and parietal bossing, prominent jaw, odontogenic keratocysts of the jaw, calcification of the falx and bifid, absent or rudimentary ribs [144]. Surveillance is by annual dermatological investigations and regular monitoring of jaw cysts which can erode locally if left untreated. Basal cell nevi and jaw cysts are present in 90% of gene carriers by age 40 [144]. Approximately 40% of cases are due to new mutations.

The causative gene is *PTCH* and mutation analysis using a variety of mutation detection techniques such as SSCP and direct sequencing has identified many mutations throughout the coding region, the majority of which are truncating (www.cybergene.se/ PTCH) [145]. Large deletions have been identified in a number of cases and mosaicism has occasionally been found. No genotype–phenotype correlations have so far been identified.

5.13   Tuberous sclerosis

Tuberous sclerosis (TS) is an autosomal dominant disorder with an incidence of 1 in 26 000 characterized by the development of tumor-like growths called hamartomas in a variety of tissues and organs including the brain, skin, kidneys and heart [146]. Diagnosis is based on combinations of clinical radiological and histopathological signs and diagnostic criteria are very stringent. Around two thirds of cases of TS are caused by *de novo* mutations.

TS shows genetic heterogeneity and two genes have been isolated, *TSC1* on chromosome 9q34 and *TSC2* on 16p13.3. Mutations in the genes have been identified by a combination of SSCP, dHPLC, direct sequencing, Southern blotting and quantitative PCR [146–148]. Mutations can be detected in over 80% of cases [148]. Several hundred different mutations have been identified in the two genes (http://expmed.bwh.harvard.edu/ts/review). These are summarized in *Table 5.4* and show the differences in the nature of mutations between *TSC1* and *2* [146]. Genotype–phenotype correlations exist. *TSC2* mutations are five times more common than *TSC1* mutations in sporadic cases of TS though this is not the case in familial disease. In general patients with *TSC1* mutations have a milder phenotype [146,148]. Intellectual disability is significantly more common in sporadic cases with *TSC2* mutations than those with *TSC1*. Patients in whom a mutation has not been found also tend to have a milder phenotype [148]. Somatic and germ-line mosaicism has been described for both *TSC1* and *2* which may complicate diagnostic testing as these mutations can easily be missed.

5.14   Familial gastric cancer

Gastric cancer is a major cause of death from cancer worldwide. Around 10% of cases show familial clustering. In 1998, linkage in a large Maori pedigree was found to chromosome 16q21 [149]. The E cadherin gene (*CDH1*) lay within this region and mutations

Table 5.4   Mutations in *TSC1* and *2*

Mutation	TSC1	TSC2
Large deletions	0%	18%
Insertions	17%	9%
Deletions	36%	23%
Nonsense	39%	19%
Splicing	7%	11%
Missense	1%	20%

were identified in the gene in some of the families. E cadherin is a cell adhesion molecule that binds to β catenin at adherens junctions preventing cell signaling. Loss of E cadherin results in increased cell mobility. A second study of non Maori families selected on the basis of either two cases of gastric cancer in first-degree relatives with one affected before age 50 years, or three or more cases of gastric cancer, found mutations in 25% of cases confirming a role for E cadherin in familial gastric cancer [150]. In addition, one of the mutation carriers in this study developed colorectal cancer at age 30, indicating that *CDH1* may also be implicated in some cases of early-onset colorectal cancer. However, only a proportion of familial gastric cancers can be accounted for by *CDH1* mutations and other genes remain to be identified. Guidelines for the management of familial gastric cancers have recently been developed [151].

5.15   Familial melanoma

Cutaneous malignant melanoma is a potentially fatal form of skin cancer which is relatively uncommon but with an incidence that is rising in many regions of the world. In Europe, the incidence of the disease is around 10 in 100 000, and the incidence has been rising at a rate of over 7% per annum. In the US, the incidence of this cancer is rising faster than any other, and in Australia the lifetime risk of melanoma is 5%. The most important risk factor is the presence of atypical nevi, lesions which are considered to be a precursor of melanoma [152]. These are variable in number, reddish brown to pink in color with an irregular border. Melanomas are associated with a number of genetic conditions such as Li–Fraumeni syndrome, HNPCC, familial retinoblastoma and most importantly familial melanoma which accounts for aproximately 10% of melanoma cases. The autosomal dominant nature of this condition has been recognized for over a century and is characterized primarily by the presence of superficial spreading-type melanomas but others such as nodular melanomas are also found.

A number of nonrandom karyotypic changes have been identified in the progression from the normal melanocyte to a malignant melanoma and have been shown to involve primarily chromosomes 1, 6, 7, 9 and 10 [153]. Initial linkage was shown to chromosome 1p but has been largely unsubstantiated. Strong evidence for a familial melanoma gene on chromosome 9p [154] was however confirmed by a number of groups. This region was identified because of consistent LOH in this area and because of a case of a cytogenetic abnormality involving chromosomes 9p and 5p in an individual with multiple melanoma. A gene within the candidate region was identified as the *p16* or cyclin-dependent kinase inhibitor 2 (*CDKN2*) gene [155]. This gene encodes a protein whose product inhibits the cyclin-dependent kinases, CDK4 and 6, and subsequently inhibits their ability to phosphorylate the retinoblastoma protein (see Section 4.4). Any inactivation of *CDKN2* would therefore lead to loss of the RB control of the G_1 checkpoint and continuous cell proliferation (see *Figure 4.3*). Germ-line mutations, both truncating and missense, have been reported in *CDKN2* in around 40% of melanoma families confirming that this is a major causative gene [156]. A further study has also identified a mutation in the *CDK4* gene itself in two unrelated families [157]. This mutation occurs in the p16-binding domain of *CDK4*, thereby generating a gene which is resistant to the inhibitory effects of p16, and acts as a dominant oncogene. However, to date mutations in this gene are relatively rare and it appears to play a relatively minor role in melanoma predisposition [158]. *CDKN2* also encodes the ARF protein in an alternate reading frame and this protein is involved in control of cell growth through the p53 pathway (see Section 4.3). The possibility that this alternate gene product plays a role in melanoma has also been investigated particularly as deletions at the *p16* locus could disable two separate pathways which control cell growth [159]. To date this alternate gene product has not been shown to have a major role but a recent report has identified a splice site mutation in exon 1β in a family with multiple melanomas. DNA from tumors showed a mutation in the normal allele and loss of the mutant allele resulting in somatic changes in both ARF and CDKN2A [160]. The importance of inactivation of both genes for the development of melanomas remains to be proven.

5.16 Hereditary prostate cancer

Prostate cancer is a major health problem accounting for over 8000 deaths per year in the UK and in the US is the second most common cause of death from cancer. A strong hereditary component to prostate cancer has been postulated. A gene for familial prostate cancer has been localized to chromosome 1q24–25 and named *HPC1* [161]. However there is clear evidence of locus heterogeneity and other possible candidate genes have been identified [161].

5.17 Conclusion

This chapter has highlighted the benefits of mutation screening and presymptomatic diagnosis in families with a strong history of cancer. Further genes are likely to be identified in the immediate future especially with the recent publication of the entire sequence of the human genome. This will mean increased benefits for patients with the ability to target screening to high-risk individuals. In addition, there will be cost benefits to health services. As well as benefiting those individuals with familial cancers it will potentially offer insight into the molecular defects in sporadic cancer and alternative chemotherapeutic options should also become available.

References

1. **Peterson, G.M., Brensinger, J.D., Johnson, K.A. and Giardello, F.M.** (1999) *Cancer*, **86**, 2540.
2. **Terdiman, J.P., Conrad, P.G. and Sleisenger, M.H.** (1999) *Am. J. Gastroenterol.*, **94**, 2344.
3. **Clarke, A.** (1998) *Genetic Testing of Children*. BIOS Scientific Publishers, Oxford.
4. **Burn, J., Chapman, P. and Delhanty, J.** (1991) *J. Med. Genet.*, **28**, 289.
5. **Eccles, D.M., Lunt, P.W., Wallis, Y.** *et al.* (1997) *Arch. Dis. Childhood*, **77**, 431.
6. **Eccles, D.M., Evans, D.G.R. and McKay, J.** (2000) *J. Med. Genet.* **37**, 203.
7. **Morton, D., Macdonald, F., Cachon-Gonzales, M.B.** *et al.* (1992) *J. Med. Genet.*, **29**, 571.
8. **Dunlop, M.G., Farrington, S.M., Carothers, A.D.** *et al.* (1997) *Hum. Mol. Genet.*, **6**, 105.
9. **Morton, D.G., Macdonald, F., Haydon, J.** *et al.* (1993) *Br. J. Surg.*, **80**, 255.
10. **Eccles, D.M., van der Luijt, R., Breukel, C.** *et al.* (1996) *Am. J. Hum. Genet.*, **59**, 1193.
11. **Bussey, H.J.R.** (1975) *Familial Adenomatous Polyposis*. The Johns Hopkins Press, Baltimore, MA, USA.
12. **Spirio, L., Olschwang, S., Groden, J.** *et al.* (1993) *Cell*, **75**, 951.
13. **Wallis, Y.L., Morton, D.G., McKeown, C.M. and Macdonald, F.** (1999) *J. Med. Genet.* **36**, 14.
14. **Van der Luijt, R.B., Khan, P.M., Vasen, H.F.A.** *et al.* (1997) *Hum. Mut.* **9**, 7.
15. **Laken, S.J., Petersen, G.M., Gruber, S.B.** *et al.* (1997) *Nat. Genet.*, **17**, 79.
16. **Frayling, I.M., Beck, N.E., Ilyas, M.** *et al.* (1998) *Proc. Natl. Acad. Sci.*, **95**, 10722.
17. **Sieber, O., Lipton, L., Heinimann, K. and Tomlinson, I.** (2003) *J. Pathol.*, **199**, 137.
18. **Wallis, Y.L., Macdonald, F., Hulten, M.** *et al.* (1994) *Hum. Genet.*, **94**, 543.
19. **Flintoff, K.J., Sheridan, E., Turner, G.** *et al.* (2001) *J. Med. Genet.*, **38**, 129.
20. **Dobbie, Z., Spycher, M., Mary, J-L.** *et al.* (1996) *J. Med. Genet.*, **33**, 274.
21. **Olschwang, S., Tiret, A., Laurent-Puig, P.** *et al.* (1993) *Cell*, **75**, 959.
22. **Caspari, R., Olschwang, S., Friedl, W.** *et al.* (1995) *Hum. Mol. Genet.* **4**, 337.

23. **Nagase, H., Miyoshi, Y., Horii, A. et al.** (1992) *Cancer Res.*, **52**, 4055.
24. **Caspari, R., Friedl, W., Mandl, M. et al.** (1994) *Lancet*, **343**, 629.
25. **Houlston, R., Crabtree, M., Phillips, R. and Tomlinson, I.** (2001) *Gut*, **48**, 1.
26. **Moser, A.R., Mattes, E.M., Dove, W.F. et al.** (1993) *Proc. Natl. Acad. Sci.*, **90**, 8977.
27. **Silverman, K.A., Koratkar, R., Siracusa, L.D. and Buchberg, A.M.** (2002) *Genome Res.*, **12**, 88.
28. **Silverman, K.A., Koratkar, R., Siracusa, L.D. and Buchberg, A.M.** (2003) *Mammalian Genome*, **14**, 119.
29. **Hawk, E., Lubert, R. and Limburg, P.** (1999) *Cancer*, **86**, 2551.
30. **Giardiello, F.M., Yang, V.W., Hylind, L.M. et al.** (2002) *N. Engl. J. Med.*, **346**, 1054.
31. **Lynch, H.T. and De la Chapelle, A.** (1999) *J. Med. Genet.*, **36**, 801.
32. **Vasen, H.F.A., Mecklin, J.P., Meera Khan, P. and Lynch, H.T.** (1991) *Dis. Colon Rectum*, **34**, 424.
33. **Rodriguez-Bigas, M.A., Boland, C.R., Hamilton, S.R. et al.** (1997) *J.N.C.I.*, **89**, 1758.
34. **Peltomaki, P.** (2001) *Hum. Mol. Genet.*, **10**, 735.
35. **Berends, M.J., Wu, Y., Sijmonds, R.H. et al.** (2002) *Am. J. Hum. Genet.*, **70**, 26.
36. **Wu, Y., Berends, M.J., Sijmonds, R.H. et al.** (2001) *Nat. Genet.*, **29**, 137.
37. **Liu, T., Yan, H., Kuismanen, S. et al.** (2001) *Cancer Res.* **61**, 7798.
38. **Trojan, J., Zeuzem, S., Randolph, A. et al.** (2002) *Gastroenterol.*, **122**, 211.
39. **Nystrom-Lahi, M., Perera, C., Raschle, M. et al.** (2002) *Genes, Chrom. Cancer*, **33**, 160.
40. **Heinen, C.D., Wilson, T., Mazurek, A. et al.** (2002) *Cancer Cell*, **1**, 469.
41. **Shin, K.H., Shin, J.H., Kim, J.H. and Park, J.G.** (2002) *Cancer Res.*, **62**, 38.
42. **Wang, Y., Friedl, W., Lamberti, C. et al.** (2003) *Int. J. Cancer*, **103**, 636.
43. **Viel, A., Petronzelli, F., Della Puppa, L. et al.** (2002) *Hum. Mut.*, **20**, 368.
44. **Nystrom Lahti, M., Kristo, P., Nicolaides, N.C. et al.** (1995) *Nat. Med.* **1**, 1203.
45. **Froggatt, N.J., Green, J., Brassett, C. et al.** (1999) *J. Med. Genet.* **36**, 97.
46. **Foulkes, W.D., Thiffault, I. and Gruber, S.B.** (2002) *Am. J. Hum. Genet.*, **71**, 1395.
47. **Holinski-Feder, E., Muller-Koch, Y., Friedl, W. et al.** (2001) *J. Biochem. Biophys. Methods*, **47**, 21.
48. **Kohonen-Corish, M., Ross, V.L., Doe. W.F. et al.** (1996) *Am. J. Hum. Genet.*, **59**, 818.
49. **Wang, Y., Friedl, W., Sengteller, M. et al.** (2002) *Hum. Mut.*, **19**, 279.
50. **Schouten, J.P., McElgunn, C.J., Waaijer, R. et al.** (2002) *Nucleic Acids Res.*, **30**, e57.
51. **Wijnen, J., Meera Khan, P., Vasen, H. et al.** (1997) *Am. J. Hum. Genet.*, **61**, 329.
52. **Vasen, H.F.A., Wijnen, J., Menko, F. et al.** (1996) *Gastroenterology*, **110**, 1020.
53. **Miyaki, M., Konishi, M., Tanaka, K. et al.** (1997) *Nat. Genet.*, **17**, 271.
54. **Bisgaard, M.L., Jager, A.C., Myrhoj, T. et al.** (2002) *Hum. Mut.*, **20**, 20.
55. **Bapat. B., Madlensky, L., Temple, L.K.F. et al.** (1999) *Hum. Genet.*, **104**, 167.
56. **Loukola, A., de la Chapelle, A. and Aaltonen, L.** (1999). *J. Med. Genet.*, **36**, 819.
57. **Wahlberg, S.S., Schmeits, J., Thomas, G. et al.** (2002) *Cancer Res.*, **62**, 3485.
58. **Hendriks, Y., Franken, P., Dierssen, J.W. et al.** (2003) *Am. J. Pathol.*, **162**, 469.
59. **Ruschoff, J., Wallinger, S. and Dietmaier, W.** (1998) *Proc. Natl. Acad. Sci.*, **95**, 11301.
60. **Howe, J.R., Roth, S., Ringold, J.C. et al.** (1998) *Science*, **280**, 1086.
61. **Lagna, G., Hata, A., Hemmati-Brivanlou, A. and Massague, J.** (1996) *Nature*, **383**, 832.
62. **Friedl, W., Kruse, R., Uhlhaas, S. et al.** (1999) *Genes, Chrom. Cancer*, **25**, 403.
63. **Howe, J.R., Blair, J.L., Sayed, M.G. et al.** (2001) *Nat. Genet.*, **28**, 184.
64. **Liaw, D., Marsh, D.J., Li, J. et al.** (1997) *Nat. Genet.*, **16**, 64.
65. **Dahia, P.L.M., Aguiar, R.S., Alberta, J.B. et al.** (1999) *Hum. Mol. Genet.*, **8**, 185.
66. **Marsh, D.J., Couolon, V., Lunetta, K.L. et al.** (1998) *Hum. Mol. Genet.*, **7**, 507.
67. **Hemminki, A., Markie, D. Tomlinson, I. et al.** (1998) *Nature*, **391**, 184.
68. **Jenne, D.E., Reimann, H., Nezu, J. et al.** (1998) *Nat. Genet.*, **18**, 38.
69. **Al-Tassan, N., Chmiel, N.H., Maynard, J. et al.** (2002) *Nat. Genet.*, **30**, 227.
70. **Jones, S., Emmerson, P., Maynard, J. et al.** (2002) *Hum. Mol. Genet.*, **11**, 2961.
71. **Sieber, O.M., Lipton, L., Crabtree, M. et al.** (2003) *N. Engl. J. Med.*, **348**, 791.
72. **DeYong, M.M., Nolte, I.M., te Meerman, G.T. et al.** (2002) *J. Med. Genet.*, 39, 225.
73. **Easton, D.F., Steele, L., Fields, P. et al.** (1997) *Am. J. Hum. Genet.* **61**, 120.
74. **Ford, D., Easton, D.F., Stratton, M. et al.** (1998) *Am. J. Hum. Genet.*, **62**, 676 .

75. **Burke, W., Daly, M., Garber, J. et al.** (1997) *J.A.M.A.,* **277**, 997.
76. **Vasen, H.F., Haites, N.E., Evans, D.G. et al.** (1998) *Eur. J. Cancer,* **34**, 1922.
77. **Easton, D.F., Ford, D. and Bishop, D.T.** (1995) *Am. J. Hum. Genet.,* **56**, 265.
78. **Eisen, A., Rebbeck, T.R., Wood, W.C. and Weber, B.L.** (2000) *J. Clin. Oncol.,* **18**, 1980.
79. **Gayther, S.A., Harrington, P., Russell, P. et al.** (1996) *Am. J. Hum. Genet,* **58**, 451.
80. **Hogervorst, F.B.L., Cornelis, R.S., Bout, M. et al.** (1995) *Nat. Genet.,* **10**, 208.
81. **The BRCA1 exon 13 Duplication Screening Group** (2000) *Am. J. Hum. Genet.,* **67**, 207.
82. **Struewing, J.P., Abeliovich, D., Peretz, T. et al.** (1995) *Nat. Genet,* **11**, 198.
83. **Neuhausen, S., Gilewski , T., Norton, L. et al.** (1996) *Nat. Genet.,* **13,** 126.
84. **Abeliovich, D., Kaduri, L., Lerer, I. et al.** (1997) *Am. J. Hum. Genet.,* **60**, 505.
85. **Thorlacius, S., Tryggvadottir, L., Olafsdottir, G.H. et al.** (1995) *Lancet,* **346**, 544.
86. **Petrij-Bosch, A., Peelen, T., van Vliet, M. et al.** (1997) *Nat. Genet.,* **17**, 341.
87. **Tonin, P.N., Mes-Masson, A-M., Futreal, P.A. et al.** (1998) *Am. J. Hum. Genet.,* **63**, 1341.
88. **Breast Cancer Linkage Consortium** (1999) *J.N.C.I.,* **91**, 1310.
89. **Gayther, S.A.,Warren, W., Mazoyer, S. et al.** (1995) *Nat. Genet.,* **11**, 428.
90. **Gayther, S.A., Mangion, J., Russell, P. et al.** (1997) *Nat. Genet.,* **15**, 103.
91. **Seitz, S., Rohde, K., Bender, E. et al.** (1997) *Oncogene,* **14**, 741.
92. **Kainu, T., Juo, S.H., Desper, R. et al.** (2000) *Proc. Natl. Acad. Sci.,* **97**, 960.
93. **Thompson, D., Szabo, C.I., Mangion, J. et al.** (2002) *Proc. Natl. Acad. Sci.,* **99**, 827.
94. **Li, F.P. and Fraumeni, J.F.** (1982) *J.A.M.A.,* **247**, 2692.
95. **Li, F.P., Fraumeni, J.F., Mulvihill, J.J. et al.** (1988) *Cancer Res.,* **48**, 5358.
96. **Garber, J.E., Goldstein, A.M., Kantor, A.F. et al.** (1991) *Cancer Res.,* **51**, 6094.
97. **Prosser, J., Elder, P.A., Condie, A. et al.** (1991) *Br. J. Cancer,* **63**, 181.
98. **Tominaga, K., Morisaki, H., Kaneko, Y. et al.** (1999) *J. Biol. Chem.,* **274**, 31463.
99. **Meijers-Heijboer, H., Van Den Ouweland, A., Klijn, J. et al.** (2002) *Nat. Genet.,* **31**, 55.
100. **Vahteristo, P., Barkova, J., Eerola, H. et al.** (2002) *Am. J. Hum. Genet,* **71**, 432.
101. **Schutte, M., Seal, S., Barfoot, R. et al.** (2003) *Am. J. Hum. Genet.,* **27**, 72.
102. **Stankovic, T., Kidd, A.M., Sutcliffe, A. et al.** (1998) *Am. J. Hum. Genet.,* **62**, 334.
103. **Thorstenson, Y.R., Roxas, A., Kroiss, R. et al.** (2001) *Am. J. Hum. Genet.,* **69**, 205.
104. **Malkin, D.** (1998) In: Vogelstein, B. and Kinzler, K. (eds) *The Genetic Basis of Human Cancer,* McGraw-Hill, New York, p. 393.
105. **Malkin, D., Li, F.P., Strong, L.C. et al.** (1990) *Science,* **250**, 1233.
106. **Varley, J.M., Evans, D.G.R. and Birch, J.M.** (1997) *Br. J. Cancer,* **76**, 1.
107. **Bell, D.W., Varley, J.M., Szydlo, T.E. et al.** (1999) *Science,* **286**, 2528.
108. **Lee, S.B., Kim, S.H., Bell, D.W. et al.** (2001) *Cancer Res.,* **61**, 8062.
109. **Villablanca, A., Hoog, A., Larsson, C. and Teh, B.T.** (2001) *J. End. Genet.* **2**, 3.
110. **Larsson, C., Skogseid, B., Oberg, K. et al.** (1988) *Nature,* **332**, 85.
111. **European Consortium on MEN1** (1997) *Hum. Mol. Genet.,* **6**, 1177.
112. **Agarwal, S.K., Guru, S.C., Heppner, C. et al.** (1999) *Cell,* **96**, 143.
113. **Marx, S.J., Agarwal, S.K., Heppner, C. et al.** (1999) *Bone,* **25**, 119.
114. **Giraud, S., Zhang, C.X., Serova-Sinilnikova, O. et al.** (1998) *Am. J. Hum. Genet.,* **63**, 455.
115. **Ponder, B.A.J.** (1998) In: Vogelstein, B. and Kinzler, K. (eds) *The Genetic Basis of Human Cancer,* McGraw-Hill, New York, p. 475.
116. **Mathew, C.G.P., Chin, K.S., Easton, D.F. et al.** (1987) *Nature,* **328**, 527.
117. **Donis Keller, H., Dou, S., Chi, D. et al.** (1993) *Hum. Mol. Genet.,* **2**, 851.
118. **Mulligan, L.M., Kwok, J.B.J., Healey, C.S. et al.** (1993) *Nature,* **363**, 458.
119. **Eng, C., Clayton, D., Schuffenecker, I. et al.** (1996) *J.A.M.A.,* **276**, 1575.
120. **Clifford, S.C. and Maher, E.R.** (2001) *Adv. Cancer Res.,* **82**, 85.
121. **Moore, A.T., Maher, E.R., Rosen, P. et al.** (1991) *Eye,* **5**, 723.
122. **Zbar, B., Kishida, T., Chen, F. et al.** (1996) *Hum. Mut.,* **8**, 348.
123. **Stolle, C., Glenn, G., Zbar, B. et al.** (1998) *Hum. Mut.,* **12**, 417.
124. **Maher, E.R., Webster, A.R., Richards, F.M. et al.** (1996) *J. Med. Genet.,* **33**, 84.
125. **Bender, B.U., Eng, C., Olschwski, M. et al.** (2001) *J. Med. Genet.,* **38**, 50.
126. **Crossey, P.A., Eng, C., Ginalska-Malinowska, M. et al.** (1995) *J. Med. Genet.,* **32**, 885.
127. **Hes, F.J., McKee, S., Taphoorn, M.J.B. et al.** (2000) *J. Med. Genet.* **37**, 939.
128. **Schmidt, L., Duh, A-M., Chen, F. et al.** (1997) *Nat. Genet.,* **16**, 68.

129. **Benn, D.E., Marsh, D.J. and Robinson, B.G.** (2002) *Curr. Opin. Endocrin. Diab.,* **9**, 79.
130. **Van der Mey, A.G., Maaswinkel-Mooy, P.D., Corneliise, C.J.** *et al.* (1989) *Lancet,* **2**, 1291.
131. **Baysal, B.E.** (2002) *J. Med. Genet.,* **39**, 617.
132. **Baysal, B.E., Ferrell, R.E., Willett-Brozick, J.E.** *et al.* (2000) *Science,* **287**, 848.
133. **Taschner, P.E.M., Jansen, J.C., Baysal, B.E.** *et al.* (2001) *Genes, Chrom. and Cancer,* **31**, 274.
134. **Niemann, S. and Muller, U.** (2000) *Nat. Genet.,* **26**, 268.
135. **Astuti, D., Latif, F., Dallol, A.** *et al.* (2001) *Am. J. Hum. Genet.,* **69**, 49.
136. **Shen, M.H., Harper, P.S. and Upadhyaya, M.** (1996) *J. Med. Genet.,* **33**, 2.
137. **Fahsold, R., Hoffmeyer, S., Mischung, C.** *et al.* (2000) *Am. J. Hum. Genet.,* **66**, 790.
138. **Evans D.G.R., Sainio, M. and Baser, M.E.** (2000) *J. Med. Genet.,* **37**, 897.
139. **Zucman-Rossi, J., Legoix, P., der Sarkissian, H.** *et al.* (1998) *Hum. Mol. Genet.,* **7**, 2095.
140. **Evans, D.G.R., Trueman, L., Wallace, A.** *et al.* (1998) *J. Med. Genet,* **35**, 450.
141. **Friedman, J.M., Woods, R., Joe H.** *et al.* (1999) *Am. J. Hum. Genet.,* **65**, A126.
142. **Kluwe, L. and Mautner, V.F.** (1998) *Hum. Mol. Genet.,* **7**, 2051.
143. **Gorlin, R.J.** (1987) *Medicine,* **66**, 98.
144. **Evans, D.G.R., Ladusans, E.J., Rimmer, S.** *et al.* (1993) *J. Med. Genet.,* **30**, 460.
145. **Wicking, C., Shanley, S., Smyth, I.** *et al.* (1997) *Am. J. Hum. Genet.,* **60**, 21.
146. **Cheadle, J.P., Reeve, M.P., Sampson, J.R. and Kwiatkowski, D.J.** (2000) *Hum. Genet.,* **107**, 97.
147. **Jones, A.C., Shyamsundar, M.M., Thomas, M.W.** *et al.* (1999) *Am. J. Hum. Genet.,* **64**, 1305.
148. **Dabora, S.L., Jozwiak, S., Franz, D.N.** *et al.* (2001) *Am. J. Hum. Genet.,* **68**, 64.
149. **Guilford, P., Hopkins, J., Harraway, J.** *et al.* (1998) *Nature,* **392**, 402.
150. **Richards, F.M., McKee, S.A., Rajpar, M.H.** *et al.* (1999) *Hum. Mol. Genet.,* **8**, 607.
151. **Caldas, C., Carneiro, F., Lynch, H.T.** *et al.* (1999) *J. Med. Genet.,* **36**, 873.
152. **Green, M.H.** (1997) *Mayo Clinic Proc.,* **72**, 467.
153. **Fountain, J.W., Bale, S.J., Housman, D.E. and Dracopoli, N.C.** (1990) *Cancer Surv.,* **9**, 645.
154. **Cannon-Albright, L., Goldgar, D.E., Meyer, L.J.** *et al.* (1992) *Science,* **258**, 1148.
155. **Kamb, A., Gruis, N., Weaver-Feldhaus, J.** *et al.* (1994) *Science,* **264**, 436.
156. **Dracopoli, N.C. and Fountain, J.W.** (1996) *Cancer Surv.,* **26**, 115.
157. **Zuo, L., Weger, J., Yang, Q.** *et al.* (1996) *Nat. Genet.,* **12**, 97.
158. **Goldstein, A.M., Chidambraram, A., Halpern, A.** *et al.* (2002) *Melanoma Res.,* **12**, 51.
159. **Piepkorn, M.** (2000) *J. Am. Acad. Dermatol.,* **42**, 705.
160. **Hewitt, C., Lee Wu, C., Evans, G.** *et al.* (2002) *Hum. Mol. Genet.,* **11**, 1273.
161. **Eales, R., the UK Familial Prostate Study Co-ordinating Group and the CRC/BPG UK Familial Prostate Cancer Study Collaborators** (1999) *Prost. Cancer and Prost. Dis.,* **2**, 9.

Lung cancer

6

6.1 Introduction

Lung cancer is one of the most important malignancies worldwide in terms of both incidence and prevalence, and as a leading cause of cancer death [1]. Despite well-defined risk factor associations, particularly tobacco consumption, preventive strategies aimed at control of cigarette smoking, and through lung cancer screening, have met with limited success [2].

Until the 1930s, carcinoma of the lung was considered a rare disease. Throughout the remainder of the 20th century, incidence and mortality rates increased dramatically in North America and Europe. Statistics from the United States illustrate the magnitude of the problem [3]. The 2001 Cancer Progress Report prepared by National Institutes of Health estimated there would be 169 500 new cases, and 157 400 deaths from lung cancer that year [4]. Although overall lung cancer rates have declined slightly since 1992, likely reflecting reduced tobacco consumption in males, lung cancer incidence continues to climb in females, and is not expected to decline until 2020.

The incidence and mortality rates for lung cancer vary (as for other cancers) throughout the world, reflecting the prevalence of cigarette smoking over the past three to four decades. Of particular concern are current trends in tobacco consumption in developing countries in Asia, Africa and South America, which will likely account for increasing lung cancer rates throughout the early decades of the 21st century.

Numerous studies have implicated tobacco consumption as a major risk factor for the development of lung cancer. Tobacco comprises a complex mixture of potent carcinogens, predominantly polycyclic aromatic hydrocarbons derived from combustion of tars. Risk varies according to the type of tobacco smoked, amount, and duration of exposure. Other well-defined risk factors include gender, occupation, diet, exposure to radon, and air pollution, including personal exposure to passive smoke. Interactions between risk factors, and in particular with cigarette smoking, may increase lung cancer risk significantly. Hereditary factors and genetic susceptibility to lung cancer currently remain ill-defined.

This chapter will review the spectrum of molecular genetic alterations in human lung cancer, and discuss their potential clinical application for screening, early diagnosis, staging and therapy.

6.2 Pathology

Despite considerable cellular heterogeneity within individual tumors, primary lung cancer is generally classified into two major types: non-small cell lung carcinoma (NSCLC) and small cell lung cancer (SCLC). This broad stratification reflects fundamental differences in tumor biology and clinical behavior, and underlies current treatment strategies.

6.2.1 Small cell lung cancer (SCLC)

SCLC, which accounts for about 20% of all primary lung tumors, is characterized histologically by small hyperchromatic cells with almost no visible cytoplasm, absent nucleoli,

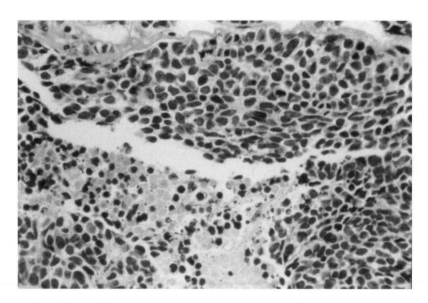

Figure 6.1

Histology of small cell lung cancer.

and variable degrees of necrosis (*Figure 6.1*). The histologic classification of SCLC is confusing, and previously included oat cell and intermediate cell types. SCLC is now generally considered a neuroendocrine tumor (with small and large cell variants), and immunohistochemical studies have consistently demonstrated characteristic biomarkers, including calcitonin, gastrin-releasing peptide, L-dopa decarboxylase, chromogranin, synaptophysin, and neuron-specific enolase. However, the precise cell of origin for lung cancer is controversial, and a mosaic of cellular elements (including NSCLC) is not uncommon in tumors with otherwise predominantly small-cell histology.

6.2.2 Non-small cell lung cancer (NSCLC)

NSCLC accounts for 80% of all primary lung cancer. Histologic subtypes of NSCLC, as recently defined by the World Health Organization (WHO) and the International Association for the Study of Lung Cancer (IASLC), are summarized in *Table 6.1*.

Adenocarcinomas are currently the most frequent histologic subtype of NSCLC. Primary lung adenocarcinomas may present as a solitary pulmonary parenchymal lesion or may be multifocal. Four histologic patterns predominate: acinar (the most common; *Figure 6.2*), solid, papillary, and bronchioloalveolar carcinoma (BAC). Although BAC of the lung has a characteristic histologic appearance (*Figure 6.3*), it is a particularly controversial tumor with respect to its pathogenesis and classification. Primary lung adenocarcinomas must be distinguished from secondary tumors which are often adenocarcinomas also, but this may actually be quite difficult in clinical practice, despite the use of special stains, immunohistochemistry and electron microscopy. Many secondary tumors tend to retain immunohistochemical profiles of the epithelium from which they arise, and the use of cytokeratin 7 and 20 immunohistochemistry has been useful to distinguish primary lung adenocarcinomas from metastatic colorectal adenocarcinomas [5,6].

Table 6.1   Histopathologic subtypes of primary non-small cell lung cancer

Squamous cell carcinoma
   Variants: Papillary, clear cell, small cell, basaloid

Adenocarcinoma
   Acinar
   Papillary
   Bronchioloalveolar
      Nonmucinous
      Mucinous
      Mixed or indeterminate
      Solid, with mucin formation
   Adenocarcinoma with mixed subtypes
      Variants: Well-differentiated fetal, mucinous (colloid), mucinous
      cystadenocarcinoma, signet ring, clear cell

Large cell carcinoma
   Variants: Large-cell neuroendocrine, combined large-cell neuroendocrine, basaloid,
   lymphoepithelioma-like, clear-cell, large-cell with rhabdoid phenotype

Adenosquamous carcinoma

Sarcomatoid carcinomas
   Variants: Carcinomas with spindle cells and/or giant cells, carcinosarcoma, blastoma,
   pleomorphic carcinoma

Neuroendocrine tumors
   Typical carcinoid
   Atypical carcinoid

Unclassified carcinomas

Modified from the 1999 World Health Organization and the International Association for the Study of
Lung Cancer Histologic Typing of Lung and Pleural Tumors.

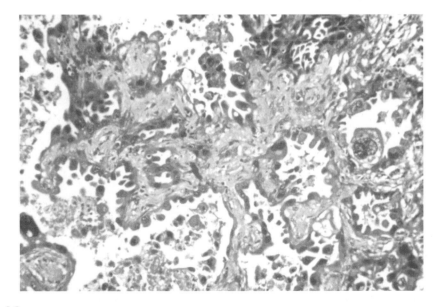

Figure 6.2

Histology of primary lung adenocarcinoma.

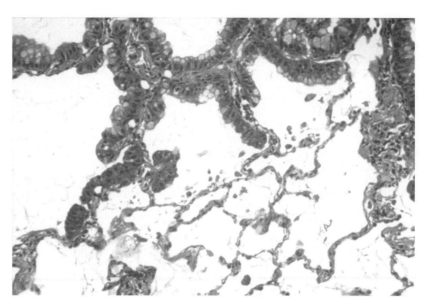

Figure 6.3

Bronchioloalveolar carcinoma of the lung.

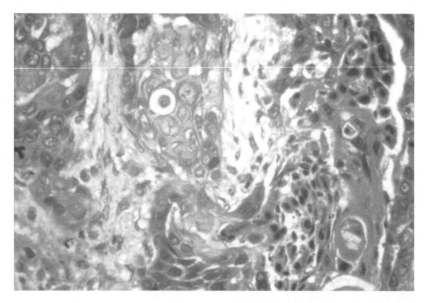

Figure 6.4

Squamous cell carcinoma of the lung.

Squamous cell carcinomas are characterized histologically by cytokeratin and intracellular bridges (*Figure 6.4*). Poorly differentiated squamous cell carcinomas may be difficult to distinguish from large-cell undifferentiated carcinomas (*Figure 6.5*), which comprise approximately 15% of all NSCLC.

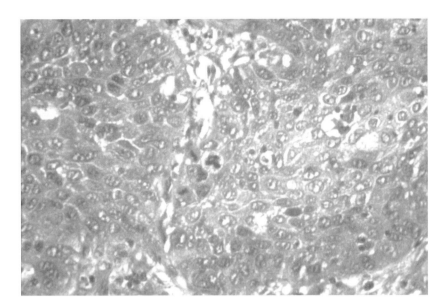

Figure 6.5

Large-cell undifferentiated lung cancer.

6.3 Clinical features

Regardless of histologic subtype, patients with NSCLC generally present with well-defined symptoms related to the primary tumor, including cough, hemoptysis, shortness of breath, unresolved pneumonia, or chest wall pain (*Figures 6.6* and *6.7*). Lung cancers may be found incidentally in up to 20% of individuals with no symptoms (e.g. during routine chest radiography for insurance purposes). Additional symptoms may arise from mediastinal regional lymph node metastases (e.g. superior vena cava syndrome from enlarged right paratracheal nodes; left recurrent laryngeal nerve paresis from subaortic nodal metastases; difficulty swallowing from extrinsic esophageal compression by enlarged subcarinal nodes), from distant metastases (e.g. jaundice from liver metastases, shortness of breath from a malignant pleural effusion, loss of consciousness from brain metastases, bone pain from skeletal metastases) or from associated paraneoplastic syndromes.

Staging the extent of the tumor permits patients to be stratified for potentially curative treatment (usually surgery), or for palliation. The overall prognosis for NSCLC is dismal with 5-year survival generally reported below 10% (for all patients), and which has not changed appreciably over the past three decades. However, patients with early-stage NSCLC have consistently improved survival (up to 80% at 5 years) following surgical resection. Several clinical trials are currently evaluating the efficacy of multimodality therapy (e.g. preoperative chemotherapy, radiotherapy or both) to improve survival of patients with locally advanced NSCLC.

SCLC may arise as a solitary peripheral pulmonary nodule, but it most frequently presents as a large central mass with extensive mediastinal lymphadenopathy, often with distant metastases. SCLC appears to have an aggressive tumor biology, and the median survival of patients with limited-stage disease is reported to be 15–18 months only. Systemic therapy (multiagent combination chemotherapy) is currently the primary treatment modality, with radiation therapy, and occasionally surgery, for highly selected patients.

Although the histologic features and clinical features of lung cancer have been well characterized, individual tumors differ considerably in their biologic behavior and

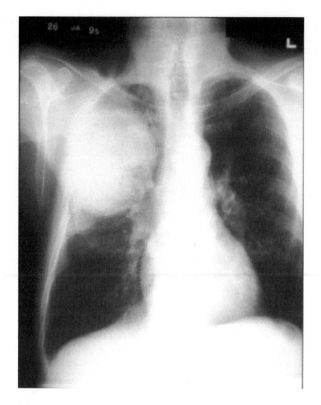

Figure 6.6

Plain chest radiograph illustrating a large right upper lobe lung cancer.

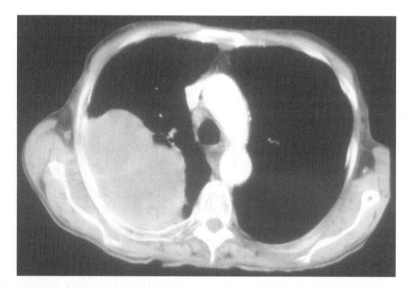

Figure 6.7

Computed tomography of the chest (same patient as *Figure 6.6*) illustrating a large right upper lobe lung cancer adjacent to the chest wall. A small right lower paratracheal lymph node is also seen in the mediastinum.

genomic profile [7]. The remainder of this chapter will summarize our current understanding of the molecular genetic alterations associated with NSCLC and SCLC, and will focus on potential clinical applications.

6.4 Genetic factors and individual susceptibility

The observation that only 15–20% of heavy smokers ever develop lung cancer, and that nonsmokers (and others who have never been exposed to known carcinogens) do develop lung malignancy, suggested underlying genetic factors, or individual susceptibility to this disease. Epidemiologic studies have suggested an increased familial risk for lung cancer, where the development of malignancy in individuals below the age of 50 years appears consistent with the Mendelian codominant inheritance of a rare autosomal gene [8].

Recent molecular epidemiologic studies have used lung cancer as a model to investigate gene–environment interactions, and individual susceptibility to tobacco carcinogen metabolism [9–11]. Phase I metabolizing enzymes, including the cytochrome P450 (CYP) enzymes, oxidize a wide range of substrates (e.g. polycyclic aromatic hydrocarbons in cigarette smoke) resulting in highly active procarcinogens. Several CYP-related enzymes display genetic polymorphisms in defined populations [12], and individual susceptibility to malignancy appears to be modified by the genotype of the enzyme [13]. Polymorphisms of CYP1A1, CYP2E1, and CYP2D6 have been implicated in the development of lung cancer [14,15]. Microsomal epoxide hydroxylase (mEH), another phase I enzyme, has dual roles in both activation and detoxification of environmental carcinogens, and has similarly been associated with lung cancer risk in nonsmokers [16].

The effect of such genetic polymorphisms is modified by further interactions with phase II enzymes (e.g. the glutathione S-transferase, GST, supergene family), which play a central role in the detoxification of toxic and carcinogenic electrophilic intermediates [17,18]. While GSTM1 polymorphisms have been associated with an increased risk of SCLC and squamous cell carcinomas, the combination of polymorphisms of CYP1A1 and GSTM1 variants has been associated with an increased risk for squamous cell carcinoma [14]. An association between GST isoenzymes and p53 mutations has recently been reported to increase risk for lung cancer development [19,20].

6.5 Molecular alterations in lung cancer

Several molecular alterations have been reported in human lung cancer over the past two decades [21]. The availability of several established SCLC cell lines has led to the extensive characterization of this cell type, but as this appears to be biologically distinct to NSCLC, important differences in the frequency of molecular alterations are noted between histologic subtypes of lung cancer.

Cytogenetic studies have reported several abnormalities in lung cancer, even within histologically similar tumors. The high frequency of chromosomal deletions (3, 11, 13, 17) seen in both SCLC and NSCLC suggests that loss of tumor suppressor function may be a critical early step in lung tumorigenesis [22]. Deletions of the short arm of chromosome 3 (p14–23) appear to be most frequent in SCLC [23], in both cell lines and fresh tissues, whereas 3p deletions are found in only 50% of NSCLC tumors [22]. Allelic loss on chromosome 3 has been reported to preceed p53 alteration in lung cancer progression [24]. A recent study of lung cancer cell lines has confirmed that 11p harbors several putative tumor suppressor genes which become inactivated at different stages of tumorigenesis [25]. Frequent loss of heterozygosity on chromosomes 13q and 17p have implicated the retinoblastoma (RB) and p53 tumor suppressor genes, respectively.

6.6 Tumor suppressor genes

The *p53* tumor suppressor gene encodes a 53 kDa polypeptide that regulates cell cycle progression, DNA repair, apoptosis, and neovascularization in normal and malignant cells via highly complex DNA and protein interactions [26]. This gene appears to have a central role in human malignancy, and has been characterized extensively in lung cancer, where mutations are reported in SCLC and NSCLC in up to 50% of cases [27]. Differences in techniques used to study *p53* (e.g. immunohistochemistry to evaluate protein accumulation vs. sequencing to characterize point mutations) account for variations in the published frequency of *p53* alterations in lung cancer. Patterns of *p53* mutations suggest a strong influence of tobacco carcinogens on the molecular pathogenesis of lung cancer [28,29]. Specifically, the high frequency of guanine (G) to thymine (T) transversions suggests a direct mutagenic effect of benzo[a]pyrene derived from cigarette smoke. By contrast, for lung cancers associated with radon exposure, G to T transversions are uncommon, with *p53* mutations comprising cytosine (C) to adenine (A) transversions, G to A transitions, and a 'hot-spot' mutation (AGG to ATG) at codon 249. Recent meta-analyses have attempted to clarify the prognostic value of *p53* in NSCLC [30,31]. Despite the limitations of selected patient populations (usually with early-stage disease who undergo surgical resection) and the technique of *p53* analysis (*Figure 6.8*), *p53* alterations appear to be associated with poor prognosis, particularly for patients with lung

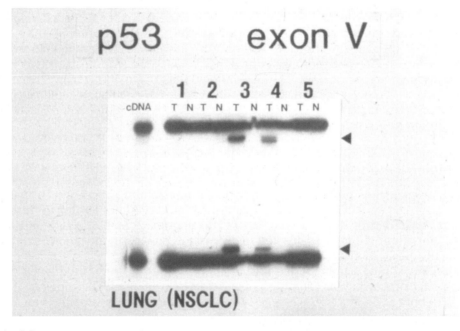

Figure 6.8

Single strand conformation polymorphism (SSCP) analysis of *p53* exon 5. Lung tumors (T) and matched normal (N) tissues from five patients were electrophoresed with a full-length *p53* cDNA marker to identify bands of interest. Electrophoretic mobility shifts were detected in two tumors (patients 3 and 4), and subsequently confirmed by sequencing to represent point mutations. An artifact of loading is present in the upper band of 3N. (A.G. Casson *et al. Journal of Surgical Oncology* © (1994) **56**, 13).

adenocarcinomas, and for *p53* missense mutations [32]. The finding of *p53* mutations in dysplastic bronchial epithelium suggests that *p53* is an early molecular alteration in lung tumorigenesis, and is further supported by the detection of *p53* mutations in cells obtained from sputum of high-risk patients preceding a diagnosis of lung cancer [33]. *p53* antibody has also been detected in serum from up to 30% of patients with lung cancer, particularly when the primary tumor has a missense mutation [34,35].

Whereas the retinoblastoma (*RB*) gene and its product are altered in almost 100% of SCLC, *RB* inactivation has been shown in only 10–30% of NSCLC. One study of 101 patients with stage I NSCLC reported reduced survival for 24 patients with absent *RB* (demonstrated immunohistochemically in surgically resected tumors), with even poorer prognosis for a subset of patients whose tumors also overexpressed *p53* protein [36]. However, in a larger study of 219 patients with NSCLC, no correlation was seen between loss of *RB* expression (32% of tumors) and survival [37]. This was also confirmed in a subsequent study, which found that loss of *p16* expression was more predictive of a poor prognosis [38]. *p16* point mutations are uncommon in lung cancer, and the most frequent mechanisms of *p16* inactivation are homozygous deletion and hypermethylation of *p16* promoter CpG islands. *p16* promoter hypermethylation is found in bronchial cell hyperplasia, and increases throughout premalignancy from squamous metaplasia to carcinoma *in situ*.

6.7 Oncogenes

Common dominant oncogene alterations in lung cancer include mutations in *RAS* and *MYC* family members. Mutational activation of the *KRAS* oncogene (codon 12) has been reported in up to one third of lung adenocarcinomas and large-cell undifferentiated carcinomas, but is rarely found in squamous cell carcinomas or SCLC [39]. *KRAS* activation has been reported to be associated with smoking history, exposure to asbestos, tumor differentiation, metastatic potential, and as an independent predictor of poor survival [40]. The prognostic value of *KRAS* mutations in NSCLC is currently being evaluated in a prospective multicenter clinical trial. *KRAS* mutations have also been found in cells obtained from sputum or bronchial lavage, and may be detected prior to the diagnosis of lung cancer [33].

Abnormal expression of all three members of the *MYC* family (*MYC*, *MYCN*, *MYCL*) has been detected in lung cancer. Amplification and overexpression of one or other of the *MYC* genes is found in up to 20% of tumors, but the significance of this finding is unclear. Amplification of *MYC* was initially described in a subset of SCLC cell lines with variant morphology [41]. These cell lines had elevated levels of bombesin isoenzymes, a rapid doubling time, increased tumorigenicity, and resistance to chemotherapy or radiation therapy. *MYC* overexpression is also seen in SCLC cell lines with intermediate neuroendocrine differentiation, but is variably expressed in classic SCLC, and NSCLC cell lines. It is thought that *MYC* expression is associated with SCLC progression, and *MYC* amplification is reported more frequently in treated than untreated SCLC, and is associated with reduced disease-free survival [42].

6.8 Growth factors and receptors

Lung tumors that produce growth factors may exhibit autocrine (and paracrine) growth by binding to specific growth factor receptors. Signal transduction usually results in phosphorylation of the receptor tyrosine kinase, resulting in a cascade of events including activation of protein kinase C, activation of oncogenes, increased cellular proliferation and tumorigenesis. This mechanism was initially described in SCLC cell lines, where autocrine growth mediated by bombesin resulted in clonal proliferation. By contrast, bombesin receptors are not expressed by NSCLC, but several other growth factors have been implicated in NSCLC progression.

Epidermal growth factor receptor (EGFR) has been demonstrated to be expressed using immunohistochemistry in a high percentage of squamous cell carcinomas and up to half of adenocarcinomas of the lung [43,44]. Southern analysis has also confirmed *EGFR* gene amplification. As NSCLC cell lines do not appear to produce EGF, the alternative ligand, transforming growth factor alpha (TGFα), is thought to function as an autocrine growth factor, and has been shown to be overexpressed in fresh tumors and cell lines [45]. The prognostic value of EGFR is unclear, but when coexpressed with TGFα, HER2 or p53, is associated with poor prognosis [46,47].

Despite initial reports of HER2 expression in up to one third of NSCLC and a clear association with reduced survival for adenocarcinomas [48], subsequent studies have reported HER2 expression in only 5% of NSCLC with no prognostic value [49,50]. Although several other growth factors and their receptors have been studied, including insulin-like growth factor (IGF), platelet-derived growth factor (PDGF), and hepatocyte growth factor (MET/HGFR), their role in lung cancer development and progression remains unclear [51].

6.9 Mesothelioma

Malignant mesothelioma is an uncommon tumor arising diffusely from the pleural surfaces. Asbestos exposure (and the synergistic effect of smoking) is recognized as a major risk factor for this disease, which has a long latency period (over 20 years). The incidence of mesothelioma is increasing and constitutes a significant public health problem. Furthermore, the prognosis for this malignancy is dismal, as conventional therapies are relatively ineffective in treating this disease. The histologic distinction between malignant mesothelioma and metastatic adenocarcinoma involving the pleura may be difficult, and currently additional immunohistochemical and electron microscopic studies are required to confirm the diagnosis.

The biology of malignant mesothelioma is poorly understood. The limited number of molecular markers studied, to date, suggest that alterations are quite different to lung cancer. Whereas *p53* and *RB* alterations are uncommon in mesothelioma, deletions of *p16*, *NF2* and *WT1* are found relatively frequently [52,53]. Several chromosomal deletions are reported to occur early, including loss of 1p, 4p, 4q, 6p, 15q and 22q, suggesting putative tumor suppressor genes at these loci [54,55].

6.10 Multiple primary tumors of the upper aerodigestive tract

Patients who have a primary lung cancer have a higher lifetime risk of developing second primary cancers of the upper aerodigestive tract (e.g. lung, esophagus, head and neck) [56–58]. This effect of field cancerization proposes that the epithelial surface of the upper aerodigestive tract shares a common carcinogen exposure, with an increased risk of second primary (vs. recurrent) carcinoma development. This hypothesis has been supported by several recent molecular studies of independent primary tumors arising from preneoplastic epithelia [59–61], and has implications for treatment and chemoprevention [62–64].

6.11 Implications for therapy

Over the next decade, the molecular alterations described above may well see increasing clinical application in the diagnosis and early detection of lung cancer, as prognostic markers and stratification factors for conventional therapies, and as intermediate markers in chemoprevention strategies [65–67].

There has been considerable interest in lung cancer chemoprevention using retinoids (vitamin A-related analogs) over the past decade [68]. Retinoic acid receptors, encoded on chromosome 3p (a region commonly deleted in lung cancer), have been shown to be altered frequently in lung cancer [69]. However, despite experimental evidence showing receptor up-regulation following 13-*cis*-retinoic acid administration [70,71], several large randomized clinical trials have failed to show any benefit for retinoids in the chemoprevention of lung cancer. Several ongoing trials are evaluating novel agents, including farnesyl transferase and tyrosine kinase inhibitors in high-risk patients with premalignant lesions.

Preliminary clinical studies have utilized molecular alterations in lung cancer as targets for innovative gene therapies [72]. Gene replacement therapy has been tested in a limited number of patients with advanced-stage unresectable NSCLC, otherwise unresponsive to conventional therapy [73,74]. Bronchoscopic injection of recombinant retrovirus or adenovirus expressing wild-type *p53* directly into an obstructing tumor with a *p53* mutation (detected by pretreatment biopsy) has resulted in tumor regression, with increased apoptosis seen in tumor biopsies taken after treatment [75,76]. Vector-related toxicity has been minimal. Additional clinical trials using antisense approaches are currently under evaluation [77].

It should be remembered, however, that the current epidemic of lung cancer could potentially be controlled through effective tobacco control and public health policies, as lung cancer appears, for the most part, to be an entirely preventable disease [78].

References

1. **Parkin, D.M., Pisani, P. and Ferlay, J.** (1999) *C.A. Cancer J. Clin.*, **49**, 33.
2. **Parkin, D.M. and Moss, S.M.** (2000) *Cancer*, **89**, 2369.
3. **Devesa, S.S., Blot, W.J., Stone, B.J.** *et al.* (1995) *J. Natl Cancer Inst.*, **87**, 175.
4. http://progressreport.cancer.gov.
5. **Wang, N., Zee, S., Zarbo, R.** *et al.* (1995) *Appl. Immunohistochem.*, **3**, 99.
6. **Chu, P., Wu, E. and Weiss, L.M.** (2000) *Mod. Pathol.*, **13**, 962.
7. **Wigle, D.A., Jurisica, I., Radulovich, N.** *et al.* (2002) *Cancer Res.*, **62**, 3005.
8. **Sellers, T.A., Bailey-Wilson, J.E., Elston, R.C.** *et al.* (1990) *J. Natl Cancer Inst.*, **82**, 1272.
9. **Hecht, S.S., Carmella, S.G., Foiles, P.G. and Murphy, S.E.** (1994) *Cancer Res.*, **54**, 1912s.
10. **Perera, F.P.** (1996) *J. Natl Cancer Inst.*, **88**, 496.
11. **Garte, S.** (2001) *Cancer Epidemiol. Biomarkers Prev.*, **10**, 1233.
12. **Garte, S., Gaspari, L., Alexandrie, A.K.** *et al.* (2001) *Cancer Epidemiol. Biomarkers Prev.*, **10**, 1239.
13. **Bartsch, H., Nair, U., Risch, A.** *et al.* (2000) *Cancer Epidemiol. Biomarkers Prev.*, **9**, 3.
14. **Le Marchand, L., Sivaraman, L., Pierce, L.** *et al.* (1998) *Cancer Res.*, **58**, 4858.
15. **MacLeod, S.L., Nowell, S., Massengill, J.** *et al.* (2000) *Clin. Chem. Lab. Med.*, **38**, 883.
16. **Zhou, W., Thurston, S.W., Liu, G.** *et al.* (2001) *Cancer Epidemiol. Biomarkers Prev.*, **10**, 461.
17. **Watson, M.A., Stewart, R.K., Smith, G.B.** *et al.* (1998) *Carcinogenesis*, **19**, 275.
18. **Malats, N., Camus-Radon, A.M., Nyberg, F.** *et al.* (2000) *Cancer Epidemiol. Biomarkers Prev.*, **9**, 827.
19. **Miller, D.P., Liu, G., De Vivo, I.** *et al.* (2002) *Cancer Res.*, **62**, 2819.
20. **Liu, G., Miller, D.P., Zhou, W.** *et al.* (2001) *Cancer Res.*, **61**, 8718.
21. **Sozzi, G., Pilotti, S., Pezzella, F.** *et al.* (1997) *Ann. Pathol.*, **17** (Suppl 5), 25.
22. **Whang-Peng, J., Knutsen, T., Gazdar, A.** *et al.* (1991) *Genes, Chrom., Cancer*, **3**, 168.
23. **De Fusco, P.A., Frytak, S., Dahl, R.J.** *et al.* (1989) *Mayo Clin. Proc.*, **64**, 168.
24. **Chung, G.T., Sundaresan, V., Hasleton, P.** *et al.* (1995) *Oncogene*, **11**, 2591.
25. **Bepler, G. and Koehler, A.** (1995) *Cancer Genet. Cytogenet.*, **84**, 39.
26. **Prives, C. and Hall, P.** (1999) *J. Pathol.*, **187**, 112.
27. **Hainaut, P., Soussi, T., Shomer, B.** *et al.* (1997) *Nucleic Acids Res.*, **25**, 151.

28. **Hainaut, P. and Vahakangas, K.** (1997) *Pathol. Biol. (Paris)*, **45**, 833.
29. **Fujita, T., Kiyama, M., Tomizawa, Y.** *et al.* (1999) *Int. J. Oncol.*, **15**, 927.
30. **Tammemagi, M.C., McLaughlin, J.R. and Bull, S.B.** (1999) *Cancer Epidemiol. Biomarkers Prev.*, **8**, 625.
31. **Mitsudomi, T., Hamajima, N., Ogawa, M. and Takahashi, T.** (2000) *Clin. Cancer Res.*, **6**, 4055.
32. **Tomizawa, Y., Kohno, T., Fujita, T.** *et al.* (1999) *Oncogene* **18**, 1007.
33. **Mao, L., Hruban, R.H., Boyle, J.O.** *et al.* (1994) *Cancer Res.*, **54**, 1634.
34. **Lubin, R., Zalcman, G., Bouchet, L.** *et al.* (1995) *Nat. Med.* **1**, 701.
35. **Wild, C.P., Ridanpaa, M., Anttila, S.** *et al.* (1995) *Int. J. Cancer*, **64**, 176.
36. **Xu, H.J., Quinlan, D.C., Davidson, A.G.** *et al.* (1994) *J. Natl Cancer Inst.*, **86**, 695.
37. **Reissmann, P.T., Koga, H., Takahashi, R.** *et al.* (1993) *Oncogene, ***8**, 1913.
38. **Kratzke, R.A., Greatens, T.M., Rubins, J.B.** *et al.* (1996) *Cancer Res.*, **56**, 3415.
39. **Rodenhuis, S., Slebos, R.J., Boot, A.J.** *et al.* (1988) *Cancer Res.*, **48**, 5738.
40. **Graziano, S.L., Gamble, G.P., Newman, N.B.** *et al.* (1999) *J. Clin. Oncol.*, **17**, 668.
41. **Carney, D.N., Gazdar, A.F., Bepler, G.** *et al.* (1985) *Cancer Res* **45**, 2913.
42. **Brennan, J., O'Connor, T., Makuch, R.W.** *et al.* (1991) *Cancer Res.*, **51**, 1708.
43. **Rusch, V., Klimstra, D., Venkatraman, E.** *et al.* (1997) *Clin. Cancer Res.*, **3**, 515.
44. **Nicholson, R.I., Gee, J.M. and Harper, M.E.** (2001) *Eur. J. Cancer, ***37** (Suppl. 4), S9.
45. **Hsieh, E.T., Shepherd, F.A. and Tsao, M.S.** (2000) *Lung Cancer, ***29**, 151.
46. **Ohsaki, Y., Tanno, S., Fujita, Y.** *et al.* (2000) *Oncol. Rep.*, **7**, 603.
47. **Brabender, J., Danenberg, K.D., Metzger, R.** *et al.* (2001) *Clin. Cancer Res.*, **7**, 1850.
48. **Kern, J.A., Schwartz, D.A., Nordberg, J.E.** *et al.* (1990) *Cancer Res.*, **50**, 5184.
49. **Pastorino, U., Andreola, S., Tagliabue, E.** *et al.* (1997) *J. Clin. Oncol.*, **15**, 2858.
50. **Cox, G., Vyberg, M., Melgaard, B.** *et al.* (2001) *Int. J. Cancer*, **92**, 480.
51. **Tsao, M.S., Yang, Y., Marcus, A.** *et al.* (2001) *Hum. Pathol.*, **32**, 57.
52. **Cheng, J.Q., Jhanwar, S.C., Klein, W.M.** *et al.* (1994) *Cancer Res.*, **54**, 5547.
53. **Kumar-Singh, S., Segers, K., Rodeck, U.** *et al.* (1997) *J. Pathol.*, **181**, 67.
54. **Lee, W.C., Balsara, B., Liu, Z.** *et al.* (1996) *Cancer Res.*, **56**, 4297.
55. **Balsara, B.R., Bell, D.W., Sonoda, G.** *et al.* (1999) *Cancer Res.*, **59**, 450.
56. **Licciardello, J.T., Spitz, M.R. and Hong, W.K.** (1989) *Int. J. Radiat. Oncol. Biol. Phys.*, **17**, 467.
57. **Fleisher, A.G., McElvaney, G. and Robinson, C.L.** (1991) *Ann. Thorac. Surg.*, **51**, 48.
58. **Massard, G., Wihlm, J.M., Ameur, S.** *et al.* (1996) *Eur. J. Cardiothorac. Surg.*, **10**, 397.
59. **Chung, K.Y., Mukhopadhyay, T., Kim, J.** *et al.* (1993) *Cancer Res.*, **53**, 1676.
60. **Braakhuis, B.J., Tabor, M.P., Leemans, C.R.** *et al.* (2002) *Head Neck*, **24**, 198.
61. **Tabor, M.P., Brakenhoff, R.H., Ruijter-Schippers, H.J.** *et al.* (2002) *Am. J. Pathol.*, **161**, 1051.
62. **Tabor, M.P., Brakenhoff, R.H., van Houten, V.M.** *et al.* (2001) *Clin. Cancer Res.*, **7**, 1523.
63. **Klaassen, I. and Braakhuis, B.J.** (2002) *Oral Oncol.*, **38**, 532.
64. **Klaassen, I., Cloos, J., Smeets, S.J.** *et al.* (2002) *Oncology*, **63**, 56.
65. **Rusch, V.W. and Dmitrovsky, E.** (1995) *Chest Surg. Clin. N. Am.*, **5**, 39.
66. **Iyengar, P. and Tsao, M.S.** (2002) *Surgical Oncology*, **11**, 167.
67. **O'Byrne, K.J., Cox, G., Swinson, D.** *et al.* (2001) *Lung Cancer*, **34** (Suppl. 2), S83.
68. **Khuri, F.R. and Lippman, S.M.** (2000) *Semin. Surg. Oncol.*, **18**, 100.
69. **Chambon, P.** (1996) *Faseb J.*, **10**, 940.
70. **Khuri, F.R., Lotan, R., Kemp, B.L.** *et al.* (2000) *J. Clin. Oncol.*, **18**, 2798.
71. **Martinet, N., Alla, F., Farre, G.** *et al.* (2000) *Cancer Res.*, **60**, 2869.
72. **Roth, J.A., Mukhopadhyay, T., Zhang, W.W.** *et al.* (1996) *Semin. Radiat. Oncol.*, **6**, 105.
73. **Nguyen, D.M., Spitz, F.R., Yen, N.** *et al.* (1996) *J. Thorac. Cardiovasc. Surg.*, **112**, 1372.
74. **Nemunaitis, J., Swisher, S.G., Timmons, T.** *et al.* (2000) *J. Clin. Oncol.*, **18**, 609.
75. **Roth, J.A.** (1996) *Hum. Gene Ther.*, **7**, 1013.
76. **Roth, J.A., Nguyen, D., Lawrence, D.D.** *et al.* (1996) *Nat. Med.*, **2**, 985.
77. **Khuri, F.R. and Kurie, J.M.** (2000) *Clin. Cancer Res.*, **6**, 1607.
78. **Kim, E.S., Hong, W.K. and Khuri, F.R.** (2000) *Chest Surg. Clin. N. Am.*, **10**, 663.

Colorectal cancer

7

7.1 Introduction

Colorectal cancer is the most common cause of death from cancer after cancer of the lung and breast. The age-adjusted death rates in the major industrialized countries of the world are shown in *Table 7.1* [1]. There has been little change in death rates over the last 50 years because, although there have been some improvements in survival, these have been masked by the increase in the incidence of the disease. The 5-year survival for colon cancer remains around 40% because of the late detection of the disease. In patients with nonresectable or disseminated disease at the time of presentation, the median survival is 7 months. In those who undergo surgery for their primary tumor and who undergo an apparently curative resection, over half will die within 5 years and 80% will have a recurrence within 2 years.

Molecular techniques are beginning to help in the understanding of the pathogenesis of the disorder and the genetic basis of the cancer is perhaps better understood than for any other cancer. There are two reasons for this. Firstly it has been recognized for over a quarter of a century that the majority of colorectal tumors develop from adenomas as described in Section 7.2 [2]. This has meant that molecular changes can be more clearly delineated than in other cancers. Secondly, there are a number of inherited forms of colorectal cancer, the genes for which have now been isolated, and many of these genes also play a role in sporadic disease. Oncogenes, tumor suppressor genes and cell cycle regulators all play a role in colorectal cancer as discussed below. However it is clear that environmental and dietary factors also have a major impact in this disease [3].

Colorectal cancers were originally classified by Dukes as A, B or C, according to depth and extent of tumor invasion. Although used widely, Dukes' classification has been modified extensively. To standardize staging of this disease, the UICC TNM system appears to provide greater precision in the identification of prognostic subgroups. The use of any of the onco-genes or tumor suppressor genes as prognostic indicators therefore has to be compared with this classification which remains the most useful method to date of determining survival.

Table 7.1 Age-adjusted death rates for colorectal cancer in the major industrialized countries of the world

Country	Age-adjusted death rates (year 2000) per 100 000 population	
	Male	Female
Germany	21.7	17.0
UK	18.7	13.8
France	18.3	12.1
Canada	16.4	11.6
USA	15.9	12.0
Japan	17.6	11.0
Russian Federation	17.5	12.7

Abstracted from [1].

7.2   The adenoma–carcinoma sequence

Histological observations led to the concept that most colorectal tumors arise from normal epithelium through increasingly more severe degrees of adenomatous dysplasia to frank carcinoma [2]. Adenomas are dysplastic but nonmalignant masses in the colon character-ized by their size, histologic type and degree of dysplasia. Several lines of evidence point to the fact that adenomas are frequently the precursor lesion to carcinomas. Early carcinomas are frequently seen in large adenomatous polyps and carcinomas can be surrounded by adenomatous areas. In addition, epidemiological data suggest that adenomas and carcino-mas are found in approximately similar distributions to carcinomas. Finally in animal models, carcinomas have been shown to develop from adenomas. The adenoma–carcinoma sequence describes this progression to carcinogenesis [2]. Since the original description of this process, an additional lesion, the aberrant crypt foci (ACF) has been described which is believed to be intermediate between normal tissue and adenomatous polyps [4].

Over 10 years ago, a number of oncogenes and tumor suppressor genes were assigned to each of these steps to produce a genetic pathway to colorectal cancer (*Figure 7.1*) [5]. Although the model suggests that each mutation occurs in a specific order, it is the accumulation of mutations which is more important than the order. Since the model was originally proposed, other genes have been identified in colorectal cancer as shown in *Figure 7.1* and some changes have been made to the pathway.

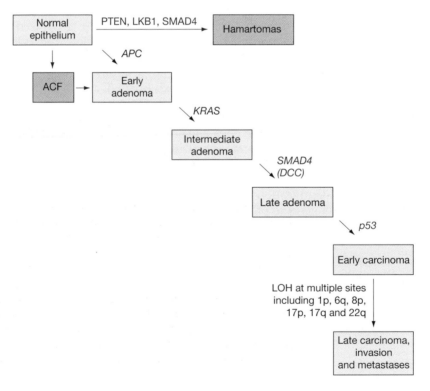

Figure 7.1

The paradigm for colorectal carcinogenesis – the adenoma–carcinoma sequence and associated oncogenes and tumor suppressor genes. Those genes shown in black are as described in the original model whereas those in purple were either identified at a later date or were incorrect in the original.

This model has been the paradigm for colorectal cancer development for many years and has led to much of our current understanding of colorectal carcinogenesis [6]. The original pathway is now however oversimplistic and several overlapping pathways seem more likely [7]. Approximately 85% of tumors show chromosomal instability (CIN) and are usually aneuploid, often left-sided, behave aggressively and show mutations in *APC*, *RAS* and *p53* [7]. The remaining 15% of tumors, which exhibit microsatellite instability (MSI), are often right-sided, diploid and behave indolently as described below in Section 7.5. In addition, mutations in *APC*, *KRAS* and *p53* are present at a much lower level in tumors showing a high degree of MSI [8]. Even within these two pathways there are subsets of tumors. Further evidence for alternative pathways come from studies which show that mutations in *APC*, *KRAS* as well as *p53* do not occur in all tumors and some tumors may only contain a mutation in one of these genes [9]. Another novel pathway has been described termed the CpG island methylator phenotype (CIMP) [10]. Two groups of tumors were identified. CIMP-positive tumors show a high degree of CpG island methylation in genes such as *p16* and *hMLH1* and are accompanied by mutations in *KRAS* and *TGFβRII*. CIMP-negative tumors, which by definition do not contain a high degree of methylation, are characterized by *p53* mutations. CIMP-positive tumors may show a degree of correlation with the MSI pathway [8]. Finally, colorectal cancers, arising from ulcerative colitis, do not develop from adenomas suggesting that they follow yet another different pathway [11]. Alternatives to the classic adenoma–carcinoma sequence are therefore likely to be described over the next few years.

7.3 Genetic changes in sporadic colorectal cancer – the role of oncogenes

7.3.1 *MYC*

Early studies of colorectal cancer using cell lines revealed amplification of *MYC* [12]. One study has found amplification in up to 26% of tumors, where it was most often found in advanced stages of tumor invasion. However because of its association with Dukes' stage, its prognostic value is not likely to be of any additional value [13].

Approximately one third of colorectal cancers show elevated expression of *MYC* but in general there are few diagnostic or prognostic implications [14]. Two studies have suggested a role for *MYC* as a marker for monitoring the transition from the benign to malignant state. Levels of *MYC* transcripts determined by northern blotting were closely related to histological type and size of colorectal polyps. In particular, adenomas containing carcinoma *in situ* or high-grade dysplasia expressed high levels of *MYC* [15,16].

Elevated expression of the *MYC* gene has also been particularly associated with tumors of the distal part of the colon [17]. On the basis of this observation it has been suggested that elevated levels of *MYC* may be a marker of a subgroup of sporadic colorectal cancers with a different etiology to those occurring on the right side of the colon. Additional studies following on from this observation have indicated loss of *MYC* regulation in carcinomas deleted for the chromosomal region containing the *APC* gene. Studies have shown that activation of *MYC* gene expression is central to signal transduction through the *APC* gene [18].

7.3.2 *RAS*

Alterations in *RAS* have been studied extensively in colorectal cancer. There has been no evidence for amplification or rearrangement of the gene, but elevated expression of *RAS* and frequent mutations in both *KRAS* and *HRAS* have been identified.

Elevated levels of the *RAS* gene family transcripts have been identified in both prema-lignant and malignant tumors of the colon and rectum, suggesting that elevated expres-sion might be critical in the process of carcinogenesis [19]. As all polyps do not progress, despite showing increased *RAS* expression, the observation is consistent with the involve-ment of other genes in the development of cancers as described above.

Analysis of point mutations in the *RAS* gene has provided more consistent results and suggests that *RAS* mutations occur early in the development of colorectal tumors. A total of 37–60% of colorectal tumors have been found to have an activated *RAS* gene with an increasing frequency in larger and more advanced lesions. The majority of the mutations occur at just three codons of *KRAS*, codons 12, 13 or 61 [5,20]. In addition, approximately 50% of adenomas contain a *RAS* gene mutation although mutations have only been found in 9% of those adenomas less than 1 cm in size [20]. This suggested that *RAS* muta-tions might be required for the conversion of small adenomas to larger ones by clonal expansion of the cell carrying the mutation. Subsequent studies have however detected *KRAS* mutations in the grossly normal mucosa of patients with colorectal cancer [21] which was shown to be due to the presence of aberrant crypts within the grossly normal-appearing mucosa indicating *RAS* involvement at an even earlier stage [22]. Over 80% of nondysplastic crypts and 63% of dysplastic crypts have been shown to contain *KRAS* mutations [23]. *APC* mutations may be required before these lesions progress further although the order in which the two mutations occurs may be different for sporadic and inherited disease [23].

For many years, the role *RAS* mutations play in colorectal carcinogenesis, particularly with respect to progression and prognosis, has been under debate. Ten years ago it was suggested that the specific position and amino acid substitution in the *KRAS* gene might influence the outcome, such that the genotype of *KRAS* could be used to identify a subset of patients with either indolent or aggressive disease [24]. Distant metastases were associ-ated with glycine to aspartic acid substitutions at codon 12 whereas tumors with valine substitutions at this position did not spread beyond the pericolonic and perirectal lymph nodes. Tumors with codon 13 mutations did not progress locally or distantly. As other studies gave conflicting results, meta-analysis of *RAS* mutations in colorectal cancer has been carried out to clarify the situation. Analysis of over 2500 patients indicated that there was a statistically significant association between the presence of a *KRAS* mutation and increased risk of recurrence and death [25]. Any mutation of a G to T but not A to C increased the risk of recurrence. When individual specific mutations were evaluated only the glycine to valine substitution conveyed an independent increased risk of recurrence. Extension of this study, on 3439 patients from 42 centers in 21 countries, has supported these findings, again highlighting that only this one amino acid substitution, found in 8.6% of all patients, had a statistically significant impact on failure-free survival and over-all survival [26]. In addition it was shown that the impact of this mutation in Dukes' C patients compared to Dukes' B patients was greater, suggesting not only that the presence of the glycine to valine mutation is important in progression but that it might predispose to a more aggressive biological behavior in patients with advanced cancer.

As *RAS* mutations have been proven to be a relatively early event in the development of colorectal cancer, the possibility of developing a molecular screening test for colorectal cancer has been considered to detect cancers at an early stage [27]. This test is dependent on the ability to detect *RAS* mutations in fecal samples. Preliminary data showed that it was possible to detect *RAS* mutations in the stools of eight out of nine patients whose can-cers had a known *RAS* mutation. Two of the positive cases had only adenomas present at the time of testing, suggesting that this technique had the potential to be used as a screening method, particularly if other genes involved in colorectal carcinogenesis could be included in the analysis. A second study confirmed the feasibility of this technique by identifying *RAS* mutations in the stools of approximately 50% of cases studied [28]. These initial studies were however limited by technical problems which have recently been overcome by the use of a more reproducible method of DNA extraction and a more robust PCR method [29]. In addition, the inclusion of three genetic markers associated with colorectal cancer, *p53*, a microsatellite instability marker BAT26, as well as *RAS*, were shown to detect up to 70% of colorectal cancers. It now remains to be determined if this approach is sufficiently sensitive to be used routinely as a screening tool given that some of the tumors in these studies were advanced, with patients already symptomatic. Larger trials of normal individuals, to evaluate the specificity of the three existing markers and studies of patients with only adenomas should prove the general usefulness of what appears to be a promising method of early detection of colorectal cancer. Further developments include the evaluation of *APC* mutations in fecal samples given that this gene is involved at an earlier stage than *RAS* (see Section 7.4.2). In addition, an alternative to screening fecal samples for mutations is to identify them in plasma DNA and an early study on a small group of patients has shown the feasibility of this approach [30].

7.3.3   Other oncogenes

The *SRC* gene has also been implicated in early colorectal cancer development with a 5–8-fold increase in activity in the majority of tumors [31]. This level of increase has been found in premalignant lesions and in small adenomatous polyps as well as benign polyps containing villous changes or severe dysplasia. *SRC* has therefore been associated with tumor progression since higher levels of expression are found in tumors compared to polyps and still higher levels can be detected in distant metastases such as those in the liver.

Overexpression of the antiapoptotic proto-oncogene *AKT* has been detected immuno-histochemically in 57% of colorectal cancers as well as in adenomas, again suggesting that this gene may be involved in early events in the development of this cancer [32]. It is less commonly overexpressed in colorectal cancers with MSI again suggesting the involvement of different pathways in the development of colorectal cancers. Early inhibition of apoptosis during colorectal carcinogenesis may be due to overexpression of *AKT*.

Finally *HER-2/neu* has been shown to be overexpressed in at least half of all colorectal cancers and has generally, though not consistently, been associated with a poor prognosis [33,34].

7.4   Genetic changes in sporadic colorectal cancer – the role of tumor suppressor genes

7.4.1   *p53*

p53 was one of the first genes to be studied extensively in colorectal cancer and close to a thousand papers have now appeared on the subject [35]. Mutation of *p53* is a relatively late event in colorectal carcinogenesis occurring at the adenoma to carcinoma transition, as shown in *Figure 7.1*. In a review of 25 studies, the incidence of *p53* mutations in colorectal cancers was shown to be between 31 and 52% [35 and www.iarc.fr/p53/ or www.umd.necker.fr:2001/P53]. Mutations are most frequently found in distal tumors, high-stage tumors, in aneuploid tumors and those without MSI or *KRAS* mutations. The frequency of mutations is lowest in mucinous tumors and tumors found in HNPCC families and they are rare in adenomas as would be expected from the pathway model for MSI-positive tumors [8]. The majority of mutations are G:C to T:A transitions primarily occurring at CpG dinucleotides [36]. Mutations at five hotspots account for 43% of all mutations in colorectal cancer as shown in *Table 7.2* [36]. Mutations in the conserved

Table 7.2   *p53* mutations in colorectal cancer

Rank	Codon number	Type of mutation	Frequency (%)
1	175	G > A	12.62
2	248	G > A	7.58
3	248	C > T	7.25
4	282	C > T	5.64
5	273	G > A	5.30
6	245	G > A	4.30
7	273	C > T	3.49
8	213	C > T	2.08
9	245	G > A	1.48
10	196	C > T	1.48

Adapted from [31].

regions of *p53* are more frequent in tumors of the distal colon as compared to the proximal colon which may reflect a different etiology [37]. Transversions are more frequent than transitions in the distal colon again perhaps reflecting a different etiology of the cancers [38].

Many different techniques have been used to look for *p53* mutations and to relate them to the patient outcome and this variation in technology may account for some of the conflicting results obtained. Immunohistochemical (IHC) studies have found widespread use and have not always shown consistent results when compared to molecular studies. The best correlation between IHC results and mutation analysis has been seen if monoclonal antibodies (D07 and Pab1801) are used rather than polyclonal ones such as CM-2 [39,40]. Variation in scoring of IHC results has also been problematic and in some studies only nuclear staining was considered to be important whereas in other studies, cytoplasmic staining was also recorded. Wild-type protein has also been shown to accumulate in the absence of a mutation, probably because of failure of the p53 degradation pathway which can also lead to false results [35]. Molecular testing is not without its own problems which primarily include the sensitivity of scanning techniques such as SSCP, leading to missed mutations or to the presence of normal cells within the tumor causing high normal background in sequencing, which can mask the presence of mutations.

The majority of reports indicate that *p53* mutations are associated with a poor prognosis but this is by no means universal as reviewed in [35] in which 14 studies showed a worse survival, eight found no association and three showed a better prognosis. This may be as a result of technological differences described above but may also reflect differences in the patient groups included in the studies such as different sites of tumors or stages of disease. For example, in a large population-based study of over 1400 patients, there were no overall prognostic implications for the group as a whole but the presence of a mutation in tumors of the proximal colon did correlate with a poorer prognosis [41]. Another study has corroborated this association between *p53* mutations, tumor location and prognosis [39]. However, in contrast, a further large study of 995 patients showed no prognostic correlations in the series overall, with tumor site or stage subgroups [42].

Different types of mutation, e.g the mutation at codon 245 [41] and mutations in the L3 zinc-binding domain [38] have been suggested to have prognostic significance in a similar manner to that described for a specific *KRAS* mutation (see Section 7.3.2). This result has now been confirmed in two studies after 5 years [43] and 10 years of follow up [44]. There is more convincing evidence that the group of patients with wild-type *p53* have a survival advantage from the use of 5-fluorouracil chemotherapy [45,46]. At present

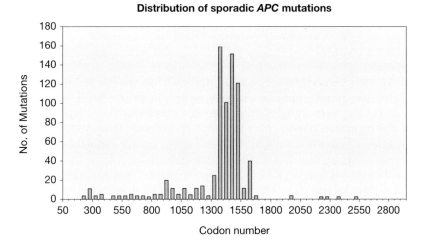

Distribution of sporadic *APC* mutations

Figure 7.2

Distribution of mutations in *APC* in sporadic colorectal cancer. The localization of the majority of mutations in the mutation cluster region (MCR: codons 1286–1513) can be seen. The codon number is shown on the x-axis.

p53 alterations have little or no value as prognostic indicators in the routine clinical setting though some associations with mutations in specific regions of the gene and in their role in response to chemotherapy may be borne out with larger studies.

7.4.2 Adenomatous polyposis coli gene (*APC*)

Loss of heterozygosity of chromosome 5q was originally found in around 30% of adenomas and up to 50% of colorectal cancers [47], suggesting the involvement of this region in the development of sporadic colorectal cancer as well as in familial adenomatous polyposis (FAP) (see Sections 3.5.1 and 5.4.1). Following the cloning of the *APC* gene, mutations have been identified in around 60–70% of sporadic colorectal cancers and adenomas [48,49]. In addition, the second copy of *APC* is either lost or can be inactivated by a second mutation [48,49]. Two 'hits' are necessary for the development of adenomas [50–52]. Mutations have been identified in adenomas as small as 3 mm in diameter and have also been found in aberrant crypt foci [53]. These observations, together with the observation that the frequency of mutations in adenomas is similar to that in carcinomas, suggests that *APC* mutations are likely to be an initiating step in colorectal carcinogenesis.

The distribution of mutations in sporadic cancers differs from that seen in FAP [49]. Two-thirds of somatic mutations are found in a region of the *APC* gene known as the mutation cluster region (MCR; *Figure 7.2*) a region which has been associated with profuse polyposis in FAP patients which represents only 8% of the coding sequence of the gene (codons 1286–1513). Three somatic mutations at codons 1309, 1450 and 1554 are the most commonly found (http://perso.curie.fr/Thierry.Soussi/APC.html). However as with germ-line mutations, the majority of sporadic mutations are truncating mutations, approximately 60% of which are small deletions or insertions causing frameshifts and the majority of the remainder are point mutations, primarily C to T transitions [49]. G to T transversions account for less than 15% of somatic mutations in *APC* suggesting that the early stages of colorectal carcinogenesis do not occur in response to a mutagen. There is strong interdependence of the two mutations in sporadic colorectal cancer as well as in

FAP. If the first mutation occurs within the MCR, the second hit is loss of heterozygosity (LOH). If the first hit is outside the MCR, the second mutation is a truncating mutation within the region [54]. A review of published data has redefined the MCR from codons 1281–1556 based on statistical theory rather than subjective assessment and examined the interdependence between the first and second mutation with respect to selective advantage [55]. This showed that two mutations within the MCR are the most likely to have a selective advantage compared to one mutation in the MCR with the second hit being LOH and these were significantly more advantageous than two mutations outside the MCR or one mutation outside plus LOH for the second. This model allows quantification of how different combinations of *APC* mutations confer different growth advantages although the mechanism requires further investigation.

As discussed in the section on *RAS*, many investigators have evaluated the possibility of developing a molecular test to screen for early cancers by detection of mutations in oncogenes and tumor suppressor genes in fecal samples. As described here, the earliest alteration in colorectal carcinogenesis is a mutation of the *APC* gene which makes it the optimal candidate to test for. The variety of different mutations in *APC* make this technically more difficult, but through the use of a technique called digital protein truncation, it has been shown that mutations can be detected in over half of the patients tested, though not in the control subjects [56]. This approach may lead to new methods for the early detection of colorectal cancers.

7.4.3 SMAD

Allele loss at chromosome 18q21 is well documented in colorectal cancers and is seen in over 65% of cases [57]. Lying within this region are three genes, *DCC* (Deleted in Colon Cancer), *SMAD4/DCP4* and *SMAD2*. For many years the *DCC* gene was thought to be the target of 18q loss but this has now been shown to be incorrect and mutations in *DCC* are very rare in colorectal cancer [58]. The two genes shown to be involved in the 18q loss are instead *SMAD4* and *SMAD2*. *SMAD4* codes for a protein which is a downstream regulator in the TGFβ signal transduction pathway and acts as a trimer with SMAD2 and SMAD3. *SMAD4* is also the gene mutated in juvenile polyposis as described in Section 5.4.3. Inactivation of *SMAD4* and *2* occurs by deletion of both alleles or by intragenic mutation of one allele and by deletion of the other making them classical tumor suppressor genes. *SMAD2* mutations only occur in about 5% of colorectal cancer [59] but *SMAD4* mutations are seen in approximately 25% of cases [60,61]. Mutations of *SMAD4* are however rarely associated with tumors with a MSI-positive phenotype [61,62]. This finding suggests that the MSI-positive pathway diverges prior to *SMAD4* inactivation [62] *SMAD4* mutations have also been associated with the presence of metastatic disease [63].

7.4.4 MCC

The 'mutated in colorectal cancer' gene (*MCC*) was originally isolated in a search for the causative gene for familial adenomatous polyposis (FAP). Although it was shown not to be involved in the inherited condition, mutations in *MCC* have been identified in a proportion of sporadic colorectal cancers [64]. However studies have shown that this gene does not appear to have a major role in prognosis or diagnosis of colorectal cancer [65].

7.4.5 NM23

NM23 was initially isolated because of its association with a low metastatic phenotype in melanoma cell lines and was localized to 17q21, a region which is also commonly deleted in colorectal tumors [66]. It has a major role in synthesis of nucleotide triphosphates other than ATP. Down-regulation of the gene has been associated with metastatic progression in many types of cancer. In colorectal cancer, although a few studies have indicated that it may be a useful marker of progression [67], the majority of recent studies, including a 5-year prospective study, have not found any evidence that it plays a major role in colorectal carcinogenesis [68].

7.4.6 Other changes

LOH at chromosome 8p has been a consistent finding in colorectal cancers and two regions on 8p have been delineated at 8p23.2-22 and at 8p21.3-8p11.22 [69]. LOH has been identified in 44–48% of cases [69,70]. LOH at this region is seen in only 10% of adenomas suggesting that this involvement is a relatively late event [71]. A novel gene has been isolated from 8p21, called *REAM* (for reduced expression associated with metastasis). Expression of this gene was reduced in over half of colorectal cancers tested, particularly those in advanced stages with liver metastasis. Somatic mutations were also found in several cases [72]. The possibility of two tumor suppressor genes within the 8p11, which confer different and independent roles in colorectal carcinogenesis, has also been described [73].

7.5 Mismatch repair defects

Microsatellite instability (MSI), also termed replication error positive or RER positive (see Section 4.6), has been found in around 15–25% of sporadic colorectal cancers as summarized in *Table 7.3* [74–81]. Tumors can be divided into MSI-high (MSI-H) where two or more markers are positive or MSI-low (MSI-L) when less than two are positive. MSI-H tumors are characterized as diploid, high-grade and mucinous tumors located primarily in the proximal colon, features which distinguish them from microsatellite stable cancers [77,80,82]. The MSI-H phenotype is associated with a good prognosis, particulary with tumors of the proximal colon [77,79,80,82] though the opposite effect is seen for the rarer distal tumors [79]. Immunohistochemical staining for MMR proteins can be used to identify those tumors with MSI and a MMR gene defect (see *Figure 5.4*) [80].

A small proportion of apparently sporadic colorectal cancers which exhibit MSI have subsequently been shown to have germ-line mutations in one of the MMR genes [81,83]. The group of MSI-positive patients, who show the highest rate of germ-line mutations, are patients who developed sporadic colon cancer at a relatively young age (<35 years). MSI-H was seen in 58% of cases and germ-line mutations have been identified in almost half of those tested [84]. These findings obviously have implications for genetic testing and subsequent management of their children. However, the majority of MSI-H sporadic cancers arise from inactivation of *MLH1*. The major cause of inactivation has been shown

Table 7.3 Microsatellite instability in sporadic colorectal cancers

Number of tumors analyzed	% with microsatellite instability	Reference
46	13	74
90	28	75
230	16	77
137	12	78
255	12	79
310	11	80
257	20	81

to be hypermethylation of the promotor of *MLH1* which has been detected in 70–90% of cases [83–87]. Promoter hypermethylation has often been shown to be biallelic [88]. Hypermethylation of the *MLH1* promoter is also present in preneoplastic lesions and in adenomas indicating that it is an early event in colorectal carcinogenesis [89,90].

7.6 TGFβ

Loss of TGFβ control is a critical event in colorectal carcinogenesis [91,92] with over 75% of cases having lost responsiveness to this control mechanism [93]. The major mechanism of loss is through mutation of both alleles of the TGFβRII as seen in the majority of MSI-positive tumors. The mutations occur within a poly-A tract in the coding region of the receptor leading to a truncated, nonfunctional receptor [94]. The inactivating mutations are a relatively late event and are associated with progression to carcinoma [95]. In MSI-negative tumors, inactivation of the TGFβ signaling pathway occurs by mutations in the SMAD proteins as described above [62]. Finally loss of TGFβ responsiveness can occur by mutation of the receptor directly [93].

7.7 Cell cycle regulators

Epigenetic changes play a major role in colorectal cancer as described above for *MLH1*. The *p16* gene is a cyclin-dependent kinase inhibitor (CKI) as described in Chapter 4 which targets cyclin D/CDK4–6 and reduces phosphorylation of RB. Aberrant methylation of CpG residues in the 5′ end of the *p16* gene has been identified in colorectal carcinomas, the result of which is transcriptional silencing of the gene [96,97]. Methylation of this gene is found in 16–42% of tumors [98–100]. In general, *p16* aberrations have been associated with a poor prognosis [98] and are more commonly found in Dukes' C and D tumors [99]. The highest levels of abnormalities have been identified in tumors of the proximal colon [100] and with MSI-positive tumors [101].

p27 is also a CKI which regulates G_1 to S progression by targeting cyclin E/CDK2. Reduced expression of p27 has been identified in colorectal cancers and correlates with poor survival [101–103]. Decreased expression has also been associated with increased likelihood of the presence of lymph node metastases independently of depth of tumor invasion [104].

References

1. **American Cancer Society**. (2003) Cancer Around the World, 2000, Death Rates per 100 000 population for 45 Countries. This can be found under Professionals, facts and Figures, Cancer Facts & Figures 2003 at http://www.cancer.org/.
2. **Morson, B.** (1974) *Cancer*, **34**, 845.
3. **Takayama, T., Katsuki, S., Tkahashi, Y. et al.** (1998) *N. Engl. J. Med.*, **329**, 1977.
4. **Gertig, D.M. and Hunter, D.J.** (1998) *Semin. Cancer Biol.*, **8**, 285.
5. **Fearon, E.R. and Vogelstein, B.** (1990) *Cell*, **61**, 759.
6. **Houlston, R.S.** (2001) *J. Clin. Pathol: Mol. Pathol.*, **54**, 206.
7. **Fodde, R.** (2002) *Eur. J. Cancer*, **38**, 867.
8. **Jass, J.R.** (2002) *Surg. Clin. N. Amer.*, **82**, 891.
9. **Smith, G., Carey, F.A., Beattie, J. et al.** (2002) *Proc. Natl Acad. Sci.*, **99**, 9433.
10. **Toyota, M., Ohe-Toyota, M., Ahuja, N. and Issa, J.-P.** (2000) *Proc. Natl Acad. Sci.*, **97**, 710.
11. **Ilyas, M. and Tomlinson, I.P.M.** (1996) *Histopathology*, **28**, 3889.
12. **Augenlicht, L.H.S., Wadler, G., Corner, C. et al.** (1997) *Cancer Res.*, **57**, 1769.
13. **Masramon, L., Arribas, R., Tortola, S. et al.** (1998) *Br. J. Cancer*, **77**, 2349.

14. **Smith, D.R., Myint, T. and Goh, H.-S.** (1993) *Br. J. Cancer*, **68**, 407.
15. **Imaseki, H., Hayashi, H., Taira, M. *et al.*** (1989) *Cancer*, **64**, 704.
16. **Pavelic, Z.P., Pavelic, L., Kuvelkar, R. and Gapany, S.R.** (1992) *Anticancer Res.*, **12**, 171.
17. **Astrin, S.M. and Costanzi, C.** (1989) *Sem. Oncol.*, **16**, 138.
18. **He, T.C., Sparks, A.B., Rago, C. *et al.*** (1998) *Science*, **281**, 1509.
19. **Field, J.K. and Spandidos, D.A.** (1990) *Anticancer Res.*, **10**, 1.
20. **Bos, J.L.** (1988) *Mut. Res.*, **195**, 255.
21. **Minamoto, T., Ronai, Z., Yamashita, N., *et al.*** (1994) *Int. J. Oncol.*, **4**, 397.
22. **Yamashita, N., Minamoto, T., Ochiai, A., Onda, M. and Esumi, H.** (1995) *Cancer*, **75**, 1527.
23. **Takayama, T., Ohi, M., Hayashi, T. *et al.*** (2001) *Gastroenterology*, **121**, 599.
24. **Finkelstein, S.D., Sayegh, R., Bakker, A. and Swalensky, P.** (1993) *Arch. Surg.*, **28**, 526.
25. **Andreyev, H.J., Norman, A.R., Cunningham, D. *et al.*** (1998) *J. Natl. Cancer Inst.*, **90**, 675.
26. **Andreyev, H.J., Norman, A.R., Cunningham, D. *et al.*** (2001) *Br. J. Cancer*, **85**, 692.
27. **Sidransky, D., Tukino, T., Hamilton, S.R. and Vogelstein, B.** (1992) *Science*, **256**, 102.
28. **Smith-Ravine, J., England, J., Talbot, I.C. and Bodmer, W.** (1995) *Gut*, **36**, 81.
29. **Dong, S., Traverso, G., Johnson, C. *et al.*** (2001) *J. Natl. Cancer Inst.*, **93**, 858.
30. **Kopreski, M.S., Benko, F.A., Borys, D.J. *et al.*** (2000) *J. Natl. Cancer Inst.*, **92**, 918.
31. **Irby, R.B. and Yeatman, T.J.** (2000) *Oncogene*, **19**, 5636.
32. **Roy, H.K., Olusola, B.F., Clemens, D.L. *et al.*** (2002) *Carcinogenesis*, **23**, 201.
33. **Ross, J.S. and McKenna, B.J.** (2001) *Cancer Invest.*, **19**, 554.
34. **McKay, J.A., Loane, J.F., Ross, V.G. *et al.*** (2002) *Br. J. Cancer*, **86**, 568.
35. **Iacopeta, B.** (2003) *Hum. Mut.*, **21**, 271.
36. **Soussi, T. and Beroud, C.** (2003) *Hum. Mut.*, **21**, 192.
37. **Jervall, P., Makinen, M., Karttunen, T., Makela, J. and Vinko, P.** (1997) *Int. J. Cancer*, **74**, 97.
38. **Borresen-Dale, A.L., Lothe, R.A., Meling, G.I. *et al.*** (1998) *Clin. Cancer Res.*, **4**, 203.
39. **Tolbert, D., Noffsinger, A.E., Miller, M.A. *et al.*** (1999) *Mol. Pathol.*, **12**, 54.
40. **Veloso, M., Wrba, F., Kaserer, K. *et al.*** (2000) *Virchows Arch.*, **437**, 241.
41. **Samowitz, W.S., Curtin, K., Ma, K.N. *et al.*** (2002) *Int. J. Cancer*, **99**, 597.
42. **Soong, R., Powell, B., Elsaleh, H. *et al.*** (2000) *Eur. J. Cancer*, **36**, 2053.
43. **Russo, A., Miglivacca, M., Zanna, I. *et al.*** (2002) *Cancer Epidemiol. Bio. Prev.*, **11**, 1322.
44. **Diez, C.B., Thorstensen, L., Meling, G.I. *et al.*** (2003) *J. Clin. Oncol.*, **21**, 820.
45. **Elsaleh, H., Powell, B., McCaul, K. *et al.*** (2001) *Clin Cancer Res.*, **7**, 1343.
46. **Barratt, P.L., Seymour, M.T., Stenning, S.P. *et al.*** (2002) *Lancet*, **360**, 1381.
47. **Vogelstein, B., Fearon, E.R., Hamilton, S.R. *et al.*** (1988) *N. Engl. J. Med.*, **319**, 525.
48. **Nagase, H. and Nakamura, Y.** (1993) *Hum. Mut.*, **2**, 425.
49. **Miyoshi, I., Nagase, H., Ando, H. *et al.*** (1992) *Hum. Mol. Genet.*, **1**, 229.
50. **Ichii, S., Hori, A., Nakatsuru, S. *et al.*** (1992) *Hum. Mol. Genet.*, **1**, 387.
51. **Ichii, S., Takeda, S., Horii, A. *et al.*** (1993) *Oncogene*, **8**, 2399.
52. **Levy, D.B., Smith, K.J., Beazer-Barclay, Y. *et al.*** (1994) *Cancer Res.*, **54**, 5953.
53. **Takayama, T., Ohi, M., Hayashi, T. *et al.*** (2001) *Gastroenterology*, **121**, 599.
54. **Rowan, A.J., Lamlum, H., Ilyas, M. *et al.*** (2000) *Proc. Natl Acad. Sci.*, **97**, 3352.
55. **Cheadle, J.P., Krawczk, M., Thomas, M.W. *et al.*** (2002) *Cancer Res.*, **62**, 363.
56. **Traverso, G., Shuber, A., Levin, B. *et al.*** (2002) *N. Engl. J. Med.*, **346**, 311.
57. **Thiagalingam, S., Lengauer, C., Leach, F.S. *et al.*** (1996) *Nat. Genet.*, **13**, 3443.
58. **Barbera, V.M., Martin, M., Marinoso, L. *et al.*** (2000) *Biochem. Biophys. Acta*, **1502**, 283.
59. **Eppert, K., Scherer, S.W., Ozeelik, H. *et al.*** (1996) *Cell*, **86**, 543.
60. **Salovaaro, R., Roth, S., Loukola, A. *et al.*** (2002) *Gut*, **51**, 56.
61. **Takagi, Y., Kohmura, H., Futamura, M. *et al.*** (1996) *Gastroenterology*, **111**, 1369.
62. **Woodford-Richens, K.L., Rowan, A.J., Gorman, P. *et al.*** (2001) *Proc. Natl Acad. Sci.*, **98**, 9719.
63. **Maitra, A., Molberg, K., Albores-Saavedra, J. and Lindberg, G.** (2000) *Am. J. Pathol.*, **157**, 1105.
64. **Kinzler, K., Nilbert, M.C., Vogelstein, B. *et al.*** (1991) *Science*, **251**, 1366.
65. **Curtis, L.J., Bubb, V.J., Gledhill, S. *et al.*** (1994) *Hum. Mol. Genet.*, **3**, 443.
66. **Purdie, C.A., Piris, J., Bird, C.C. and Wyllie, A.H.** (1995) *J. Pathol.*, **175**, 297.

67. **Cohn, K.H., Wang, F., Desoto-Paix, F.** *et al.* (1991) *Lancet*, **338**, 722.
68. **Bazan, V., Migliavacca, M., Zanna, I.** *et al.* (2002) *J. Cancer Res. Clin. Oncol.*, **128**, 650.
69. **Fukiwara, Y., Emi, M., Ohata, H.** *et al.* (1993) *Cancer Res.*, **53**, 1172.
70. **Lerebours, F., Olschwang, S., Thuille, B.** *et al.* (1999) *Genes Chrom. Cancer*, **25**, 147.
71. **Cunningham, C., Dunlop, M.G., Bird, C.C. and Wyllie, A.H.** (1994) *Br. J. Cancer*, **70**, 18.
72. **Oyama, T., Miyoshi, Y., Koyama, K.** *et al.* (2000) *Genes Chrom. Cancer*, **29**, 9.
73. **Chughtai, S.A., Crundwell, M.C., Cruickshank, N.R.** *et al.* (1999) *Oncogene*, **18**, 657.
74. **Aaltonen, I.A., Peltomaki, P., Leach, F.S.** *et al.* (1993) *Science*, **260**, 812.
75. **Thibodeau, S.N., Bren, G. and Schaid, D.** (1993) *Science*, **260**, 816.
76. **Peltomaki, P.** (2003) *J. Clin. Oncol.*, **21**, 1174.
77. **Lothe, R.A., Peltomaki, P., Meling, G.I.** *et al.* (1993) *Cancer Res.*, **53**, 5849.
78. **Ionov, Y., Peinado, M.A., Malkbosyan, S.** *et al.* (1993) *Nature*, **363**, 558.
79. **Jernvall, P., Makinen, M.J., Karttunen, T.J.** *et al.* (1999) *Eur. J. Cancer*, **35**, 197.
80. **Ward, R., Meagher, A., Tomlinson, I.** *et al.* (2001) *Gut*, **48**, 821.
81. **Cunningham, J.M., Kim, C.Y., Christensen, E.R.** *et al.* (2001) *Am. J. Hum. Genet.*, **69**, 780.
82. **Gryfe, R., Kim, H., Hsieh, E.T.K.** *et al.* (2000) *N. Engl. J. Med.*, **342**, 69.
83. **Kuismanen, S.A., Holmberg, M.T., Salovaara, R.** *et al.* (2000) *Am. J. Pathol.*, **156**, 1773.
84. **Liu, B., Farrington, S.M., Petersen, G.M.** *et al.* (1995) *Nature Med.*, **1**, 348.
85. **Miyakura, Y., Sugano, K., Konishi, F.** *et al.* (2003) *Genes, Chrom. Cancer*, **36**, 17.
86. **Menigatti, M., Gregorio, G., Borghi, F.** *et al.* (2001) *Genes, Chrom. Cancer*, **31**, 357.
87. **Wheeler, J.M.D., Beck, N.E., Kim, H.C.** et al. (1999) *Proc. Natl Acad. Sci.*, **96**, 10296.
88. **Veigl, M.L., Kasturi, L., Lechnowicz, J.** *et al.* (1998) *Proc. Natl Acad. Sci.*, **95**, 8698.
89. **Toyota, M., Ahuja, N., Ohe-Toyota, M.** *et al.* (1999) *Proc. Natl Acad. Sci.*, **96**, 8681.
90. **Kuismanen, S.A., Holmberg, M.T., Salovara, R.** *et al.* (1999) *Proc. Natl Acad. Sci.*, **96**, 12661.
91. **Wong, S.F. and Lai, L.C.** (2001) *Pathology*, **33**, 85.
92. **Derynck, R., Akhurst, R.J. and Balmain, A.** (2001) *Nat. Genet.*, **29**, 117.
93. **Grady, W.M., Myeroff, L.L., Swinler, S.E.** *et al.* (1999) *Cancer Res.*, **59**, 320.
94. **Markovitz, S., Wang, J., Myeroff, L.** *et al.* (1995) *Science*, **268**, 1336.
95. **Grady, W.M., Rajput, A., Myeroff, L.** *et al.* (1998) *Cancer Res.*, **58**, 3101.
96. **Herman, J.G., Merlo, A., Mao, L.** *et al.* (1995) *Cancer Res.*, **55**, 4525.
97. **Ahuja, N., Mohan, A.L., Li, Q.** *et al.* (1997) *Cancer Res.*, **57**, 3370.
98. **Esteller, M., Gonzalez, S., Risques, R.A.** *et al.* (2001) *J. Clin. Oncol.*, **19**, 299.
99. **Yi, J., Wang, Z.W., Cang, H.** *et al.* (2001) *World J. Gastroenterol.*, **7**, 722.
100. **Wienke, J.K., Zheng, S., Lafuente, A.** *et al.* (1999) *Cancer Epid. Bio. Prev.*, **8**, 501.
101. **Hawkins, N., Norrie, M., Cheong, K.** *et al.* (2002) *Gastroenterology*, **122**, 1376.
102. **Loda, M., Cukor, B., Tam, S.W.** *et al.* (1997) *Nat. Med.*, **3**, 231.
103. **Tenjo, T., Toyoda, M., Okuda, J.** *et al.* (2000) *Oncology*, **58**, 45.
104. **Liu, D.F., Ferguson, K., Cooper, G.S.** *et al.* (1999) *J. Clin. Lab. Anal.*, **13**, 291.

Gastrointestinal cancer

8

8.1 Introduction

Cancers of the upper gastrointestinal tract (esophagus and stomach) are among the most frequent malignancies worldwide, with a characteristic geographical distribution [1]. Over the past several decades, the epidemiology of esophagogastric cancers has altered dramatically, particularly in Western or developed countries [2–5]. Although the reasons for this change are unknown, several risk factors, including tobacco exposure, obesity, diet, *Helicobacter pylori* infection and gastroesophageal reflux have been implicated [6–11]. Esophageal and gastric cancers are associated with high mortality rates [12], suggesting an aggressive tumor biology.

This chapter will focus primarily on recently described molecular genetic alterations in esophagogastric tumors in the context of epidemiologic trends, and will discuss potential clinical applications for such molecular markers. Molecular alterations in small bowel, pancreatic, and hepatobiliary malignancy will also be summarized.

8.2 Pathology of esophagogastric cancer

8.2.1 Esophageal cancer

Patients with esophageal cancer present clinically with difficulty swallowing as a result of esophageal obstruction from the tumor mass (*Figures 8.1* and *8.2*). There are two principal histologic subtypes of esophageal cancer: squamous cell carcinoma and adenocarcinoma. Arising from squamous epithelium lining the tubular esophagus, squamous cell carcinomas account for the majority of esophageal tumors worldwide. Squamous cell carcinomas are characterized histologically by keratinization and/or the presence of intercellular bridges, and are often associated with dysplastic change. Primary esophageal adenocarcinomas, which are increasing in incidence in North America and Europe, are thought to arise from the columnar epithelium-lined (Barrett's) esophagus [13,14]. Dysplasia is widely regarded as the precursor of invasive cancer, and high-grade dysplasia in Barrett's epithelium (*Figure 8.3*) is frequently associated with esophageal adenocarcinoma [15,16].

8.2.2 Barrett's esophagus

Barrett's esophagus is currently defined as a change in the esophageal epithelium of any length that can be recognized at endoscopy, and confirmed to have intestinal metaplasia by biopsy [17]. Goblet cells are the hallmark for intestinal metaplasia, diagnosed histologically using hematoxylin and eosin staining of tissue sections (*Figure 8.4*). Goblet cells also contain acidic mucin, which shows an intense blue color with Alcian blue stain at pH 2.5 [18]. Immunohistochemical studies using monoclonal antibodies to cytokeratin (CK) 7 and 20, have recently been used to differentiate intestinal metaplasia of esophageal versus gastric cardia origin [19,20]. CK 7 positivity in superficial and deep glands, with CK 20 positivity limited to superficial glands only, is referred to as the Barrett's CK 7/20 pattern,

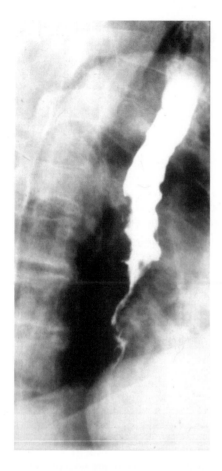

Figure 8.1

Barium swallow illustrating an obstructing carcinoma of the esophagus.

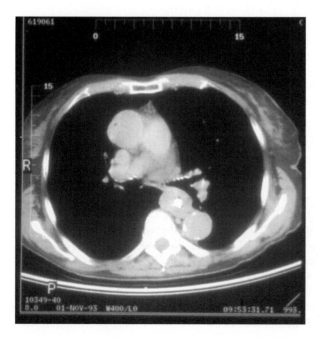

Figure 8.2

Computed tomography of the chest illustrating an esophageal carcinoma adjacent to the left mainstem bronchus.

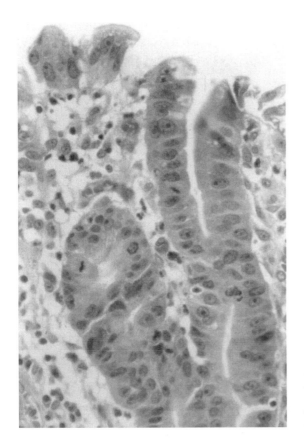

Figure 8.3

Histologic section of the columnar epithelium-lined esophagus showing high-grade dysplasia.

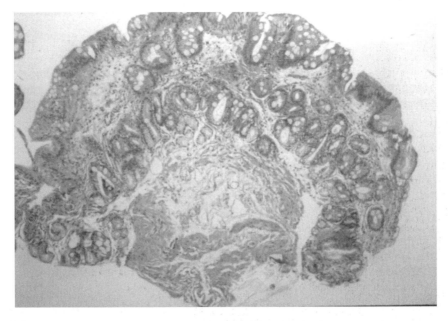

Figure 8.4

Histologic section of an esophageal biopsy illustrating intestinal metaplasia.

whereas gastric intestinal metaplasia is characterized by CK 20 immunopositivity in both superficial and deep glands, but with absent (or weak or patchy) CK 7 immunoreactivity.

8.2.3 Gastric cancer

Histologically, adenocarcinomas of the upper gastrointestinal tract are characterized by glandular differentiation. Adenocarcinomas are therefore the principal epithelial tumor arising from the gastric mucosa, but have distinct histopathologic variability (tubular, papillary, mucinous, signet-ring). Lauren classified gastric cancers into two histologic subtypes, intestinal or diffuse, to reflect differing etiology and clinical course [21]. Precursor lesions for gastric cancer include chronic atrophic gastritis (often associated with *H. pylori*) and intestinal metaplasia.

8.3 Epidemiology of esophagogastric malignancy

Squamous cell carcinoma of the esophagus shows considerable geographic diversity, with mortality rates ranging from 30–140/100 000 population in high-risk regions (i.e. China, South Africa, Iran, South America and northern France). The remarkable variability of incidence of esophageal squamous cell carcinoma worldwide led to numerous epidemiologic studies attempting to define risk factors for the disease. Heavy alcohol and tobacco consumption, diet and nutritional factors have been consistently implicated in the pathogenesis of this disease.

Over the past three decades, there has been a dramatic change in the epidemiology of esophageal malignancy in North America and Europe, which are still considered low-incidence regions globally, with mortality rates generally below 10/100 000. While the incidence of squamous cell carcinoma of the esophagus has remained stable, the incidence of esophageal adenocarcinoma has increased steadily, at a rate exceeding any other solid tumor. Although the reasons for this change are largely unknown and remain controversial, several lifestyle risk factors, including alcohol consumption, tobacco exposure, obesity, and dietary factors have been proposed. The increased use of nitrate-based fertilizers in agriculture over the past three to four decades has recently been suggested to account for these striking epidemiological trends [22,23]. It is thought that increased dietary nitrate, derived principally from green leafy vegetables, in association with gastric and ascorbic acid, results in potentially carcinogenic levels of nitric oxide in epithelia of the lower esophagus. Recent studies have implicated gastroesophageal reflux disease (GERD) as the most significant individual risk factor for esophageal adenocarcinoma [10]. It is hypothesized that GERD results in acute mucosal injury (esophagitis), promotes cellular proliferation, and induces columnar metaplasia of the normal squamous epithelium lining the esophagus. The resulting columnar epithelium-lined (Barrett's) esophagus appears predisposed to develop malignancy at an estimated risk of at least 30–40-fold higher than that of the general population. However, the absolute risk of an individual patient with Barrett's esophagus developing invasive adenocarcinoma is low, and has recently been estimated at 0.5% per patient year [24].

In North America, current estimates of the prevalence of Barrett's esophagus are approximately 1–2% for all patients undergoing endoscopy [25]. This increases to over 10% for patients who undergo endoscopy for upper gastrointestinal symptoms (predominantly reflux-related). Autopsy series indicate the prevalence of Barrett's esophagus to be much higher (approximately 20 times), suggesting that a large percentage of the general population have undiagnosed Barrett's esophagus [26]. This may not be unreasonable considering that up to 40% of the general population in North America experience reflux-related symptoms monthly. The mean age at which Barrett's esophagus is currently diagnosed is around 60 years, with a male predominance (2–3:1).

Gastric cancer is the second most frequent malignancy worldwide. High-incidence regions (with rates >40/100 000) include Japan and East Asia, Eastern Europe and South America. Steadily decreasing incidence and mortality rates for all regions have been observed over the past several decades [27,28], particularly for the intestinal type, and have been attributed to improved diet and better food storage. The association between *H. pylori* infection and gastric cancer [29,30] led the International Agency for Research on Cancer to classify *H. pylori* as a carcinogen in humans.

8.4 Cellular and molecular alterations in esophagogastric cancer

Esophagogastric tumors, as for other human cancers, are thought to arise as a multistep process, modulated by both genetic and environmental factors. The accumulation of genetic alterations leads to genomic instability and through complex interactions between stimulatory oncogenes and regulatory tumor suppressor genes. This results in widespread clonal outgrowth of cells exhibiting aberrant cell cycle regulation, with capacity for invasion. In general, genomic instability precedes the appearance of histologic changes.

The multistep process for esophageal adenocarcinoma is recognized histologically as the metaplasia–dysplasia–adenocarcinoma sequence [31]. However, few studies have investigated critical early molecular genetic events, and despite the plausibility of a link between GERD, Barrett's epithelium and esophageal adenocarcinoma, no convincing biological mechanism to explain these associations has yet been established.

It is proposed that the intestinal form of gastric cancer develops as a stepwise progression from chronic to atrophic gastritis (induced by *H. pylori*), intestinal metaplasia, dysplasia and finally invasive adenocarcinoma [32]. No obvious sequential pattern is seen for the diffuse type of gastric cancer.

8.4.1 Cell cycle regulation

Progression through each stage of the cell cycle is modulated by complex interactions between stimulatory and inhibitory signals mediated by oncogenes and tumor suppressor genes, respectively. Quiescent cells (stage G_0) enter the cell cycle in the G_1 phase and progress under external mitogenic stimulation to S phase, in which DNA synthesis occurs. The cell is now committed to divide irrespective of exogenous factors, and after proceeding through G_2, enters M phase where mitosis occurs. The cell cycle is regulated at key checkpoints (G_1/S and G_2/M) by cyclins and cyclin-dependent kinases (CDK), which interact with cellular proteins to activate transcription factors having positive or negative regulatory effects.

Cyclin D1/CDK-4 (or CDK-6) complex regulates the early to mid-G_1 phase. Following phosphorylation of the retinoblastoma (RB) tumor suppressor protein, the E2F transcription factor activates genes required for DNA synthesis in S phase, driving the cell to G_1/S transition. Whereas the *p16* tumor suppressor gene inhibits the association of CDK-4 (and -6) with cyclin D1, mutational activation of the *RAS* oncogene induces cyclin D1 expression. This regulatory process is further modulated by interactions with other upstream and downstream genes. The *p53* tumor suppressor gene may induce cell cycle arrest by transcriptional activation of *p21* (*WAF-1*), which sequesters various cyclin-dependent kinases, including CDK-4 and CDK-6, at several points in the cell cycle. Therefore, progression through the G_1/S checkpoint may result from the loss of *RB*, *p16*, or *p53*, or by overexpression of *RAS* or cyclin D1. A reciprocal relationship between RB, cyclin D1 and *p16* expression has been reported in esophageal cancer [33,34]. In general, tumors which retain RB expression typically exhibit overexpression of cyclin D1, *p16* inactivation, or both, often in the context of *p53* mutations.

8.4.2 Epithelial proliferation markers

Immunohistochemistry and flow cytometry have been used to study cell proliferation in esophageal tissues, evaluating distribution of proliferating cell nuclear antigen (PCNA) and Ki67 (a cell nuclear proliferation-associated antigen for G_1/S and G_2/M). PCNA immunostaining is normally seen in the basal layer of metaplastic Barrett's epithelium, but extends to superficial layers with high-grade dysplasia [35]. Immunohistochemical studies using the monoclonal antibody MIB-1 (against Ki67) demonstrated a higher percentage of proliferating cells in metaplastic Barrett's mucosa compared to normal gastric epithelium. Staining patterns for low-grade and high-grade dysplasia were similar to PCNA, suggesting a greater turnover of differentiated cells in the surface epithelium by immature proliferating cells arising from basal layers [36,37]. Increased proliferative activity and altered cell cycle kinetics have also been shown using flow cytometry in esophageal cancer [38] and Barrett's epithelium [39]. An increased G_1 fraction appears to be the earliest finding, progressing to increased S phase fractions with aneuploidy, high-grade dysplasia and carcinoma. These findings suggest a functional instability of Barrett's mucosa, predisposing to increasing dysplasia and malignancy.

8.4.3 Aneuploidy

Flow cytometry has also been used to study cell nuclear DNA content (ploidy). Progression of normal esophageal epithelium to Barrett's metaplasia is clearly associated with abnormal DNA content (aneuploidy) [40]. Furthermore, the prevalence of aneuploidy appears to increase with the degree of dysplasia determined histologically. In a prospective study of 62 patients, nine of 13 patients with abnormal DNA content (aneuploidy or increased 4N fraction) progressed to high-grade dysplasia or carcinoma [39]. Of 49 patients with normal flow cytometry at initial endoscopy, none progressed over the 34-month study. The exact significance of the 4N fraction is unclear, but is thought to represent an unstable intermediate stage in progression to aneuploidy. Mapping studies have shown that most aneuploid populations are localized to a single region of esophageal mucosa, suggesting clonal expansion of single progenitor cells to involve large regions of esophageal mucosa [41]. Although multiple aneuploid cell populations are occasionally encountered in Barrett's epithelium, only one aneuploid cell population typically is found in the primary tumor. Although several studies have reported that abnormal DNA content of Barrett's adenocarcinomas to be predictive of advanced-stage disease, lymph node metastases and reduced overall survival, these findings have been inconsistent.

8.4.4 Chromosomal aberrations

Conventional cytogenetic techniques (karyotypic analysis) are limited by both the size of chromosome to be visualized (at least 5 megabases of DNA) and the requirement that cells be cultured and stimulated into mitosis for banding. Recent molecular cytogenetic techniques, such as fluorescence *in situ* hybridization, will resolve smaller structural chromosomal changes (less than 1 kilobase of DNA), and may be applied to detect multiple chromosomal changes in heterogeneous cell populations. Comparative genomic hybridization has recently become the most popular cytogenetic tool to analyze the entire genome [42]. The following chromosomal aberrations and allelic losses have consistently been reported in esophageal cancer [43–45]: 2q, 3p, 5q, 9p, 11p, 12q, 13q, 17q, 17p, 18q, Xq and loss of the Y-chromosome, which increased with high-grade dysplasia. For gastric cancers, common chromosomal targets for loss or gain include 3p, 4q, 5q, 6q, 9p, 17p, 18q, and 20q [46–48].

8.4.5 Loss of heterozygosity

Frequent loss of chromosomal regions or loci from one or more alleles (loss of hetero-zygosity; LOH) is reported frequently in Barrett's esophagus and associated esophageal adenocarcinomas for the following loci (and associated tumor suppressor genes): 5q (ade-nomatous polyposis coli; *APC*), 17p (*p53*), 18q (deleted in colorectal cancer; *DCC*), 9p (*p16, p15*), 3p (fragile histidine triad; *FHIT*), 13q (retinoblastoma; *RB*). LOH is reported in several additional loci (1p, 3q, 4q, 5p, 6q, 9q, 11p, 12p, 12q, 17q and 18q) [49], but many of the associated candidate suppressor genes have yet to be characterized [50]. Particularly interesting is a recent report suggesting that deletion of a locus (31–32.1) on chromosome 14q can be used to differentiate adenocarcinomas of esophageal versus gastric cardia origin [51]. Studies of 5q and 17p allelic losses in aneuploid cell populations derived from patients with Barrett's esophagus and adenocarcinomas suggest that 17p loss precedes 5q loss in esophageal tumorigenesis [52]. This is in contrast to colorectal malignancy where 5q losses are found earlier in neoplastic progression. This finding further supports the hypothesis that alterations of the *p53* tumor suppressor gene (localized to 17p) occur as an early molecular event in esophageal tumorigenesis.

8.5 Tumor suppressor genes

8.5.1 *p53*

The *p53* tumor suppressor gene encodes a 53 kDa polypeptide that regulates cell cycle pro-gression, DNA repair, apoptosis, and neovascularization in normal and malignant cells via highly complex DNA and protein interactions [53]. *p53* mediates cell cycle arrest in part by inducing the expression of *p21* (*WAF-1*) which sequesters a variety of cyclin-dependent kinases facilitating G_1 as well as G_2/M arrest. Over 90% of *p53* mutations locate in the conserved DNA-binding domain (exons 5 to 8), with 'hot spots' at codons 175, 176, 245, 248, 249, 273 and 282 in many human tumors [54]. This gene appears to have a central role in human malignancy, and has been characterized extensively over the past decade. Furthermore, *p53* would appear to have potential clinical application for novel therapeutic strategies.

p53 gene mutations were first reported in esophageal squamous cell carcinoma [55], and subsequently in primary esophageal adenocarcinoma and associated Barrett's epithelium in 1991 [56]. These findings have now been confirmed by several investigators, and the spectrum of *p53* alterations in Barrett's esophagus and esophageal adenocarcinomas has been studied extensively (*Figures 8.5* and *8.6*). The finding of *p53* mutations in nondysplastic Barrett's epithelium suggests that *p53* is altered early in the metaplasia–dysplasia–carcinoma sequence, and may therefore be a clinically useful early molecular diagnostic marker.

In the most comprehensive study of primary esophageal adenocarcinomas (defined by strict clinico-pathologic criteria) reported to date, *p53* mutations were associated with poorly differentiated tumors, and reduced disease-free and overall survival following surgical resection [57,58]. Biologically, one of the most interesting findings derived from the analysis of *p53* relates to patterns of mutations, which differ between squamous cell carcinomas and adenocarcinomas of the esophagus. *p53* mutations in esophageal squamous tumors are pre-dominantly transitions or transversions occurring at A:T base pairs (suggesting a relation-ship to metabolites of ethanol, a well-defined risk factor for squamous cell carcinoma), or G

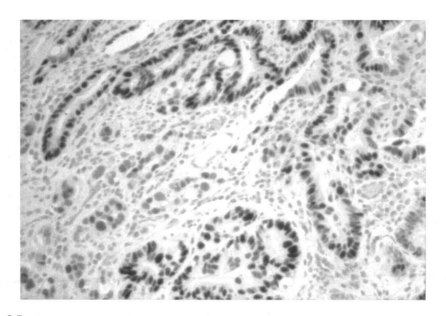

Figure 8.5

Immunohistochemical section of a primary esophageal adenocarcinoma illustrating heterogeneous cell nuclear immunoreactivity for *p53*.

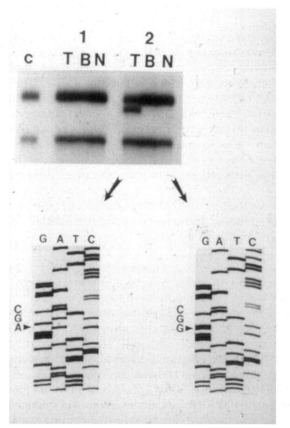

Figure 8.6

Single-strand conformation polymorphism (SSCP) analysis (top panel) and sequence analysis (below) for exon 7 of the *p53* tumor suppressor gene. No electrophoretic mobility shifts were found for patient 1 in the tumor (T), Barrett's epithelium (B) or matched histologically normal esophageal epithelium (N). An electrophoretic mobility shift is seen in the tumor (T) of patient 2, which was confirmed by sequencing (below, left) to represent a point mutation (AGC) at codon 245. The wild-type p53 sequence (GGC) was found in matched normal (N) esophageal epithelium (below, right).

to T transversions (a characteristic mutation attributed to benzo[a]pyrine, suggesting an association with tobacco). However, for esophageal adenocarcinoma, predominant mutations are G:C to A:T transitions at CpG dinucleotides. A review of all published human *p53* mutations (see IARC *p53* mutation database at www.iarc.fr/p53/index.html) indicates that esophageal adenocarcinoma is the cancer type with the highest proportion of such mutations.

There is now solid molecular evidence that these mutations primarily arise through endogenous mechanisms, involving the spontaneous deamination into thymine of the 5' methylated cytosine that frequently occurs at CpG dinucleotides. This mechanism is enhanced by exposure to oxy- and nitro-radicals. In colon cancer, a direct correlation was demonstrated between the levels of expression of inducible nitric oxide synthase 2 (*iNOS; NOS-2*) and the rate of formation of G:C to A:T transition at CpG dinucleotides, incriminating overproduction of nitric oxide as a major causative factor [59]. Exposure of human fibroblasts to nitric oxide results in *p53* protein accumulation, whereas overexpression of wild-type *p53* in human tumor cell lines results in down-regulation of *NOS-2* expression through promotor inhibition [60]. These studies suggest that *p53* provides a novel mechanism for DNA protection, reducing the potential for nitric oxide-induced DNA damage. Nitrotyrosine, a stable reaction product of nitric oxide and a marker for cellular protein damage from high levels of nitric oxide, has recently been reported in esophageal squamous cell carcinomas [61]. Furthermore, *NOS-2* mRNA expression was recently shown to be increased in up to 80% for both Barrett's epithelia and esophageal adenocarcinoma tissues [62].

Recent data on patterns of *p53* mutations in esophageal adenocarcinoma support the hypothesis that chronic stress, in particular resulting from GERD-induced esophagitis, may result in local overproduction of nitric oxide in normal and metaplastic esophageal mucosa, thus enhancing the rate of formation of 'spontaneous' mutations, including *p53*. Furthermore, based on earlier observations of *p53* mutations in nondysplastic Barrett's epithelia and the subsequent demonstration of p53 protein accumulation in near-normal esophageal epithelia with mild basal cell hypertrophy, it is proposed that *p53* alterations occur relatively early during multistep esophageal tumorigenesis. This notion is further supported by recent observations of *p53* mutations in other preneoplastic epithelia of the upper aerodigestive tract (head and neck, foregut and lung) [63,64]. Abnormalities in DNA repair (see following sections) would initiate and/or exacerbate this process.

Compared to esophageal adenocarcinomas, *p53* alterations are generally reported to be less frequent in gastric antral carcinomas [65], although *p53* mutations appear to be found in association with intestinal-type tumors. Similar *p53* mutation patterns are seen for esophageal and gastric adenocarcinomas, particularly at CpG dinucleotides [66], and a polymorphism at codon 72 is reported to be associated with gastric malignancy [67].

8.5.2 Retinoblastoma (*RB*)

The retinoblastoma (*RB*) gene, located on 13q14, encodes a 105 kDa nuclear phosphoprotein which is intimately involved in regulation of the G_1 restriction point. A variety of proteins, including cyclin-dependent kinases, transcription factors, and viral oncoproteins, interact with RB and related proteins (p107 and p130) to regulate cell cycle progression. Mutations involving the retinoblastoma gene frequently disrupt RNA splicing resulting in the loss of RB protein expression. *RB* mutations have been observed in Barrett's epithelium, and in approximately 20–40% of esophageal adenocarcinomas [68]. *RB* mutations occur more frequently in tumors with *p53* mutations [69]. Recent studies suggest that loss of RB expression correlates with advanced stage of disease, nodal metastases, and reduced survival [70].

8.5.3 *p16*

The *p16* tumor suppressor gene, located on 9p21, encodes a protein product which inhibits the activity of CDK4 and CDK6, preventing cyclin D-dependent phosphorylation of the RB protein. Inactivation of *p16* by allelic deletion or point mutation has been detected in approximately 20% of esophageal squamous cell cancers. LOH involving 9p21 was reported in 24 of 32 (75%) aneuloid cell populations derived from esophageal adenocarcinomas, as well as 7/7 samples of Barrett's epithelium whose tumors had *p16* mutations [71]. Allelic loss involving *p16* preceded the onset of aneuploidy in 13 of 15 specimens. Somatic mutation silenced the remaining *p16* allele in 23% of the aneuploid cell samples. Methylation contributes to *p16* inactivation in an additional 50% of squamous cell or adenocarcinomas of the esophagus [72,73]. Recent reports suggest that *p16* inactivation correlates with cyclin D_1 overexpression and poor prognosis [74]. Restoration of *p16* expression by gene therapy techniques profoundly inhibits the proliferation and tumorigenicity of esophageal cancer cells [34], implicating *p16* as a key molecular event during esophageal carcinogenesis.

8.5.4 *p14*

The *p14* gene product is encoded by an alternate reading frame (*ARF*) at the *p16* locus. Recent studies indicate that *ARF* prevents interaction of p53 with MDM2, a protein which facilitates p53 degradation. *ARF* also mediates *p53*-dependent apoptosis in response to activated oncogenes such as *MYC* or *RAS*. The *p14/ARF* gene locus is unique in that it encodes two separate protein products that link the *RB* and *p53* tumor suppressor pathways. Inactivation of *ARF* occurs by allelic deletion as well as methylation mechanisms which simultaneously inactivate *p16*. Expression of *ARF* in Barrett's epithelium and esophageal adenocarcinoma has not been studied in detail, to date. However, aberrant promoter methylation involving *ARF* or *p16* was observed in 15% and 40% of esophageal squamous cell carcinomas, respectively [72]. Whereas most of the *p16* methylations were exclusive, nearly all of the methylations involving *ARF* also involved *p16*. Homozygous deletions involving *ARF* or *p16* were seen in 33% and 18% of specimens, respectively. These results suggest that in esophageal squamous cell cancer, *p14/ARF* is a primary target for homozygous deletion, whereas *p16* appears to be more frequently silenced by hypermethylation. It is therefore possible that inactivation of *p14/ARF* is an alternative molecular pathway in esophageal tumors with wild-type *p53*.

8.5.5 Fragile histidine triad (*FHIT*) gene

Deletions involving 3p have been detected in 60–100% of esophageal adenocarcinomas and Barrett's epithelium, and have been proposed as an early molecular event [75]. Although the tumor suppressor gene associated with 3p LOH has not been identified conclusively, one major target appears to be the fragile histidine triad (*FHIT*) gene. *FHIT* acts by hydrolyzing di-nucleotide triphosphates, modulation of cell cycle progression and apoptosis. Point mutations and promoter methylation are mechanisms of *FHIT* inactivation, resulting in the loss of FHIT protein expression. Aberrant expression of *FHIT* has been observed in 50–90% of esophageal tumors, and 85% of specimens derived from Barrett's esophagus examined, to date [76,77].

8.5.6 Adenomatous polyposis coli (*APC*) gene

LOH of chromosome 5q, in the region of the *APC* gene, has been observed frequently in gastric cancer. Mutations in the *APC* gene appear to have been found primarily in well-differentiated gastric adenocarcinomas, particularly the signet-ring type, but less frequently in poorly differentiated tumors [78]. The pattern of *APC* mutations differs from those seen in colorectal cancers, with missense mutations being a frequent finding in gastric cancer, but rarely found in colorectal cancer. *APC* mutations are also reported in up to 20% of gastric adenomas, suggesting *APC* alterations to be an early event in gastric carcinogenesis [79,80].

Although a high frequency of allelic deletion of chromosome 5q is also reported for esophageal cancers, detailed mapping studies have suggested that *APC* may not be the target tumor suppressor gene inactivated in this region [50]. As *APC* mutations are also seen rarely in esophageal tumors, *APC* promoter hypermethylation has recently been suggested as an alternative epigenetic mechanism of *APC* gene suppression.

8.6   Growth factors and receptors

Epidermal growth factor (EGF), which stimulates epithelial cell proliferation, has been shown to be overexpressed in esophageal tumors. The epidermal growth factor receptor (EGFR) is a 170 kDa tyrosine kinase glycoprotein which is overexpressed in approximately 80% and 30% of esophageal squamous cell cancers and adenocarcinomas, respectively [81]. Furthermore, EGFR overexpression appears to correlate with the degree of dysplasia. Similarly, transforming growth factor alpha, structurally and functionally related to EGF, which binds to the EGFR to stimulate growth via autocrine mechanisms, is also overexpressed in dysplastic esophageal epithelium [82]. Several recent studies have suggested that growth factor/receptor overexpression may have prognostic significance following surgical resection of esophageal cancer, associated with both regional nodal and distant metastasis [83].

The *ERBB2* gene (*HER2/NEU*) encodes a 185 kDa tyrosine kinase receptor molecule which is structurally related to EGFR. Using immunohistochemistry, ERBB2 protein was reported to be overexpressed in up to 70% of Barrett's epithelia and esophageal adenocarcinomas [84]. In one recent study, overexpression of ERBB2 was seen in 24% of esophageal adenocarcinomas, and was associated with an advanced tumor stage [85]. However, the prognostic significance of ERBB2 for patients with Barrett's esophagus or esophageal adenocarcinoma is unknown.

Gastric cancer is associated with abnormalities in a broad spectrum of growth factors, including EGF, transforming growth factor-alpha, cripto, platelet-derived growth factor and Ksam [86]. The *KSAM* gene, a member of the fibroblast growth factor receptor family, has been found to be amplified in poorly differentiated tumors [87]. By contrast, *ERBB2* amplification, which is reported in up to 25% of gastric tumors, has been associated with well-differentiated tumors [88]. Correlations between *ERBB2* amplification or overexpression and prognosis remain conflicting.

8.7   Oncogenes

8.7.1   Cyclin D1

Cyclin D1, a protein derived from the *CCND1* gene located on chromosome 11q13, was recently reported to be overexpressed in Barrett's epithelium [89]. The cyclin D1/CDK-4 (or CDK-6) complex regulates the early to mid-G_1 phase of the cell cycle. Cyclin D1 expression may be induced by mutational activation of the *RAS* oncogene, whereas the *p16* tumor suppressor gene inhibits the association of CDK-4 (and -6) with cyclin D1. This regulatory process is further modulated by interactions with other upstream and downstream genes.

The *p53* tumor suppressor gene may induce cell cycle arrest by transcriptional activation of *p21* (*WAF-1*), which sequestors various cyclin-dependent kinases, including CDK-4 and CDK-6, at several points in the cell cycle. Progression through the G_1/S checkpoint may result from the loss of RB, p16, or p53, or by overexpression of *RAS* or cyclin D1. A reciprocal relationship between Rb, cyclin D1 and p16 expression has been reported in esophageal cancer. In general, tumors which retain RB expression typically exhibit over-expression of cyclin D1, *p16* inactivation, or both. In a large case–control study of patients with Barrett's esophagus, cyclin D1 immunopositivity was associated with a statistically significant increased risk of progression to adenocarcinoma, suggesting this molecular marker may be clinically useful for early diagnosis in future endoscopic surveillance studies [90]. Additionally, overexpression of cyclin D1 has also been observed in 40–60% of esophageal adenocarcinomas [70] and squamous cell carcinomas [91], and is associated with poor prognosis following esophageal resection.

8.7.2 *RAS*

In contrast to other gastrointestinal tumors, this oncogene is rarely mutated in human esophageal adenocarcinoma, although it is reported in carcinogen-induced animal models of esophageal cancer. However, overexpression of *RAS* and *RAS*-regulated genes (osteopontin, cathepsin L) has been reported in 58% of human primary esophageal adenocarcinomas [92].

Early immunohistochemical studies reported overexpression of *RAS* p21 protein in over 60% of gastric adenocarcinomas, particularly for the intestinal type [93]. Variations in the frequency of *RAS* mutations in gastric cancer have been described, with high frequencies in Asia and lower frequencies in Europe and North America, but the significance of these early observations is not known.

8.8   DNA repair-associated genes

Decreased efficiency of DNA repair is viewed as a crucial event in carcinogenesis, as such defects accelerate the rate of genetic change. Numerous links have been identified between oncogenesis and acquired or inherited faulty genome guardians that cause genomic instability, highlighting a key role of DNA protection. The role of DNS repair-associated genes in esophagogastric malignancy is currently under investigation.

The xeroderma pigmentosum group C (*XPC*) gene is integral to nucleotide excision repair (NER), and an intronic poly(AT) insertion/deletion polymorphism has been associated with increased risk for squamous cell carcinoma of head and neck. Similar associations have been reported for polymorphisms (C22541A and A35931C) of the xeroderma pigmentosum group D (*XPD*) gene, which is essential for transcription during NER. The X-ray repair cross-complementing 1 (*XRCC1*) gene encodes a protein which complexes with DNA ligase to repair DNA gaps resulting from base excision repair (BER). Three polymorphisms of *XRCC1* (codons 194, 280 and 399) have been associated with increased risk of gastric carcinoma [94].

The role of DNA mismatch repair (MMR) genes has been studied extensively in colorectal adenocarcinoma. Hereditary nonpolyposis colorectal cancer is due to a germ-line mutation in one of four MMR genes leading to genomic instability. *hMSH2* and *hMLH1* are reported to be altered (mutations, promoter hypermethylation) most frequently. To date, genomic/microsatellite instability caused by defective function of MMR genes has not been studied in esophageal adenocarcinoma progression, but is of particular interest following the report of *hMSH2* expression in the replicating component of esophageal and intestinal epithelium

[95]. Hypermethylation in the promoter region of *hMLH1* has been reported at high frequency in gastric adenocarcinomas [96].

8.8.1   Microsatellite instability

Microsatellite instability results from defects in MMR during DNA replication, leading to accumulation of single nucleotide mutations and alterations in the length of microsatellites in affected regions. Microsatellites are tracts of simple sequence oligonucleotide repeats (usually di- and tri-nucleotide repeats) found throughout the genome. Microsatellite instability at from one to five chromosomal loci was initially reported in 7% of Barrett's epithelia, and 22% of esophageal adenocarcinomas. This frequency was in contrast to only 2% of esophageal squamous cell carcinomas. In a subgroup of esophageal adenocarcinomas with diploid nuclei, microsatellite instability was found in 50%, suggesting this may be an early event in the metaplasia–dysplasia–adenocarcinoma sequence [97]. However, the number of unstable chromosomal loci necessary to consider a tumor to have microsatellite instability remains controversial. Using more stringent criteria (two or more of five loci), subsequent studies reported microsatellite instability to be infrequent (less than 10%) in esophageal adenocarcinomas [98]. These findings have recently been confirmed in a detailed analysis of 27 esophageal adenocarcinomas using a panel of 15 markers, where low levels of microsatellite instability were found in 67% of tumors, suggesting an inherent baseline genomic instability [99].

The frequency of microsatellite instability in gastric cancer is reported to range from 10% to 40%, particularly for intestinal-type tumors [100]. Loss of hMLH1 and hMSH2 protein is reported in all tumors with high levels of microsatellite instability, suggesting biallelic inactivation [101].

8.9   Hereditary factors

The possibility of inherited factors underlying esophageal tumorigenesis arose from observations of an association between esophageal squamous cell carcinoma and the rare autosomal dominant disease, tylosis. The tylosis esophageal cancer gene was recently mapped to a small region on chromosome 17q25, and recent loss of heterozygosity studies have further implicated this gene in sporadic esophageal cancers [102]. Although families with Barrett's esophagus have been reported, these are quite uncommon and a genetic locus for a familial syndrome of Barrett's esophagus has not been confirmed, to date.

Several inherited cancer syndromes (i.e. Li–Fraumeni, Peutz–Jeghers, hereditary nonpolyposis colon cancer, familial adenomatous polyposis) are associated with gastric cancer [103]. Excluding these rare associations, it is estimated that up to 10% of gastric cancers (usually diffuse-type) have an inherited familial component, and germ-line mutations of *CDH1*, which encodes E-cadherin, have recently been implicated [104].

8.10   Cancers of the small intestine

Despite the large surface area of the small intestine, small bowel tumors (usually adenocarcinomas) are extremely rare. Chronic inflammation (i.e. Crohn's disease) appears to be the most significant risk factor identified, to date. Mutations of *KRAS* and *p53* have been reported in small bowel adenocarcinomas (43% and 57%, respectively) [105]. Patients with hereditary nonpolyposis colorectal cancer (with germ-line *hMSH2* or *hMLH1* mutations) are also at increased lifetime risk for small bowel malignancy [106]. Similarly, patients with Peutz–Jeghers syndrome, a rare autosomal dominant syndrome characterized by epithelial hyperpigmentation and intestinal polyposis, appear predisposed to intestinal adenocarcinomas, possibly as a consequence of associated germ-line *LKB1* gene mutations [107].

8.11 Pancreatic cancer

Infiltrating adenocarcinoma of the exocrine pancreatic duct is the most frequent histologic subtype of pancreatic cancer. Adenocarcinoma of the pancreas is a highly lethal malignancy with dismal survival. Pancreatic cancer is the fourth leading cause of cancer death in males and females, with mortality and incidence rates below 10/100 000, which have remained stable from 1980 onwards. Associations between cigarette smoking, diet and pancreatitis have been reported for pancreatic cancer.

Pancreatic duct hyperplasia is now recognized as a distinct precursor lesion, although this organ is relatively inaccessible to diagnostic imaging and biopsy. These early premalignant lesions have been shown to have activating point mutations of the *KRAS* oncogene in up to 40% of cases, with increasing frequency of *KRAS* mutations (above 90%) with increasing cellular and histologic atypia [108,109]. Overexpression of *ERBB2* is also believed to be an early molecular alteration, and is reported in 70% of tumors [110]. Inactivation of *p16* (LOH, homozygous deletion/mutation, promoter hypermethylation) inactivation, is reported in up to 50% of higher-grade pancreatic duct lesions [108], and is thought to precede loss of *p53, DPC4* and *BRCA2* tumor suppressor gene function [111].

Up to 10% of individuals have a family history of pancreatic cancer, and although an autosomal recessive pattern of inheritance has been identified for some families, the precise genetic basis for inherited pancreatic malignancy is not defined. Pancreatic cancer may feature as one of the tumors seen in Lynch type 2, ataxia telangiectasia, or Li–Fraumeni syndromes. About 4% of familial pancreatic cancers have been reported to have microsatellite instability with germ-line mutations in DNA mismatch repair genes (*hMLH1* and *hMSH2*), wild-type *KRAS* and a distinct medullary histology [112].

8.12 Hepatobiliary cancer

Primary tumors of the liver are among the most frequent malignancies worldwide. In adults, hepatocellular carcinoma (liver cell carcinoma) and intrahepatic cholangiocarcinoma (bile duct carcinoma) account for the majority of liver tumors, with extrahepatic bile duct and gallbladder carcinomas seen infrequently. Hepatoblastoma is seen most often in children, and is associated with a number of congenital genetic disorders.

8.12.1 Hepatocellular carcinoma

Hepatocellular carcinomas exhibit a marked geographic variation in incidence, and epidemiologic associations with risk factors are well established in high-incidence regions: chronic hepatitis B and C infection and aflatoxin B1 exposure in developing countries (Asia, Africa); tobacco exposure and alcohol-induced liver injury in developed countries.

Integration of the hepatitis virus, resulting in altered gene expression, is the major cause of hepatocellular carcinoma in countries where this viral agent is endemic. The pattern of integration appears to be random, although insertion of the virus into intron 2 of the cyclin A gene leads to altered cyclin expression and cell cycle deregulation [113]. Viral infection may also account for overexpression of transforming growth factor beta, which is seen in 82% of tumors, and may play a role in carcinogenesis through both autocrine and paracrine pathways [114]. Binding of the hepatitis virus to the C-terminus domain of *p53* inhibits sequence-specific DNA binding and transcriptional activation, resulting in suppression of *p53*-induced apoptosis [115]. The virus may also modulate *p53* function by inhibiting base nucleotide excision repair [116]. *p53* may also be altered by aflatoxin B1 exposure, where a signature mutation at codon 249 (G to T transversion) results in a change of amino acid arginine to serine in p53 protein [117]. Epidemiologic studies have shown that this *p53* mutation is specific to regions where contaminated food is eaten, and a dose-dependent relationship between dietary aflatoxin B1 exposure, *p53* mutations at codon 249 and hepatocellular carcinoma has been demonstrated.

While mutational activation of oncogenes appears to be uncommon in hepatocellular carcinogenesis, frequent allelic losses are reported at 1p, 4q, 5q, 11p, 13q, 16p and 16q, implicating associated tumor suppressor genes [118]. LOH at chromosome 16 is seen in over 50% of tumors, and is associated with poorly differentiated tumors and advanced-stage disease [119]. DNA hypermethylation is also found at various loci on chromosome 16, even in chronic hepatitis and cirrhosis, suggesting this may be an early molecular mechanism in disease progression [120].

8.12.2 Intrahepatic cholangiocarcinoma

Intrahepatic cholangiocarcinomas, which arise from large bile ducts within the liver parenchyma, are the second most frequent primary liver cancer. High-incidence regions include Laos and northern Thailand, where chronic parasitic infection is reported as the most significant risk factor. In tumors from low-incidence regions, *KRAS* mutations (particularly at codon 12) are reported for the majority of sporadic tumors, and are associated with the periductal infiltrating form of the disease [121]. Missense *p53* mutations, predominantly G to A transitions, are reported frequently in cholangiocarcinoma types which are slow growing and noninvasive [122].

8.12.3 Extrahepatic bile duct and gallbladder cancer

Although closely related anatomically, these rare tumors have quite different risk factors and associations. Unlike bile duct carcinoma, cancers of the gallbladder have a characteristic geographic and ethnic variation in incidence, and are seen most frequently in native and Hispanic Americans. Gallstones have been considered a risk factor for gallbladder cancer (despite the high prevalence of gallstones in the general population), and sclerosing cholangitis, choledochal cysts, and chronic infection with liver flukes have been implicated in the pathogenesis of extrahepatic bile duct carcinoma. Differences in the frequency of molecular alterations are reported for each tumor subtype. Whereas *p53* mutations are found in the majority of gallbladder cancers, *RAS* oncogene mutations appear to be uncommon [123,124]. By contrast, bile duct cancers are reported to have an increased frequency of *KRAS* mutations, and a reduced frequency of *p53* mutations [125].

8.12.4 Hepatoblastoma

Most frequently seen in children, this embryonal tumor is associated with many clinical syndromes and congenital anomalies. Cytogenetic anomalies include deletions of chromosome 11p (of maternal origin), implicating the *WT1* gene at 11p13 and possibly *WT2* at 11p15, involved with the Beckwith–Wiedemann syndrome [126]. Inconsistent losses of the long and short arms of chromosome 1 have also been reported, in addition to germ-line mutations of the *APC* gene. *p53* alterations appear to be infrequent [127].

References

1. **Parkin, C.M., Pisani, P. and Ferlay, J.** (1999) *Ca. Cancer J. Clin.*, **49**, 33.
2. **Devesa, S.S., Blot, W.J. and Fraumeni, J.F.** (1998) *Cancer*, **83**, 2049.
3. **Blot, W.J. and McLaughlin, J.K.** (1999) *Semin. Oncol.*, **26** (Suppl. 15), 2.
4. **Botterweck, A.A.M., Schouten, L.J., Volovics, A. et al.** (2000) *Int. J. Epidemiol.*, **29**, 645.
5. **Powell, J., McConkey, C.C., Gillison, E.W. and Spychal, R.T.** (2002) *Int. J. Cancer*, **102**, 422.
6. **Chow, W-H., Finkle, W.D., McLaughlin, J.K. et al.** (1995) *J.A.M.A.*, **274**, 474.
7. **Zhang, Z-F., Kurtz, R.C., Sun, M. et al.** (1996) *Cancer Epidemiol. Biomark. Prev.*, **5**, 761.
8. **Gammon, M.D., Schoenberg, J.B., Ahsan, H. et al.** (1997) *J. Natl Cancer Inst.*, **89**, 1277.
9. **Lagergren, J., Bergstrom, R. and Nyren, O.** (1999) *Ann. Intern. Med.*, **130**, 883.
10. **Lagergren, J., Bergstrom, R., Lindgren, A. and Nyren, O.** (1999) *N. Engl. J. Med.*, **340**, 825.
11. **Mayne, S.T., Risch, H.A., Dubrow, R. et al.** (2001) *Cancer Epidemiol. Biomark. Prev.*, **10**, 1055.
12. **Farrow, D.C. and Vaughan, T.L.** (1996) *Cancer Causes & Control*, **7**, 322.
13. **Dent, J., Bremner, C.G., Collen, M.J. et al.** (1991) *J. Gastroenterol. Hepatol.*, **6**, 1.
14. **DeMeester, S.R. and DeMeester, T.R.** (2000) *Ann. Surg.*, **231**, 303.
15. **McArdle, J.E., Lewin, K.J., Randall, G. and Weinstein, W.** (1992) *Hum. Pathol.*, **32**, 479.
16. **Haggitt, R.C.** (1994) *Hum. Pathol.*, **25**, 982.
17. **Sampliner, R.E.** (1998) *Am. J. Gastroenterol.*, **93**, 1028.
18. **Antonioli, D.A. and Wang, H.H.** (1997) *Gastroenterol. Clin. N. Am.*, **26**, 495.
19. **Ormsby, A.H., Vaezi, M.F., Richter, J.E. et al.** (2000) *Gastroenterology*, **119**, 683.
20. **Couvelard, A., Cauvin, J-M., Goldfain, D. et al.** (2001) *Gut*, **49**, 761.
21. **Lauren, T.** (1965) *Acta Pathol. Microbiol. Scand.*, **64**, 34.
22. **Iijima, K., Moriya, A., Wirz, A. et al.** (2002) *Gastroenterology*, **122**, 1248.
23. **Moriya, A., Grant, J., Mowat, C. et al.** (2002) *Scand. J. Gastroenterol.*, **3**, 253.
24. **Shaheen, N. and Ransohoff, D.F.** (2002) *J.A.M.A.*, **287**, 1972.
25. **Sharma, P. and Sampliner, R.E.** (2001) *Curr. Opin. Gastroenterol.*, **17**, 381.
26. **Cameron, A.J., Zinsmeister, A.R. and Ballard, D.J.** (1990) *Gastroenterology*, **99,** 918.
27. **Thompson, G.B., van Heerden, J.A. and Sarr, M.G.** (1993) *Lancet*, **342**, 713.
28. **Fuchs, C.S. and Mayer, R.J.** (1995) *N. Engl. J. Med.*, **333**, 32.
29. **Nomura, A., Stammermann, G.N., Chyou, P.H. et al.** (1991) *N. Engl. J. Med.*, **325**, 1132.
30. **Parsonnet, J., Hansen, S., Rodriguez, L. et al.** (1994) *N. Engl. J. Med.*, **330**, 738.
31. **Jankowski, J.A, Wright, N.A., Meltzer, S.J. et al.** (1999) *Am. J. Pathol.*, **154**, 965.
32. **Rugge, M., Cassaro, M., Leandro, G. et al.** (1996) *Dig. Dis. Sci.*, **41**, 950.
33. **Jiang, W., Zhang, Y.J., Kahn, S.M. et al.** (1993) *Proc. Natl Acad. Sci. USA*, **90**, 9026.
34. **Schrump, D.S., Chen, G.A., Consuli, U. et al.** (1996) *Cancer Gene Ther.*, **3**, 357.
35. **Hong, M.K., Laskin, W.B., Herman, B.E. et al.** (1995) *Cancer*, **75**, 423.
36. **Lauwers, G.Y., Kandemir, O., Kubilis, P.S. and Scott, G.V.** (1997) *Mod. Pathol.*, **10**, 6917.
37. **Whittles, C.E., Biddlestone, L.R., Burton, A. et al.** (1999) *J. Pathol.*, **187,** 535.
38. **Chanvitan, A., Nekarda, H. and Casson, A.G.** (1995) *Int. J. Cancer*, **63,** 381.
39. **Reid, B.J., Blount, P.L., Rubin, C.E. et al.** (1992) *Gastroenterology*, **102**, 1212.
40. **Reid, B.J., Levine, D.S., Longton, G. et al.** (2000) *Am. J. Gastroenterol.*, **95**, 1669.
41. **Raskind, W.H., Norwood, T.H., Levine, D.S. et al.** (1992) *Cancer Res.*, **52**, 2946.
42. **Forozan, F., Karhu, R., Kononen, J. et al.** (1997) *Trends Genet.*, **13**, 405.
43. **Aoki, T., Mori, T., Du, X. et al.** (1994) *Genes Chromosomes Cancer*, **10**, 177.
44. **Menke-Pluymers, M.B., van Drunen, E., Vissers, K.J. et al.** (1996) *Cancer Genet. Cytogenet.*, **90**, 109.
45. **Pack, S.D., Karkera, J.D., Zhuang, Z. et al.** (1999) *Genes Chromosomes Cancer*, **25**, 160.
46. **Sano, T., Tsujino, T., Yoshida, K. et al.** (1991) *Cancer Res.*, **51**, 2926.
47. **Moskaluk, C.A., Hu, J. and Perlman, E.J.** (1998) *Genes Chromosomes Cancer*, **22**, 305.
48. **Yustein, A.S., Harper, J.C., Petroni, G.R. et al.** (1999) *Cancer Res.*, **59**, 1437.
49. **Dolan, K., Garde, J., Gosney, J. et al.** (1998) *Br. J. Cancer*, **78**, 950.
50. **Peralta, R.C., Casson, A.G., Wang, R. et al.** (1998) *Int. J. Cancer*, **78**, 600.
51. **van Dekken, H., Geelen, E., Dinjens, W.N. et al.** (1999) *Cancer Res.*, **59**, 748.
52. **Blount, P.L., Meltzer, S.J., Yin, J. et al.** (1993) *Proc. Natl Acad. Sci. USA*, **90**, 3221.
53. **Prives, C. and Hall, P.A.** (1999) *J. Pathol.*, **187**, 112.

54. **Hainaut, P. and Hollstein, M.** (2000) *Adv. Cancer Res.*, **77**, 81.
55. **Hollstein, M.C., Metcalf, R.A., Welsh, J.A.** *et al.* (1990) *Proc. Natl Acad. Sci. USA*, **87**, 9958.
56. **Casson, A.G., Mukhopadhyay, T., Cleary, K.R.** *et al.* (1991) *Cancer Res.*, **51**, 4495.
57. **Casson, A.G., Tammemagi, M., Eskandarian, S.** *et al.* (1998) *Mol. Pathol.*, **51**, 71.
58. **Casson, A.G., Evans, S.E., Gillis, A.** *et al.* (2003) *J. Thorac. Cardiovasc. Surg.*, **125**, 1121.
59. **Ambs, S., Bennett, W.P., Merrimam, W.G.** *et al.* (1999) *J. Natl Cancer Inst.*, **91**, 86.
60. **Chazotte-Aubert, L., Hainaut, P. and Ohshima, H.** (2000) *Biochem. Biophys. Res. Comm.*, **267**, 609.
61. **Sepehr, A., Taniere, P., Martel-Planche, G.** *et al.* (2001) *Oncogene*, **20**, 7368.
62. **Wilson, K.T., Fu, S., Ramanujam, K.S. and Meltzer, S.J.** (1998) *Cancer Res.*, **58**, 2929.
63. **Boyle, J.O., Hakim, J., Koch, W.** *et al.* (1993) *Cancer Res.*, **53**, 4477.
64. **Shi, S.T., Yang, G.Y., Wang, L.D.** *et al.* (1999) *Carcinogenesis*, **20**, 591.
65. **Flejou, J-F., Gratio, V., Muzeau, F. and Hamelin, R.** (1999) *Mol. Pathol.*, **52**, 263.
66. **Taniere, P., Martel-Planche, G., Maurici, D.** *et al.* (2001) *Am. J. Pathol.*, **158**, 33.
67. **Shepherd, T., Tolbert, D., Benedetti, J.** *et al.* (2000) *Gastroenterology*, **118**, 1039.
68. **Montesano, R., Hollstein, M. and Hainaut, P.** (1996) *Int. J. Cancer (Pred. Oncol.)*, **69**, 225.
69. **Xing, E.P., Yang, G.Y., Wang, L.D.** *et al.* (1999) *Clin. Cancer Res.*, **5**, 1281.
70. **Roncalli, M., Bosari, S., Marchetti, A.** *et al.* (1998) *Lab. Invest.*, **78**, 1049.
71. **Barrett, M.T., Sanchez, C.A., Galipeau, P.C.** *et al.* (1996) *Oncogene*, **13**, 1867.
72. **Xing, E.P., Nie, Y., Song, Y.** *et al.* (1999) *Clin. Cancer Res.*, **5**, 2704.
73. **Klump, B., Hsieh, C.J., Holzmann, K.** *et al.* (1998) *Gastroenterology*, **115**, 1381.
74. **Takueshi, H., Ozawa, S., Ando, N.** *et al.* (1997) *Clin. Cancer Res.*, **3**, 2229.
75. **Krishnadath, K.K., Tilanus, H.W., van Blankenstein, M.** *et al.* (1995) *Cancer Res.*, **55**, 1971.
76. **Chen, Y.J., Chen, P.H., Lee, M.D. and Chang, J.G.** (1997) *Int. J. Cancer*, **72**, 955.
77. **Michael, D., Beer, D.G., Wilke, C.W.** *et al.* (1997) *Oncogene*, **15**, 1653.
78. **Nakatsuru, S., Yanagisawa, A., Ichii, S.** *et al.* (1993) *Hum. Mol. Genet.*, **1**, 559.
79. **Nakatsuru, S., Yanagisawa, A., Furukawa, Y.** *et al.* (1993) *Hum. Mol. Genet.*, **2**, 1463.
80. **Tamura, G., Maesawa, C., Suzuki, Y.** *et al.* (1994) *Cancer Res.*, **54**, 1149.
81. **Al-Kasspooles, M., Moore, J.H., Orringer, M.B. and Beer, D.G.** (1993) *Int. J. Cancer*, **54**, 213.
82. **Iihara, K., Shiozaki, H., Tahara, H.** *et al.* (1993) *Cancer*, **71**, 2902.
83. **Shimada, Y., Imamura, M., Watanabe, G.** *et al.* (1999) *Br. J. Cancer*, **80**, 1281.
84. **Hardwick, R.H., Shepherd, N.A., Moorghen, M.** *et al.* (1995) *J. Clin. Pathol.*, **48**, 129.
85. **Polkowski, W., van Sandick, J.W., Offerhaus, G.J.** *et al.* (1999) *Ann. Surg. Oncol.*, **6**, 290.
86. **Tahara, E., Semba, S. and Tahara, H.** (1996) *Semin. Oncol.*, **23**, 307.
87. **Tahara, E.** (1995) *Cancer*, **75**, 1410.
88. **David, L., Seruca, R., Nesland, J.M.** *et al.* (1992) *Mod. Pathol.*, **5**, 384.
89. **Arber, N., Lightdale, C., Rotterdam, H.** *et al.* (1996) *Cancer Epidemiol. Biomark. Prev.*, **5**, 457.
90. **Bani-Hani, K., Martin, I.G., Hardie, L.J.** *et al.* (2000) *J. Natl Cancer Inst.*, **92**, 1316.
91. **Research Committee on Malignancy of Esophageal Cancer, Japanese Society for Esophageal Diseases** (2001) *J. Am. Coll. Surg.*, **192**, 708.
92. **Casson, A.G., Wilson, S.E., McCart, J.A.** *et al.* (1997) *Int. J. Cancer*, **72**, 739.
93. **Tahara, E., Yasui, W., Taniyami, K.** *et al.* (1986) *Cancer Res.*, **77**, 517.
94. **Shen, H., Xu, Y., Qiao, Y.** *et al.* (2000) *Int. J. Cancer*, **88**, 601.
95. **Leach, F.S., Polyak, K., Burrell, M.** *et al.* (1996) *Cancer Res.*, **56**, 235.
96. **Ogata, S., Tamura, G., Endoh, Y.** *et al.* (2001) *J. Pathol.*, **194**, 334.
97. **Meltzer, S., Yin, J., Manin, B.** *et al.* (1994) *Cancer Res.*, **54**, 3379.
98. **Gleeson, C., Sloan, J., McGuigan, J.A.** *et al.* (1996) *Cancer Res.*, **56**, 259.
99. **Evans, S.E., Gillis, A., Malatjalian, D.** *et al.* (2003) *Proc. AACR* (abstract), **44**, 62.
100. **Seruca, R., Santos, N.R., David, L.** *et al.* (1995) *Int. J. Cancer*, **64**, 32.
101. **Leung, S.Y., Yuen, S.T., Chung, L.P.** *et al.* (1999) *Cancer Res.*, **59**, 159.
102. **Kelsell, D.P., Risk, J.M., Leigh, I.M.** *et al.* (1996) *Hum. Mol. Genet.*, **5**, 857.
103. **LaVecchia, C., Negri, E., Franceschi, S. and Gentile, A.** (1992) *Cancer*, **70**, 50.
104. **Guiford, P., Hopkins, J., Grady, W.** *et al.* (1999) *Hum. Mutat.*, **14**, 249.
105. **Rashid, A. and Hamilton, S.R**. (1997) *Gastroenterology*, **113**, 127.

106. **Vasen, H.F., Wijnen, J.T., Menko, F.H. *et al.*** (1996) *Gastroenterology,* **110**, 1020.
107. **Hizawa, K., Iida, M., Matsumoto, T. *et al.*** (1993) *Dis. Colon Rectum,* **36**, 953.
108. **Moskaluk, C.A., Hruban, R.H. and Kern, S.E.** (1997) *Cancer Res.,* **57**, 2140.
109. **Terhune, P.G., Phifer, D.M., Tosteson, T.D. and Longnecker, D.S.** (1998) *Cancer Epidemiol. Biomark. Prev.,* **7**, 515.
110. **Day, J.D., DiGiuseppe, J.A., Yeo, C.J. *et al.*** (1996) *Hum. Pathol.,* **27**, 119.
111. **Yamano, M., Fujii, H., Takagaki, T. *et al.*** (2000) *Am. J. Pathol.,* **156**, 2123.
112. **Goggins, M., Offerhaus, G.J., Hilgers, W. *et al.*** (1998) *Am. J. Pathol.,* **152**, 1501.
113. **Wang, J., Zindy, F., Chenivesse, X. *et al.*** (1992) *Oncogene,* **7**, 1653.
114. **Hsia, C.C., Axiotis, C.A., DiBisceglie, A.M. and Tabor, E.** (1992) *Cancer,* **70**, 1049.
115. **Wang, X.W., Forrester, K., Yeh, H. *et al.*** (1994) *Proc. Natl Acad. Sci. USA,* **91**, 2230.
116. **Jia, L., Wang, X.W. and Harris, C.C.** (1999) *Int. J. Cancer,* **80**, 875.
117. **Hsu, I.C., Metcalf, R.A., Sun, T. *et al.*** (1991) *Nature,* **350**, 427.
118. **Fujimori, M., Tokino, T., Hino, O. *et al.*** (1991) *Cancer Res.,* **51**, 89.
119. **Tsuda, H., Zhang, W.D., Shimosato, Y. *et al.*** (1990) *Proc. Natl Acad. Sci. USA,* **87**, 6791.
120. **Kanai, Y., Ushijima, S., Tsuda, H. *et al.*** (1996) *Jpn J. Cancer Res.,* **87**, 1210.
121. **Levi, S., Urbano, I.A., Gill, R. *et al.*** (1991) *Cancer Res.,* **51**, 3497.
122. **Kang, Y.K., Kim, W.H., Lee, H.W. *et al.*** (1999) *Lab. Invest.,* **79**, 477.
123. **Wistuba, I.I. and Albores-Saavendra, J.** (1999) *J. Hepatobiliary Pancreatic Surg.,* **6**, 237.
124. **Hanada, K., Itoh, M., Fujii, K. *et al.*** (1996) *Cancer,* **77**, 452.
125. **Watanabe, M., Asaka, M., Tanaka, J. *et al.*** (1994) *Gastroenterology,* **107**, 1147.
126. **Albrecht, S., von Schweinitz, D., Waha, A. *et al.*** (1994) *Cancer Res.,* **54**, 5041.
127. **Ohnishi, H., Kawamura, M., Hanada, R. *et al.*** (1996) *Genes Chromosomes Cancer,* **15**, 187.

Breast cancer

9

9.1 Introduction

Breast cancer is one of the most common cancers in women in the developed countries of the world and it is the cause of death in approximately 20% of all females who die from cancer in these countries. Up until fairly recently the mortality figures had not altered significantly over the years. The age-adjusted death rates for women in the major industrialized countries are shown in *Table 9.1*.

There are marked differences in the incidence of breast cancer in different places, the predominant impression being that the disease is more common among Caucasians living in the colder climates and more highly industrialized countries of the Western hemisphere [2]. The Western diet is associated both with earlier age at menarche and with post-menopausal obesity, which increases estrogen production and hence breast cancer risk. The higher breast cancer incidence in most Western countries than in many developing countries is partly (and perhaps largely) accounted for by these dietary effects combined with later first childbirth, lower parity and shortest breastfeeding [3 and references therein].

In the 1997 edition of this book [4] a survey was cited which showed that, with the possible exception of China, breast cancer incidence rates had been increasing over the previous 20 years in all age groups in all countries of the world for which rates were obtainable [5]. Despite this, it is encouraging to be able to report that recent global breast cancer statistics indicate that mortality is currently declining in the USA and other countries where historically the disease has been a leading cause of death [6]. This has been due, in part, to the use of endocrine therapy that reduces the levels of circulating estrogens or competes with estrogen for binding to its receptor [7].

Breast tumors are classified as noninvasive or invasive, the majority (76%) belonging to the group of invasive ductal breast cancers. Patients can be staged using the clinical characteristics of tumor diameter (T), lymph node involvement (N) and the presence of distant metastases (M), hence the nomenclature TNM staging. Several staging systems have been used for patients with breast cancer and the International Union Against Cancer (UICC) and

Table 9.1 Age-adjusted death rates for breast cancer in the major industrialized countries of the world (2000)

Country	Age-adjusted death rates per 100 000 population
United Kingdom	26.8
Germany	23.7
Canada	22.7
France	21.4
USA	21.2
Russian Federation	16.7
Japan	7.7

Abstracted from [1].

the American Joint Committee on Cancer (AJCC) have now agreed on a single staging system [8,9]. Prognostic factors include lymph node status, degree of differentiation of the tumor and the presence of estrogen and/or progesterone receptors. The DNA content (ploidy) of tumors has been suggested as a prognostic indicator but its use is limited at present. However, DNA analysis of *in situ* ductal carcinomas of the breast by flow cytometry has indicated that major genetic alterations and DNA heterogeneity are early events in carcinogenesis and that they are already established at the preinvasive stage [10]. Recent consensus conferences on the treatment of breast cancer in Europe and the USA have developed guidelines on the eligibility of patients to receive adjuvant chemotherapy based on histological and clinical characteristics [11,12]. When these guidelines are applied, up to 90% of lymph node-negative young breast cancer patients are candidates for adjuvant systemic treatment.

A number of risk factors have been associated with development of the disease, including the dose and duration of estrogen exposure, early menarche, late menopause, age at first childbirth, nulliparity, fat in the diet, postmenopausal hormone replacement therapy, alcohol consumption, cigarette smoking and ionizing radiation [3,13 and references therein]. Several models have been developed to estimate breast cancer risk, e.g. the Gail model and the Claus model. The Gail model was developed to address the challenge of combining epidemiological risk factors and the Claus model estimates the effect of family breast cancer history alone on age-specific breast cancer risks [13].

9.2 Genetic changes in breast cancer

Although relatively little is known about the molecular mechanisms leading to breast cancer development, breast cancers have probably been studied more than any other tumor type with regard to oncogene expression. *MYC*, *ERBB2* or one of the *RAS* family have been found to be expressed in over 60% of cases.

Mutations in highly penetrant susceptibility genes are estimated to underlie only 5–10% of breast cancers yet these are important as individuals carrying such mutations are at very high risk of developing breast cancers, of having more than one primary cancer, and of passing these altered genes to their offspring [13]. Following the genomic localization and subsequent identification of the breast cancer susceptibility genes *BRCA1* and *BRCA2* (considered to be tumor suppressor genes), the basic pattern of cancer risk associated with these genes has been defined [14].

9.2.1 *MYC*

Very few examples of rearrangements of the *MYCN* or *MYCL* genes have been found in breast cancer. On the other hand, there is considerable evidence for *MYC* amplification although the reported incidence varies from 4% to 52% [15,16]. Three large studies of *MYC* amplification in breast cancer have produced different conclusions as to the value of this marker in prognosis [17].

Some reports indicate that *MYC* amplification is predictive for shortened relapse-free and/or overall survival. Depending on the study, *MYC* amplification has also been correlated with highly proliferative tumors, high tumor grade, nodal status, steroid receptor status, tumor size, age, cathepsin D expression, aneuploidy and a particularly aggressive form of breast cancer, inflammatory carcinoma [16 and references therein]. Some studies have found that co-amplification *MYC* and *ERBB2* is relatively rare, while one study found a positive correlation for amplification of the two loci and still others have found no correlation. The variation in results with these studies, as with such studies in general, is not surprising considering the broad range of sample size, composition, follow up, as well as inconsistencies in experimental and statistical methodology [16].

In one study, including 80 primary breast carcinomas as well as benign tissues and nodal metastases, amplification was the major alteration in the *MYC* gene occurring in 18% of tumors, and one example of a rearrangement was also seen. Amplification was primarily associated with infiltrating ductal carcinomas and poorly differentiated tumors. No stage-I tumors showed any abnormalities in *MYC*. A significant correlation was seen between amplification of *MYC* and poor short-term prognosis, implicating this gene in the progression of breast cancer. In a second study of tumors from 121 patients, over 30% had abnormalities in *MYC*. There was a significant correlation between tumors of postmenopausal patients and amplification of *MYC*, but no correlation with tumor grade, receptor status or presence of metastatic disease was found. No survival rates were presented in this study. A prospective study of 125 patients found amplification of *MYC* in 18% of cases but no clinical correlations were detected, except a highly significant association between *MYC* amplification and inflammatory carcinomas, suggesting that *MYC* might contribute to the rapid progression of this subtype of breast cancer. Increased levels of *MYC* mRNA have also been found in breast cancer and have not always correlated with gene amplification. In one study almost 50% of tumors had elevated mRNA levels and this correlated with the presence of lymph node metastases. Amplification was seen in only 11% of tumors [17].

Immunocytochemistry using monoclonal antibodies Myc1-6E10 and Myc1-9E10 has been used to study p62 levels in breast cancers but has not clarified the role of *MYC* in this disease. Seventy per cent of cases in one study, using antibody Myc1-9E10, showed expression of p62 in both the nucleus and occasionally in the cytoplasm, but there was no correlation with histopathological grade, presence of metastases in the lymph nodes, or receptor status [17]. This antibody may not be of value for detecting abnormalities in the *MYC* gene because discrepancies have been demonstrated between the levels of MYC protein, as determined immunohistochemically, and the levels of mRNA measured by *in-situ* hybridization. Neither of these techniques could be used to indicate the presence of amplification [18]. Another study of 85 breast tumors showed that *MYC* overexpression is less common than found in previous studies, possibly because a different scoring system was used [16]. These results suggest that there may be problems in using measurement of oncoprotein levels with monoclonal antibodies in routinely processed tissues as a measure of abnormalities in the oncogene itself.

A sensitive enzyme immunoassay has also been used as a means of measuring MYC oncoprotein expression in tissue extracts. All tumors had elevated levels of p62 compared with normal tissue, but no correlation with age, nodal status, receptor status, histopathological grade or survival was observed. There was, however, an association between tumor size (T) and high levels of MYC oncoprotein expression [17].

The data on the prognostic value of *MYC* amplification are therefore conflicting with some studies suggesting it is an independent powerful prognostic factor, particularly in node-negative and steroid receptor-positive cancers [19], and others suggesting that it is not a suitable prognostic marker for routine purposes [20]. Thus, amplification, rearrangement and overexpression of MYC oncoprotein are present in a significant number of breast cancers, but the predictive value and pathophysiological consequences of amplification are not clear at present. A study of benign breast disease, which has been associated with breast cancer risk, found none of the genetic alterations commonly identified in breast cancer, including *MYC* amplification [16]. Taken together these results do not show any clearly defined diagnostic or prognostic role for *MYC* in breast cancer at the present time, and any role it does have is likely to be complex. However, a recent meta-analysis of 29 published studies showed significant associations between *MYC* and tumor grade, lymph node metastasis, negative progesterone receptor status and postmenopausal status. Amplification was significantly associated with risk of relapse and death. It was stated that *MYC* amplification is relatively common in breast cancer and that it may provide independent prognostic information. Despite these findings the authors concluded that more rigorous studies with consistent methodology are required to validate these associations and to investigate its potential as a predictor of specific therapy response [21].

As indicated above, there is substantial evidence that *MYC* is frequently amplified and overexpressed in breast cancer, and that such overexpression can contribute to breast neoplasia. However, the evidence from *in vitro* and *in vivo* studies indicates that MYC alone is not sufficient for tumorigenesis. Recent reports suggest that *MYC* overexpression will not be advantageous for a breast tumor cell line unless it is counterbalanced by additional signals which promote cell survival, such as growth factor stimulation, or other genetic changes such as loss of Bax expression or elevated *BCL2* expression. Clinical studies which examine *MYC* expression in conjunction with a survey of apoptosis regulators and growth factors, or markers of signal transduction pathway activation, may therefore provide new prognostic information for breast cancer and may identify new targets for therapy [16]. There is also evidence that the related family member *NMYC* is widely overexpressed in breast cancer in the absence of amplification [16].

9.2.2 *ERBB2* and *RAS*

More useful information has been obtained from measurements of *ERBB2* (*HER-2/neu*) and the *RAS* gene family.

The protein encoded by *ERBB2* is a transmembrane tyrosine kinase receptor with extensive homology to the epidermal growth factor receptor (EGFR) and is considered to be one of its family members. It has been estimated that 20–30% of breast cancers overexpress the *ERBB2* gene product at a sufficiently high level that the gene acts as an oncogene [22]. *ERBB2* amplification in breast cancer has been demonstrated by many investigators, in frequencies ranging from 10% to 30%. In a preliminary study, amplification of *ERBB2* correlated with poor prognosis in lymph node-positive patients. Several subsequent studies confirmed this finding, but others have been unable to show any correlation. The workers who produced the initial results corroborated their findings in a large study of 526 patients [23] and two other groups found similar results in lymph node-negative breast cancer. In one study co-amplification of *ERBB2* and *ERBA1* was found in 23% of breast tumors and was shown to be a strong indicator of metastatic potential.

Antibodies to the ERBB2 protein have also been used to study overexpression of *ERBB2* in breast cancer and there are several reports which correlate the levels of ERBB2 protein with gene amplification. As with direct analysis of gene amplification, immunostaining for ERBB2 protein as a prognostic marker has shown both positive and negative correlations. Strong staining was found in nine, 14 and 17% of cases of breast cancer in three different studies. Several groups found no relationship between staining and stage, node status, receptor status or size of tumor, but others have shown a significant association between reactivity and mortality or recurrence [24]. One study has demonstrated a correlation between *ERBB2* overexpression and the presence of hematogenous metastases.

These conflicting results are likely to be due to variations in the numbers of patients examined, differences in antibodies used, length of follow-up and methods of scoring immunoreactivity. The immunohistochemical technique has potential as a routine method for the analysis of expression of *ERBB2* in tumors, once all the parameters have been carefully evaluated and standardized. Data obtained from such studies should confirm or refute a role for *ERBB2* as a prognostic indicator in this disease. In a review of 47 published studies comprising more than 15 000 patients it was concluded that the preponderance of evidence indicates that *ERBB2* gene amplification and protein overexpression are associated with an adverse outcome in breast cancer [25]. The reported clinical therapeutic efficacy of anti-ERBB2 monoclonal antibodies in breast cancer (see Chapter 13) highlights the importance of understanding the biology of ERBB2 [26].

The range of downstream signaling pathways activated by ERBB2 is wide, but ultimately these function not only to increase the levels of cyclin-D-CDK4 complexes leading to a loss of the G_1-S checkpoint but also to disrupt the delicate balance between cell survival and death signals. An understanding of these molecular abnormalities is leading to the design of novel therapies and in this context it has been recently demonstrated that antisense ERBB2 produces a significant enhancement of doxorubicin-induced apoptosis in breast cancer cell lines and xenografts [27 and references therein].

RAS has been the subject of intense research activity over the last two decades and although considerable insight has been obtained into what it does it has been pointed out recently that many more questions than answers have been generated [28]. Consequently, its exact role in breast cancer has still to be determined. That it is of potential importance has been suggested in a recent study in which primary human mammary epithelial cells were transformed into invasive, malignant cells after the introduction of genes encoding the SV40 large T antigen, the telomerase catalytic subunit and HRAS. The requirement for all three genetic changes was demonstrated in this study, as well as that the latency and frequency of tumor formation were influenced by the stromal microenvironment [29].

The majority of studies have detected elevated levels of HRAS mRNA in primary breast carcinomas. Elevated expression was associated with advanced histological types. In contrast, a further study detected elevated expression of NRAS and KRAS, but only one tumor showed any increase in HRAS [17].

Increased expression of RAS p21 has been detected in 63–83% of malignant breast tumors compared with low levels in benign tissues. A single study was unable to detect differences in RAS p21 expression between malignant and benign breast tumors. In general there are no significant clinical correlations, although one study demonstrated higher RAS p21 expression in carcinomas from postmenopausal women. A direct-binding quantitative competition radioimmunoassay was developed for RAS p21 using monoclonal antibody Y13 259 to determine absolute levels of the protein. Again, approximately two-thirds of carcinomas demonstrated high levels of the protein compared with lactating breast and benign tissues. Recent studies using this technique demonstrated higher levels of p21 in postmenopausal patients. Levels of RAS p21 have also been determined in breast tissue using Western blotting techniques and the results have confirmed the immunohistochemical findings of high levels of the protein in malignant tissues. An association was also found between high levels of RAS p21 and extent of tumor and a significant correlation was seen between high levels and short disease-free interval.

Studies of the RAS family in breast cancer have therefore found little evidence of point mutations or structural alterations in the genes, but high levels of RAS appear to be associated with progression of breast cancer and poor prognosis [17].

As in lung cancer, polymorphisms associated with HRAS have been investigated in breast cancer. In one study, four common alleles and 16 rare alleles were found in the normal and cancer-bearing populations. The frequency of two particular common alleles was diminished in the breast cancer patients with concomitant increases in two rare alleles. One of these rare alleles was significantly associated with breast cancer and is potentially of use in risk analysis. In a second study, HRAS allele loss has been found in breast tumors and correlates significantly with grade-III tumors, lack of estrogen/progesterone receptors and the presence of distant metastases. A third study has confirmed that allele loss correlates with low levels of estrogen receptors [17]. The association of HRAS polymorphisms and breast cancer has been submitted to meta-analysis and this has shown that women with a single copy of the rare allele have a 1.7-fold increased risk of breast cancer compared to the general population, while those who are homozygous for the rare allele have a 4.6-fold increased risk [30].

9.2.3 *p53*

Numerous studies have shown the importance of the *p53* tumor suppressor gene in cancer, including breast cancer [31,32 and references therein]. In addition to mutations, a second mechanism is believed to inactivate wild-type p53. In some types of breast cancer, the wild-type p53 can lose its tumor-suppressive function by being sequestered in the cytoplasm and prevented from entering the nucleus, its normal site of action [33]. This has been suggested as a possible explanation for the development of breast cancer in cases that do not show any *p53* mutations.

Most breast cancer cases are believed to be sporadic, the results of a spontaneously arising mutation. However, a small proportion of cancers result from an inherited predisposition. Evidence for inherited disease includes early age of onset and extensive bilateral breast tumors clustered within families. For example, the Li–Fraumeni syndrome is a familial predisposition to the early childhood development of soft tissue sarcomas as well as early-onset breast cancer in parents and relatives [31].

Breast tumors from Li–Fraumeni cancer patients are associated with inherited mutations within exon 7 of the *p53* gene. These mutations are primarily CG–TA transitions at CG dinucleotides [31,34]. Sporadic breast tumors are associated with mutations clustered within exons 5, 6, 7 and 8. However, most other inherited breast cancer cases are rarely caused by germ-line point mutations in *p53*. Several studies have shown the possible involvement of a number of other tumor suppressor genes and several oncogenes in the development of the disease [34 and references therein].

Although mutations within *p53* are significantly associated with breast cancer development, several other factors can play a role in the disease process. These include estrogen receptor status, accumulation of p53 protein and the *p53*-binding murine double minute-2 (*MDM2*) gene product which has been identified as a negative regulator of *p53*. MDM2 overexpression may be another mechanism by which cancer cells overcome *p53*-regulated growth control without selecting for *p53* mutations *per se* [35].

The recent discovery of two paralogs of p53 (p63 and p73) has greatly complicated interpretation of the p53 literature [36 and references therein]. Functional p53 protein is directly or indirectly compromised in most sporadic cancers. In contrast, p63 and p73 are only rarely mutated in human tumors. p73 is located at 1p36, a region that is frequently deleted in a variety of tumors including breast cancer [36].

Familial breast cancer

Many studies have shown that, with the exception of the Li–Fraumeni syndrome, *p53* gene mutations are not common in breast cancer patients with a family history of malignancy. In a study of five families with a history of breast cancer none showed any mutations of the *p53* gene [37]. The eight cases that demonstrated a *p53* mutation did not have any family history of the disease. The authors suggested, therefore, that a mutation in another tumor suppressor gene on chromosome 17p is involved in heritable breast cancer.

In a related study, 126 patients with early-onset breast cancer were screened and only one of the affected patients was identified with an inherited germ-line *p53* mutation [38]. A study of 25 families with a strong family history of breast cancer demonstrated a complete absence of any germ-line *p53* mutations [31]. An independent study obtained similar findings and these authors concluded that germ-line tumor suppressor *p53* mutations occur in extremely few breast cancer patients that do not have family histories suggestive of the Li–Fraumeni syndrome [39].

Five to ten per cent of breast cancers have been suggested to be caused by highly penetrant autosomal dominantly inherited susceptibility genes and around one in 200 women in the general population will develop breast cancer as a result of inheriting one of these genes. There are three main syndromes which show an increased risk of breast cancer: (1) Li–Fraumeni syndrome, associated with mutations in *p53* as discussed above; (2) site-specific breast cancer; and (3) breast–ovarian cancer syndrome. Genes for these last two syndromes have now been identified which means that accurate presymptomatic testing of women in at-risk families can be carried out (see Chapter 5).

Sporadic breast cancer

Sporadic breast cancer cases tend to show a greater incidence of *p53* mutations. In one study 13% of sporadic breast tumors were shown to exhibit *p53* mutations [40]. However, this analysis only included exons 5 and 6, thus the authors speculated that further studies, especially with exons 7 and 8, would reveal a greater mutation frequency.

A mutation frequency of 46% amongst sporadic breast tumors has been obtained in other studies [34,41]. GC–TA transversions and CG–TA transitions are common in tumors caused by mutagenic factors, suggesting that some exogenous carcinogens also play a role in breast cancer development [34].

Thus from a comparison of both sporadic and hereditary breast cancer studies, it can be concluded that *p53* mutation is a rare etiological factor in hereditary breast cancer but the mutation rate is significantly higher in sporadic tumors.

Prognostic significance of *p53*

When assessing the relevance of alterations in *p53* to breast cancer, and particularly to prognosis, it is important to recognize the limitations common to many of the studies [35]. The two techniques most often used to study *p53* are immunohistochemical staining, to detect protein accumulation, and nucleic acid sequencing. With regard to immunohistochemical studies there are variations at every step, from patient selection, through specimen processing, actual immunohistochemical technique used, interpretation of the data and statistical analyses [42]. As a consequence of this variation, standardization of the immunohistochemical methods being used, so that a more meaningful analysis of results can be obtained, has been proposed by several groups [35,42,43].

With the nucleic acid approach, many investigations have only concentrated on the four highly conserved domains encompassed by exons 5–8, however, *p53* mutations have been detected outside this region and, thus, the number of *p53* mutations may be underestimated [35]. Another area of concern is the independence of *p53* as a prognostic variable from other better-characterized predictors as well as the striking correlation between the number of cases examined in a particular study and the likelihood of *p53* being found to be a prognostic factor [42]. It should be noted that these limitations also apply to studies of *p53* in other cancers as well as to the study of other oncogenes.

With these limitations in mind, several researchers have demonstrated that the detection of mutant *p53* in patients with breast cancer has significant prognostic value [44–47]. Patients with node-negative breast cancer that have overexpressed mutant p53 proteins and/or mutations within the *p53* gene have a much lower chance of long-term survival and a greater incidence of recurrence following surgery. Thus p53 accumulates within the cell nucleus and can be detected by immunohistochemistry [46,47]. Similarly, point mutations within the *p53* gene which can be detected by SSCP analysis are also indicators of poorer prognosis. This is not surprising considering that such a mutation can result in an altered protein conformation resulting in prolongation of its half-life [48,49].

A study in which the entire coding region of *p53* was sequenced in 316 consecutive cases of breast cancer showed that mutations in the evolutionary conserved regions II and IV of the *p53* protein (codons 117–142 and 270–286 respectively) were associated with a significantly worse prognosis [50]. In another study of 392 breast carcinomas analyzed immunohistochemically, abnormal expression of p53 protein was found to be an independent prognostic indicator, although it was only a weak one [51]. A recent immunohistochemical analysis of p53 expression in 192 cases of infiltrating ductal carcinomas of the breast showed prognostic significance between p53 expression and shorter survival time, and disease-free interval, for all patients in general as well as those who presented with lymph node metastases at the time of diagnosis [52].

Responsiveness to therapy and *p53*

Using the nuclear accumulation of p53 as a surrogate marker of alteration in *p53* function, no statistically significant evidence was found that *p53* status, as assessed by immunohistochemistry, can predict the response to combination chemotherapy in lymph node-negative breast cancer patients. There was, however, a trend for patients who were *p53*-negative by immunohistochemistry to receive greater benefit from chemotherapy, but the sample size was too small for this observation to be statistically meaningful [53]. In another study, therapy, especially adjuvant tamoxifen, was found to be of less value to patients with *p53* mutations and lymph node-positive tumors [50].

A comparison of four anti-*p53* antibodies in an immunohistochemical analysis of 245 breast cancers showed a significant relationship between immunodetection of p53 with each of the four antibodies and poor responsiveness to endocrine therapy [54]. However, as pointed out by these authors, the evaluation of immunohistochemical detection of p53 protein should be carried out with great care as the observed tumor phenotypes do not always reflect the underlying biochemical and biological cellular changes and hence it is no longer sufficient to designate cases as '*p53* positive'.

The use of adjuvant chemotherapy and toxic drugs for women with node-negative breast tumors has been hotly debated by clinicians. Since node-negative tumors have a much lower malignant potential (than axillary lymph node-positive tumors), postsurgical therapy along with its associated risks is not always necessary. However, 20–30% of these cancer patients will relapse [44]. It is possible that *p53* status may provide a mechanism for screening patients at greater risk of relapse or recurrence of disease. Thus more aggressive postsurgical therapy would be warranted for those women showing overexpression of mutant p53 protein and/or *p53* gene mutations and this continues to provide a major impetus for clarification of the role of *p53* in breast cancer (see also Section 9.5).

9.2.4 *BRCA1* and *BRCA2*

The breast cancer susceptibility genes *BRCA1* and *BRCA2* encode multifunctional proteins, the mutant phenotype of which predisposes to both breast and ovarian cancer [14, 55 and references therein]. The combined contribution to overall breast cancer incidence of strongly predisposing mutations in *BRCA1* and *BRCA2* which confer individual risks of about 60% by age 70, is less than 5%. By contrast, a predisposing allele with a relative risk of 2 and a frequency of 20% could account for up to 20% of breast cancer incidence [56]. The risk to close relatives of a case, averaged across all ages, is about twofold. Most of this familial risk is probably genetic in origin [3]. The risk is about the same for the mother, sisters or daughters of a case, suggesting dominant rather than recessive effects. Large population-based studies indicate that only 15–20% of overall familial risk is attributable to mutations in *BRCA1* and *BRCA2* and that *BRCA1* and *BRCA2* mutations are rare in the general population [56, 57 and references therein]. A full discussion of *BRCA1* and *BRCA2* is provided in Chapter 3 and their contributions to familial breast cancer have been covered in Chapter 5, therefore only a brief review of the structure and functions of these genes will be provided here.

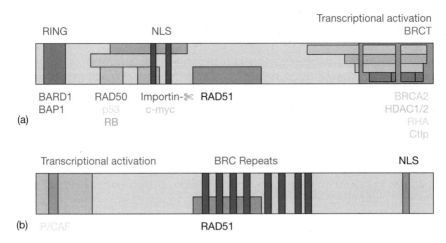

Figure 9.1

Structural and functional features of (a) BRCA1 (1863 amino acids) and (b) BRCA2 (3418 amino acids) proteins. The N-terminal RING finger of BRCA1 interacts with BRCA1-associated RING domain 1 protein (BARD1) and with the deubiquitinating enzyme, BRCA1-associated protein-1 (BAP1). Two nuclear localization signals (NLS) bind importin-α; nearby regions interact with p53, RB1, RAD50 and MYC. A domain within BRCA1 (amino acids 758–1064) interacts with RAD51. The BRCA1 C-terminal (BRCT) repeats interact with BRCA2, histone deacetylase (HDAC) 1 and 2, the CtBp-interacting protein (CtIP) and RNA helicase A (RHA). The BRCA2 N-terminus interacts with p300/CBP-associated factor (P/CAF) which has histone acetylase activity. Four of the eight BRC repeats of BRCA2 interact with RAD51. Putative transcriptional activation domains lie at the C-terminus of BRCA1 and the N-terminus of BRCA2. Adapted from [55].

BRCA1 and *BRCA2* are large genes which span approximately 80 kb of genomic DNA. Both have large central exons encoding >50% of the protein and both encode large proteins of 1863 (BRCA1) and 3418 (BRCA2) amino acids [55] (*Figure 9.1*). The molecular mechanism by which BRCA1 loss promotes tumor formation is unclear but it has been recently established that the BRCA1 RING finger region heterodimerizes with BARD1 and functions as an enzymatic mediator of protein ubiquitization [58].

In normal cells, BRCA1 and BRCA2 are nuclear proteins and *BRCA1* and *BRCA2* mRNA and proteins are preferentially expressed during the late G1–early S phase of the cell cycle. BRCA1 and BRCA2 are expressed ubiquitously and in mouse embryos they are highly expressed during mammary epithelial proliferation and differentiation. In the mammary gland the expression of both messages is developmentally regulated and is induced during puberty and pregnancy and reduced during lactation. Null mutations in both genes result in embryonic lethality [55].

A role for BRCA1 and BRCA2 in homologous recombination and DNA repair is suggested by the strong biochemical interaction of BRCA1 and BRCA2 with proteins known to be involved in these processes. A striking connection between double-strand break repair defects and tumorigenesis has been shown for BRCA1 and BRCA2. Both proteins interact with RAD51 protein which catalyzes structural change as an early step in homologous recombination that results in the formation heteroduplex molecules [59 and references therein]. BRCA1 and BRCA2 are essential for normal development as patients in whom both alleles of *BRCA1* or *BRCA2* are mutated have not been found [60]. It appears that the role of BRCA1 in DNA repair, which could involve the integration of several pathways, is broader than that of BRCA2. BRCA1 functions in the signaling of DNA damage and its repair by homologous recombination, nucleotide-excision repair and possibly non-homologous end-joining. BRCA2 regulates the activity of RAD51 which is required for homologous recombination [61]. A recent study has established that BRCA2 functions directly in homologous recombination and provides a structural and biochemical basis for understanding the loss of recombination-mediated double-strand break (DSB) repair seen in BRCA2-associated cancers [62]. These findings have prompted searches for alterations of RAD51 in patients with breast tumors. In one such study with patients with early-onset breast cancer no germ-line mutations were identified although in another study a single nucleotide polymorphism (SNP) was found in the 5′ UTR region of RAD51 which was associated with an increased breast cancer risk in *BRCA2* carriers [59 and references therein].

An association between a heterozygous *BRCA1* or *BRCA2* mutation in human cells and radiosensitivity has been demonstrated recently. The authors comment that their data are not sufficient evidence to prove a causal relationship between heterozygous *BRCA* mutations and radiosensitivity but they are important as this phenotype could possibly predispose to an increased rate of radiation-induced mutagenesis and carcinogenesis [63].

Somatic mutation of *p53* is frequent in human *BRCA1* and *BRCA2* breast tumors, and many *BRCA*-associated *p53* mutations are novel. Genetic instability caused by the loss of *BRCA1* or *BRCA2* could trigger mutations, including mutations in checkpoint genes such as *p53*. In this scheme, mutant p53 would overcome incomplete proliferation and lead instead to uncontrolled proliferation and invasive growth [55].

Rapid proliferation of breast epithelial cells during puberty and pregnancy offer an ideal opportunity for somatic mutation, including the loss of a *BRCA1* or a *BRCA2* allele. Tissue specificity of *BRCA1* and *BRCA2* tumorigenesis might be attributable to both the hormonal environment of breast (and ovarian) cells, as well as to tissue-specific expression of as-yet-unknown genes [55].

BRCA1- and *BRCA2*-associated human breast cancers have different morphological and immunohistochemical characteristics, including different gene expression profiles [64,65]. Importantly, from a therapeutic viewpoint, restoring BRCA1 or BRCA2 function would not be useful after checkpoint mutations had occurred [66].

An important recent finding has been that an apparent *BRCA1* homozygous knockout was found to be an artefact of the PCR. In addition to challenging the results of an earlier study it suggests that mispriming due to mismatched primers at the site of SNPs, leading to preferential amplification of one allele, may represent a significant proportion of instances of mutation detection insensitivity. These results have important cautionary implications for all PCR-based mutation detection methods used in research and in particular in clinical genetic testing laboratories [60].

Finally, a recent review of preclinical and clinical studies relevant to the prognostic associations of *BRCA1-* and *BRCA2*-associated breast carcinoma concluded that the current data are inadequate to support the use of *BRCA1* or *BRCA2* status to counsel individuals regarding their prognosis or to select treatment [64] (see also Chapter 5).

9.2.5 *CCND1*

Amplification of the 11q13 region on chromosome 11 is associated with a number of cancers, including breast carcinoma. The gene for cyclin D1 (*CCND1, PRAD1*) has been located at this region, near to the *BCL1* breakpoint, and its association with a number of cancers, especially B-cell tumors, qualifies it as an oncogene [67]. In a study of 226 breast carcinomas 11q13 amplification was associated with a significantly shorter relapse-free survival and a higher breast cancer-specific mortality suggesting that 11q13 amplification identifies a subgroup of node-positive patients at high risk [68]. At the present time it is still not resolved whether amplification of the 11q13 region reflects a selection for *CCND1*, or other genes which have been also identified there (*HST1, INT2, BCL1*). As a result, there is no direct evidence that the *CCND1* gene has a causative role in breast cancer [69].

However, up to 50% of all cases of breast cancer have elevated levels of cyclin D1 [70]. Tissue-specific transgenic expression of cyclin D1 in mice results in mammary hyperplasia and adenocarcinomas and cyclin D1 knockout mice show a marked defect in the development of breast epithelium during pregnancy, both of which indicate a prominent role for cyclin D1 in the development of breast cancer [70].

It has been recently proposed that for oncogenesis, the role that constitutive cyclin D kinase activity plays in driving cell growth and metabolism in early G_1 may be of equal importance to its role in promoting G_0 to early G_1 progression [71].

9.2.6 *RB1*

RB1 has been found to be disrupted by deletion or chromosomal rearrangement in some breast cancer cell lines leading to loss of wild-type RB1 expression. However, the correlation between loss of heterozygosity (LOH) at *RB1* and the loss of RB1 protein is uncertain as the published literature is conflicting on this point [72 and references therein].

The protein RB1CC1 (retinoblastoma 1 (RB1)-inducible coiled-coil 1) has been identified as a key regulator of the *RB1* gene. RB1CC1 is proposed to be a transcription factor and the *RB1CC1* gene is localized to a region of chromosome 8q11. It has been recently demonstrated that 20% of primary breast cancers contain mutations in *RB1CC1* and that this gene has all the characteristics of a tumor suppressor gene [73].

9.2.7 Other chromosomal abnormalities and potential tumor suppressors

In addition to the genetic alterations indicated above there are a number of other reported alterations associated with breast cancer. LOH on the long arm of chromosome 6 has been detected in 48% of primary breast cancers, confirming an earlier study indicating that this is a frequent occurrence in primary breast carcinomas [74]. LOH for 7q31 may also be an early event in breast cancer development [75].

Chromosome 1 abnormalities have also been reported to be a frequent occurrence and rearrangements of 1p correlate with a poor prognosis [76]. Using DNA fingerprinting followed by LOH analysis it has been shown that regions at 2q21-24 and 6p21-23 may harbor novel tumor suppressor genes in breast cancer [77]. These authors also showed that LOH at 2q21-24 was more frequent in high-grade than in low-grade tumors.

Chromosome 3p contains several tumor suppressor genes and 21 polymorphic and 2 nonpolymorphic 3p markers were recently investigated by allelotyping and real-time PCR in a variety of epithelial malignancies, including breast cancer. A high frequency of LOH was found in all the types of cancer that were studied, including 41/51 of the breast cancer cases [78].

It has been shown that the *E-cadherin* gene, responsible for the cell–cell adhesion molecule E-cadherin, was not mutated in 42 cases of infiltrative ductal or medullary breast carcinomas but was mutated in four out of seven infiltrative lobular breast cancers using a PCR/SSCP method. Although this study suffers from the sample size problem alluded to earlier, if confirmed, the results could offer a molecular explanation for the typical scattered tumor cell growth seen in infiltrative lobular breast carcinoma [79].

Recent data support the involvement of sigma 14-3-3 (a member of a class of molecular chaperones known to interact with cyclin-dependent kinases in breast cancer cells) in the neoplastic transformation of breast epithelial cells by virtue of its role as a tumor suppressor [80]. Finally, a growing body of evidence supports the hypothesis that the retinoic acid receptor beta 2 (*RARβ2*) is a tumor suppressor. Biallelic inactivation of *RARβ2* by epigenetic changes has recently been reported in breast cancer [81].

9.3 Estrogens and cell cycle regulation

Clinical and experimental data have indicated that exposure to estrogens is one of the leading causes of sporadic female breast cancer and in December 2002 estrogen was declared to be a known human carcinogen by the National Toxicology Program of the National Cancer Institute in the USA. It is becoming apparent that estrogen has separate hormonal and DNA-damaging cancer-promoting effects.

Estrogen has been found to control several key G_1 phase cell cycle regulators (e.g. cyclin D1, MYC, Cdk2, Cdk4, Cdk inhibitors and Cdc25A) [82]. Cyclin D1 overexpression, for example, is seen preferentially in ER-positive breast cancers [70]. Several models have been proposed for the mechanism of ER activation by cyclin D1. One of these postulates an autostimulatory loop in which autoactivation of ER by hormone leads to transcriptional induction of cyclin D1. This newly synthesized cyclin D1 protein can then bind to ER to cause further ER activation. In this way cyclin D1 and estrogen may act synergistically in stimulating breast cancer growth. Such an autostimulatory loop could explain why coexpression of ER and cyclin D1 is seen in many breast cancers [70].

Tamoxifen, the primary drug used for breast cancer treatment and prevention blocks the ability of estrogen to promote breast cancer hormonally. Recent research indicates that the anticancer activity of tamoxifen may be determined in part by tumor levels of AIB1 (SRC-3, an ER co-activator that has been shown to reduce the antagonistic activity of tamoxifen-bound ERs) and ERBB2. Thus, AIB1 may be an important diagnostic and therapeutic target [83].

Mammalian cells convert estrogen into related metabolites that generate free radicals capable of damaging DNA and which can also bind to DNA and remove a nucleotide base (depurination). Although there are mechanisms to repair this damage they are error prone resulting in incorporation of an incorrect base and a mutation that can lead to cancer. It is anticipated that this new understanding of the direct cancer-causing potential of estrogen will provide additional drug targets in the treatment of breast cancer.

9.4 Male breast cancer

Male breast cancer is rare and represents approximately 1% of all cancers in men, causing approximately 0.1% of male cancer deaths per year [84]. In a study of 23 male breast cancers, LOH on chromosome 11q13 was detected in 68% of informative cases [85]. Mutations in the androgen receptor gene, primarily at codons 607 and 608 which are in the second zinc finger of the DNA-binding domain of the gene, have been detected [86]. In addition, the risk of breast cancer in males carrying *BRCA2* mutations, though small, is probably greater than in men carrying *BRCA1* mutations [87].

9.5 Gene expression profiling

The strongest predictors for metastatic spread (e.g. lymph node status and histological grade) fail to accurately classify breast cancers according to their clinical behavior. Chemotherapy or hormonal therapy can reduce the risk of distant metastasis by about 30%, however, 70–80% of patients receiving these treatments would survive without it. In a recent landmark study by Friend and colleagues, the results of gene expression profiling of breast tumors, after they had been surgically removed, was used to predict which patients would develop clinical metastases [88]. These authors used gene expression microarrays to analyze 25 000 genes in 98 primary breast cancers from young women. The gene expression data were analyzed using statistical methods to identify gene clusters that correlated with the patients' known clinical outcome. Seventy-eight sporadic, lymph node-negative patients were selected to search for a prognostic signature in their expression profiles. An optimum of 70 marker genes 'the prognosis classifier' correctly predicted the later appearance or absence of clinical metastasis in 65/78 patients. The authors then determined an 'optimized threshold' that would result in no more than 10% of poor-prognosis patients being misclassified and they applied this threshold to a similar set of 19 breast cancer patients. They found that the predictive power of the poor-prognosis gene expression profile for metastasis was much greater than that of currently used methods. The study also developed an expression signature that defines ER status that can be used to decide on adjuvant hormonal therapy, and an expression signature that reveals *BRCA1* status that may be of value in improving the diagnosis of hereditary breast cancer [88].

In an extension of this study to 295 patients, the same research team was able to use the 70 predictor marker genes to assign 115 patients to the good-prognosis category and 180 patients to the poor-prognosis category. The molecular prognosis profile was associated with age, histologic grade and ER status but, surprisingly, not with lymph node status as node-negative and node-positive patients were uniformly distributed in the two prognostic categories [89]. The ability of the profile to identify patients with lymph node-positive cancer with an unexpectedly good prognosis implies that the development of lymph node metastases and the development of distant metastases are independent properties of the tumor. A clear implication of the study is that the gene expression profile is a better predictor of clinical outcome than any of the currently used criteria [90].

These results have very important implications for the future treatment of patients with breast cancer (and potentially for other cancers if similar results are obtained) [91]. These authors point out that 70–90% of patients cured by surgery and radiotherapy would be advised unnecessarily to receive adjuvant therapy using the current predictive criteria [11,12]. Importantly, the number of patients with poor prognosis wrongly assigned to the 'no treatment' group would not differ significantly using conventional or molecular criteria [91]. Despite the very clear implications for patient management a note of caution is raised by these authors who point out that the study [88]: was a pilot study with a small sample size; used a retrospective tumor bank; and only tumors where cancer cells constituted more than 50% of the overall cell composition were analyzed [91]. Many questions have yet to be answered before molecular profiling can be integrated into the routine management of breast cancer, including the reproducibility of profiling results. The studies reported to date have been applied to a specific group of women (those less than 55 years old with small tumors and stage I or II disease). The promising results need to be validated in larger groups of patients including older women and those with advanced disease [90]. Nonetheless, adaptation of this approach to high-throughput automated analysis has the potential to greatly improve cancer treatment by better predicting the response of individual tumors to therapy [91].

References

1. **American Cancer Society.** (2003) Cancer Around the World, 2000, Death Rates per 100,000 Population for 45 Countries. This can be found under Professionals, Facts and Figures, Cancer Facts & Figures 2003 at http://www.cancer.org/
2. **Baum, M., Saunders, C. and Meredith, S.** (1994) *Breast Cancer. A Guide for Every Woman.* Oxford University Press, Oxford.
3. **Peto, J.** (2001) *Nature*, **411**, 390.
4. **Macdonald, F. and Ford, C.H.J.** (1997) *Molecular Biology of Cancer.* BIOS Scientific Publishers, Oxford.
5. **Ursin, G., Bernstein, L. and Pike, M.C.** (1994) In *Cancer Surveys, Vol. 19: Trends in Cancer Incidence and Mortality* (Doll, R., Fraumeni, J.F., Jr. and Muir, C.S., eds). Imperial Cancer Research Fund. Cold Spring Harbor Laboratory Press, p. 241.

6. **Mettlin, C.** (1999) *CA-Cancer J. Clin.*, **49**, 138.
7. **Ali, S. and Coombes, C.** (2002) *Nature Rev. Cancer*, **2**, 101.
8. **Brower, S.T., Cunningham, J.D. and Tartter, P.I.** (1999) In *Breast Cancer* (Roses, D.F., ed.). Churchill Livingstone, New York, Chapter 17, p. 273.
9. **Harris, J.R.** (2000) In *Diseases of the Breast, 2nd Edition* (Harris, J.R., Lippman, M.E., Morrow, M. and Osborne, C.K., eds.). Lippincott Williams & Wilkins, Philadelphia, Chapter 28, p. 403.
10. **Ottesen, G.L., Christensen, I.J., Larsen, J.K.** *et al.* (1995) *Cytometry*, **22**, 168.
11. **Goldhirsch, A., Glick, J.H., Gelber, R.D. and Senn, H-J.** (1998) *J. Natl. Cancer Inst.*, **90**, 1601.
12. **National Institutes of Health Consensus Development Panel** (2001) *J. Natl. Cancer Inst.*, **93**, 979.
13. **Ang, P. and Garber, J.E.** (2001) *Semin. Oncol.*, **28**, 419.
14. **Rahman, N. and Stratton, M.R.** (1998) *Annu. Rev. Genet.*, **32**, 95.
15. **Bieche, I. and Lidereau, R.** (1995) *Genes Chromosom. Cancer*, **14**, 227.
16. **Nass, S.J. and Dickson, R.B.** (1997) *Breast Cancer Res. Treat.*, **44**, 1.
17. **Field, J.K. and Spandidos, D.A.** (1990) *Anticancer Res.*, **10**, 1.
18. **Walker, R.A., Senior, P.V., Jones, J.L.** *et al.* (1989) *J. Pathol.*, **158**, 97.
19. **Berns, E.J.M.M., Klijn, J.G.M., van Putten, W.L.J.** *et al.* (1992) *Cancer Res.*, **52**, 1107.
20. **Borg, A., Baldertorp, B., Ferno, M.** *et al.* (1992) *Int. J. Cancer*, **51**, 687.
21. **Deming, S.L., Nass, S.J., Dickson, R.B. and Trock, B.J.** (2000) *Br. J. Cancer*, **83**, 1688.
22. **Dickson, R.B. and Lippman, M.E.** (2000) In *Diseases of the Breast, 2nd Edition* (Harris, J.R., Lippman, M.E., Morrow, M. and Osborne, C.K., eds.). Lippincott Williams & Wilkins, Philadelphia, Chapter 20, p. 281.
23. **Slamon, D.J., Godolphin, W., Jones, L.A.** *et al.* (1989) *Science*, **244**, 707.
24. **Walker, R.A., Gullick, W.J. and Varley, J.M.** (1989) *Br. J. Cancer*, **60**, 426.
25. **Ross, J.S. and Fletcher, J.A.** (1998) *Stem Cells*, **16**, 413.
26. **Yarden, Y.** (2001) *Oncology*, **66** (suppl. 2), 1.
27. **Makin, G. and Dive, C.** (2001) *Trends Cell Biol.*, **11**, s22.
28. **Shields, J.M., Pruitt, K., McFall, A.** *et al.* (2000) *Trends in Cell Biol.*, **10**, 147.
29. **Elenbaas, B., Spiro, L., Koerner, F.** *et al.* (2001) *Genes Dev.*, **15**, 50.
30. **Krontiris, T.G., Devlin, B., Karp, D.D.** *et al.* (1993) *New Engl. J. Med.*, **329**, 517.
31. **Warren, W., Eeles, R.A., Ponder, B.A.J.** *et al.* (1992) *Oncogene*, **7**, 1043.
32. **Delmolino, L., Band, H. and Band, V.** (1993) *Carcinogenesis*, **14**, 827.
33. **Moll, U.M., Riou, G. and Levine, A.J.** (1992) *Proc. Natl Acad. Sci. USA*, **89**, 7262.
34. **Coles, C., Condie, A., Chetty, U.** *et al.* (1992) *Cancer Res.*, **52**, 5291.
35. **Ozbun, M.A. and Butel, J.S.** (1995) *Adv. Cancer Res.*, **66**, 71.
36. **Irwin, M.S. and Kaelin, W.G.** (2001) *Apoptosis*, **6**, 17.
37. **Prosser, J., Porter, D., Coles, C.** *et al.* (1992) *Br. J. Cancer*, **65**, 527.
38. **Sidransky, D., Tokino, T., Helzlsouer, K.** *et al.* (1992) *Cancer Res.*, **52**, 2984.
39. **Borresen, A.L., Andersen, T.I., Garber, J.** *et al.* (1992) *Cancer Res.*, **52**, 3234.
40. **Prosser, J., Thompson, A.M., Cranston, G. and Evans, H.J.** (1990) *Oncogene*, **5**, 1573.
41. **Osborne, R.J., Merlo, G.R., Mitsudomi, T.** *et al.* (1991) *Cancer Res.*, **51**, 6194.
42. **Dowell, S.P. and Hall, P.A.** (1995) *J. Pathol.*, **177**, 221.
43. **Silvestrini, R., Rao, S., Benini, E.** *et al.* (1995) *J. Natl Cancer Inst.*, **87**, 1020.
44. **Callahan, R.** (1992) *J. Natl Cancer Inst.*, **84**, 826.
45. **Isola, J., Visakorpi, T., Holli, K. and Kallioniemi, O-P.** (1992) *J. Natl Cancer Inst.*, **84**, 1109.
46. **Thor, A.D., Moore, D.H., Edgerton, S.M.** *et al.* (1992) *J. Natl Cancer Inst.*, **84**, 845.
47. **Allred, D.C., Clark, G.M., Elledge, R.** *et al.* (1993) *J. Natl Cancer Inst.*, **85**, 200.
48. **Runnebaum, I.B., Nagarajan, M., Bowman, M.** *et al.* (1991) *Proc. Natl Acad. Sci. USA*, **88**, 10657.
49. **Elledge, R.M., Fugua, S.A.W., Clark, G.M.** *et al.* (1993) *Breast Cancer Res. Treat.*, **26**, 225.
50. **Bergh, J., Norberg, T., Sjogren, S.** *et al.* (1995) *Nature Med.*, **1**, 1029.
51. **Pietilainen, T., Lipponen, P., Aaltoma, S.** *et al.* (1995) *J. Pathol.*, **177**, 225.
52. **Sirvent, J.J., Fortuño-Mar, A., Olona, M. and Orti, A.** (2001) *Histol. Histopath.*, **16**, 99.
53. **Elledge, R.M., Gray, R., Mansour, E.** *et al.* (1995) *J. Natl Cancer Inst.*, **87**, 1254.
54. **Horne, G.M., Anderson, J.J., Tiniakos, D.G.** *et al.* (1996). *Br. J. Cancer*, **73**, 29.

55. **Welsch, P.L., Owens, K.N. and King, M-C.** (2000) *Trends Genet.*, **16**, 69.
56. **Ponder, B.A.J.** (2001) *Nature,* **411**, 336.
57. **Anglia Breast Cancer Study Group** (2000) *Br. J. Cancer*, **83**, 1301.
58. **Baer, R. and Ludwig, T.** (2002) *Curr. Opin. Genet. Develop.*, **12**, 86.
59. **Pierce, A.J., Stark, J.M., Araujo, F.D. et al.** (2001) *Trends Cell Biol.*, **11**, s52.
60. **Kuschel, B., Gayther, S.A., Easton, D.F. et al.** (2001) *Genes, Chromosom. Cancer*, **31**, 96.
61. **Tutt, A. and Ashworth, A.** (2002) *Trends Mol. Med.*, **8**, 571.
62. **Yang, H., Jeffrey, P.D., Miller, J. et al.** (2002) *Science*, **297**, 1837.
63. **Buchholz, T.A., Wu, X., Hussain, A. et al.** (2002) *Int. J. Cancer*, **97**, 557.
64. **Phillips, K-A., Andrulis, I.L. and Goodwin, P.J.** (1999) *J. Clin. Oncol.*, **17**, 3653.
65. **Hedenfalk, I., Duggan, D., Chen, Y.D. et al.** (2001) *New Engl. J. Med.*, **344**, 539.
66. **Brugarolas, J. and Jacks, T.** (1997) *Nature Med.*, **3**, 721.
67. **Gillet, C., Fantl, V., Smith, R. et al.** (1994) *Cancer Res.*, **54**, 1812.
68. **Schurring, E., Verhoeven, E., van Tinteren, H. et al.** (1992) *Cancer Res.*, **52**, 5229.
69. **Dickson, C., Fantl, V., Gillet, C. et al.** (1995) *Cancer Letters*, **90**, 43.
70. **Bernards, R.** (1999) *Biochim. Biophys. Acta.*, **1424**, M17.
71. **Ho, A. and Dowdy, S.F.** (2002) *Curr. Opin. Genet. Develop.*, **12**, 47.
72. **Devilee, P. and Cornelisse, C.J.** (1994) *Biochim. Biophys. Acta*, **1198**, 113.
73. **Chano, T., Kontani, K., Teramoto, K. et al.** (2002) *Nature Genet.*, **31**, 285.
74. **Zheng, Z.M., Marchetti, A., Buttita, F. et al.** (1996) *Br. J. Cancer*, **73**, 144.
75. **Champeme, M.H., Bieche, I., Beuzelin, M. and Lidereau, R.** (1995) *Genes, Chromosom. Cancer*, **12**, 304.
76. **Hainsworth, P.J., Raphael, K.L., Stillwell, R.G. et al.** (1992) *Br. J. Cancer*, **66**, 131.
77. **Piao, Z., Lee, K.S., Kim, H. et al.** (2001) *Genes Chromosom. Cancer,* **30**, 113.
78. **Braga, E., Senchenko, V., Bazov, I. et al.** (2002) *Int. J. Cancer*, **100**, 534.
79. **Berx, G., Cleton-Jansen, A-M., Nollet, F. et al.** (1995). *EMBO J.*, **14**, 6107.
80. **Vercoutter-Edouart, A.S., Lemoine, J., Le Bourhis, X. et al.** (2001) *Cancer Res.*, **61**, 76.
81. **Yang, Q.F., Mori, I., Shan, L. et al.** (2001) *Am. J. Pathol.*, **158**, 299.
82. **Foster, J.S., Henley, D.C., Ahamed, S. and Wimalasena, J.** (2001) *Trends Endocrin. Metab.*, **12**, 320.
83. **Osborne, C.K., Bardou, V., Hopp, T.A. et al.** (2003) *J. Natl. Cancer Inst.*, **95**, 353.
84. **Thomas, D.B.** (1993) *Epidemiol. Rev.*, **15**, 220.
85. **Sanz-Ortega, J., Chuaqui, R., Zhuang, Z. et al.** (1995) *J. Natl. Cancer Inst.*, **87**, 1408.
86. **Wooster, R., Mangion, J., Eeles, R. et al.** (1992) *Nature Genet.*, **2**, 132.
87. **Wooster, R., Neuhausen, S.L., Mangion, J. et al.** (1994) *Science*, **265**, 2088.
88. **van't Veer, L., Dal, H., van de Vijver, M.J. et al.** (2002) *Nature*, **415**, 530.
89. **van de Vijver, M.J., He, Y.D., van't Veer et al.** (2002) *New Engl. J. Med.*, **347**, 1999.
90. **Kallioniemi, A.** (2002) *New Engl. J. Med.*, **347**, 2067.
91. **Caldas, C. and Aparicho, S.A.J.** (2002) *Nature,* **415**, 485.

Genitourinary cancer

10

10.1 Ovarian and endometrial cancers

It is difficult to summarize international trends in the incidence and mortality of ovarian cancer since they differ between countries and between age groups [1]. Incidence rates are high in most of the industrial countries where women have relatively few children and lower in Asian and African countries with higher fertility rates. In recent years the mortality rate has started to decline in the younger age groups (under 55 years) but this is not reflected in the incidence data. Reproductive and hormonal factors are the main determinants of risk, with a decline in risk associated with increasing parity, oral contraceptive use, hysterectomy and sterilization by tubal ligation [2]. Ovarian cancer differs from many other human tumors in displaying considerable disease heterogeneity, a poorly understood progression pathway, and a few good tumor markers. In addition, ovarian cancers are usually diagnosed at a later stage, and the survival rate, therefore, is poor [3].

There are several similarities between ovarian and breast cancers, for example, both often express steroid hormone receptors and it is suggested that they have common etiological factors. Several prognostic factors have been identified in ovarian cancer, including stage of disease, presence or absence of ascites and volume of residual disease following surgery. Histological grading and ploidy are also strongly associated with clinical outcome. However, stage at diagnosis is the major prognostic factor with stage I patients having a 5-year survival rate of 80–90% compared with 15–20% for women with stage II and IV disease [4]. Lack of early symptoms and reliable screening tests result in only ~25% of cases being diagnosed as stage I and disease heterogeneity, poorly understood progression and the absence of a definite precursor lesion confound attempts to improve early detection and therapy [5]. Molecular genetic studies of ovarian cancer have been limited by this lack of a well-defined precursor lesion, the inaccessible location of the ovary and the advanced stage of tumors available for analysis [5,6].

A high frequency of loss of heterozygosity (LOH) has been detected on chromosomes 6p, 6q, 9q, 13q, 17p and 17q and together with mutation analysis this suggests that chromosome 17 plays a significant role in ovarian cancer development. On the short arm, LOH and mutations of the *p53* locus as well as LOH at a more distal locus (17p13.3) have been observed in a high percentage of tumors. Similarly, on the long arm, losses in the *BRCA1* region and a more distally located locus (17q22-23) are frequently seen [6 and references therein].

It has been estimated that 5–10% of ovarian cancers are the result of an inherited predisposition. The susceptibility to ovarian cancer is inherited through one of three distinct patterns: (1) the breast–ovarian cancer syndrome, which is typified by multiple cases of early age of onset breast and ovarian cancer; (2) hereditary nonpolyposis colorectal cancer, in which excess cancer cases are primarily cancers of the proximal colon, endometrium, and ovary; and (3) site-specific ovarian cancer, which is the least common. As with retinoblastoma and other inherited cancers, the susceptibility to cancer in families with a history of site-specific ovarian cancer is transmitted as an autosomal dominant trait with an early age of onset [7]. In families with a history of breast–ovarian cancer, the cumulative risk for breast cancer or ovarian cancer in women with a mutant *BRCA1* allele is estimated to be 76% by age 70, thus indicating that *BRCA1* is a highly penetrating predisposing gene for both malignancies [6,8]. At present it is unclear what fraction of hereditary ovarian cancers are due to *BRCA1* mutations (see Chapter 5).

Complementary DNA (cDNA) microarrays have been used to examine the role of mutations in *BRCA1* and *BRCA2* in ovarian carcinogenesis by comparing gene expression patterns in ovarian cancers that are associated with germ-line *BRCA1* and *BRCA2* mutations with expression patterns in sporadic ovarian cancers. Approximately 6500 genes were analyzed and 110 showed statistically significant expression levels [9]. It was found that each sporadic sample had a molecular profile similar to that of either the *BRCA1*- or the *BRCA2*-associated tumors, suggesting that the *BRCA*-associated molecular pathways are different at the RNA level for the two genes and may be altered in a similar way in sporadic tumors. However, as pointed out in a subsequent report, the labeling of sporadic cases was based on a negative test for only three founder mutations and a larger number of sporadic cases are required to strengthen the validity of the results by allowing the demonstration of more heterogeneity in the expression profiles [10]. The observation that *BRCA*-associated gene expression profiles occur in subsets of sporadic tumors indicates that molecular mechanisms common to both hereditary and sporadic ovarian carcinogenesis exist but additional studies are needed to establish the underlying mechanism linking mutations in these genes to the evolution of their distinct gene-expression subtypes [3].

Despite the apparent difference between *BRCA1*- and *BRCA2*-like profiles indicated above, a recent report using the protein truncation test (PTT) to study *BRCA2* mutations in a group of sporadic ovarian cancers previously evaluated for *BRCA1* mutations, found that *BRCA2* dysfunction was usually accompanied by simultaneous *BRCA1* dysfunction [11]. There is clearly still much to learn of the role of *BRCA1* and *BRCA2* in ovarian carcinogenesis.

In addition to *BRCA1* (on the long arm at 17q22-23) the *p53* locus is located on the short arm of chromosome 17 as well as two other tumor suppressor genes, *OVCA1* and *OVCA2* at 17p13.3 [12]. Mutation of *p53* with resultant overexpression of mutant p53 protein occurs in 50% of stage III/IV, 15% of stage I/II and 4% of borderline ovarian cancers. Most mutations are transitions, which suggests that they arise spontaneously rather than being due to exogenous carcinogens [8]. Mutations of *p53* in ovarian cancer have been shown to be associated with advanced stage of disease [13,14], poorly differentiated carcinomas [15] and a poor prognosis [16]. In addition, a significant correlation between p53 accumulation, type of mutation (missense) and pathological response to cisplatin-based chemotherapy [17] and the fact that transfection of a wild-type *p53* construct into an ovarian cancer cell line with a mutant *p53* gene can inhibit proliferation [18] have been reported. A study of 19 ovarian carcinomas and 17 borderline tumors has confirmed that loss or inactivation of tumor suppressor gene function by chromosome 17p allelic deletions, or *p53* mutations, are important genetic changes in ovarian cancer [19]. However, the role of this gene in prognosis and in determining response to chemotherapy has yet to be determined [20].

Although amplification of *MYC* occurs in approximately 30% of ovarian cancers and is more frequently seen in advanced stage serous cancers [8] this does not appear to correlate with prognosis [21].

In a series of 74 ovarian cancers, *KRAS* mutations were detected in 45% of mucinous, compared to only 16% of serous, cancers. In the mucinous tumors, *KRAS* mutations were present in half of the borderline and in an equal number of invasive tumors. *KRAS* mutations were seen in eight of 31 mucinous tumors that lacked allele loss for more than 30 polymorphic markers spanning the human genome. Borderline ovarian tumors constitute a unique subgroup characterized by an unusual degree of epithelial cell proliferation and atypia compared with benign tumors, but they lack the stromal invasion characteristic of invasive ovarian tumors. The clinical course of these tumors is between that of benign and malignant tumors and they metastasize within the peritoneal cavity but rarely result in death. The frequent finding of *KRAS* mutations in borderline mucinous tumors suggests that these mutations are an early event in the development of mucinous cancers of the ovary and that borderline tumors are precursors of invasive tumors [6 and references therein].

Like breast cancers, approximately 30% of ovarian cancers carry amplified *ERBB2* (*HER-2/neu*), but here the presence of the amplified oncogene appears to correlate with poor survival. In one study, the survival times for patients with one, two to five, or over ten copies of *ERBB2* were 1879, 959 and 243 days, respectively [8 and references therein].

The alpha 110 kDa subunit of phosphatidyl inositol 3 kinase (p13 kinase) is amplified in approximately 80% of ovarian cancers. The AKT serine-threonine kinases are activated by the products of p13 kinase and AKT2 is amplified in 12% of ovarian cancers which are associated with a poorly differentiated histology [21 and references therein]. Since amplification of p13 kinase and AKT2 are frequently seen in ovarian cancer, but rarely in breast or epithelial cancers from other sites, these authors speculate that the p13 kinase pathway may be particularly important in ovarian cancer.

With the recent advent of newer molecular techniques a variety of approaches have been used to identify genes and their protein products which are of importance in the development and progression of ovarian cancer. Serial analysis of gene expression (SAGE) has been used to compare the global gene expression of ovarian cancer cell lines, ovarian carcinomas and nontransformed ovarian epithelium enabling differentially expressed genes to be identified. Many of the genes up-regulated in ovarian cancer were found to code for surface or secreted proteins such as claudin-3 and -4, HE4, mucin-1, epithelial cellular adhesion molecule and mesothelin [5]. An interesting observation from this study was that apolipoprotein E (ApoE) and ApoJ, two proteins involved in lipid homeostasis, were among the genes highly up-regulated in ovarian cancer. ApoJ, claudin-3 and -4 have never previously been linked to ovarian cancer.

Using an alternative strategy, two-dimensional gel electrophoresis (2-DE/PDQUEST) was used to characterize the expression of multiple proteins in ovarian tissue from 22 patients (five benign, six borderline and 11 carcinomas). Large variations were observed between tumors in the expression of various polypeptides, indicating heterogeneity in gene expression. However, the pattern of expression of nine protein markers permitted discrimination between malignant and benign tumors [22]. Although there was apparently no overlap in terms of the findings in both these reports, studies such as these and the public availability of expression profiles through SAGEmap should facilitate and stimulate further research [5,23 and references therein].

An immunohistochemical study of the prognostic and predictive relevance of apoptosis-related proteins Bax, Bcl-XL and Mcl-1 in 185 stage III ovarian cancer patients concluded that Bax and BCL2 expression data, when combined, were of independent prognostic significance [24].

Using proteomic spectra generated by mass spectroscopy (surface-enhanced laser desorption and ionization) a preliminary 'training' set of spectra derived from the sera of 50 unaffected women and 50 patients with ovarian cancer were analyzed by an iterative searching algorithm that identified a proteomic pattern that completely discriminated cancer from noncancer. The pattern was then used to classify an independent set of 116 matched serum samples (50 from ovarian cancer patients and 66 from unaffected women or those with nonmalignant conditions). The discriminating pattern correctly identified all 50 ovarian cancer cases and 63/66 of the nonmalignant cases were recognized as not being cancer. The very encouraging results of studies such as this justify a prospective population-based assessment of proteomic pattern technology as a screening tool for all stages of ovarian cancer in high-risk and general populations [25].

It is known that estrogen replacement therapy (ERT) increases a woman's risk for uterine endometrial cancer and epithelial ovarian cancer (EOC) and a recent large study in the Swedish population has demonstrated that ever users of ERT and hormone replacement therapy with sequentially added progestins (HRTsp), but not hormone replacement therapy with continuously added progestins (HRTcp), may be at increased risk of EOC [26].

The expression level of several matrix metalloproteinases (MMPs), including MMP-2 and MMP-9, in ovarian cancer cells has been shown to be directly associated with their invasiveness and metastatic potential. Using a nude mouse xenograft model in which the mice either possessed or lacked the *MMP-9* gene, it was found that host-derived *MMP-9* expression appeared to play a critical role in angiogenesis and progressive growth of human ovarian cancer xenografts in the mice [27].

The *MUC1* gene has been shown to be expressed by a number of epithelial cancers, including ovarian cancer. A comparison of the pattern of expression of MUC1 splice variants in malignant and benign ovarian tumors demonstrated that expression of MUC1 splice variants A, D, X, Y, Z and REP was associated with malignancy whereas MUCI/SEC expression was associated with the absence of malignancy. However, no significant association of the expression of MUC1 splice variants and response to chemotherapy or patient survival was found [28].

The first mouse model of ovarian cancer with defined genetic lesions has been reported recently [29]. Trp53$^{+/+}$ and Trp53$^{-/-}$ ovarian cells from transgenic mice that express an avian virus receptor were transfected *in vitro* with avian retroviral vectors carrying oncogenes that are frequently amplified or mutated in ovarian cancer (see earlier): *KRAS*, *MYC* and *AKT*. The authors investigated which combinations of genetic lesions could induce tumor formation *in vivo* by injecting these ovarian cells subcutaneously into nude mice and monitoring for tumor growth. Tumors arose only from cells deficient for Trp53 and which had at least two oncogenes by infection. Implantation of infected cells under the ovarian capsule of nude mice resulted in ovarian tumors within 2 weeks which metastasized to the same sites as human ovarian cancer within 4 weeks. When infected ovarian cells were implanted under the ovarian capsule of the immunocompetent mice from which the cells had been previously removed, ovarian tumors developed after 3 months [29]. This model should help unravel the molecular basis of ovarian cancer development and will provide a system for testing new therapeutic approaches. In addition, it may also provide a valuable opportunity for evaluating the mechanisms underlying the epidemiological observation that ovarian cancer recurs at a lower frequency in women exposed to non-steroidal anti-inflammatory drugs (NSAIDS) [30].

Interpretation of international trends in the incidence and mortality of endometrial cancers is complicated by variation over time and, in different countries, of the coding of such tumors [1]. Despite this caveat there is an international downward trend in mortality.

Endometrial cancers can be divided into endometrioid adenocarcinomas, which account for the majority of endometrial cancers and are typified by developing from atypical endometrial hyperplasia in the setting of excess estrogenic stimulation. In contrast, serous carcinomas are representative of endometrial cancers in older women who have endometrial atrophy and lack the typical endometrial cancer risk factors reflecting unopposed estrogen exposure [31].

Overexpression of *HER-2/neu* (*ERBB2*) occurs in 10% of endometrial adenocarcinomas and correlates with intraperitoneal spread of disease and poor survival. *MYC* is amplified in 10% of cases. Point mutations in codon 12 of *KRAS* have been reported to occur in 10–20% of cases. *KRAS* mutations have also been reported in some endometrial hyperplasias, which may represent an early event in the development of some endometrial cancers. Mutations of *p53* occur in 20% of endometrial adenocarcinomas and overexpression of mutant p53 protein is associated with advanced stage and poor survival. Since *p53* mutations have not been seen in endometrial hyperplasias, this is thought to be a relatively late event in endometrial carcinogenesis. However, serous carcinomas are frequently associated with *p53* abnormalities and appear to develop from a surface lesion termed endometrial intraepithelial carcinoma. Although serous carcinomas are rare, these highly aggressive tumors account for a disproportionate number of endometrial cancer deaths [31].

Samples from a group of patients with endometrial carcinomas with retroperitoneal lymph node metastases were studied by immunohistochemistry to evaluate the prognostic significance of the tumor suppressor gene *PTEN*. It was found that PTEN-positive staining was a significant favorable prognostic indicator of survival for patients with advanced endometrial carcinoma who underwent postoperative chemotherapy. In contrast, PTEN expression did not relate to survival for those patients who underwent radiation therapy [32].

Microsatellite instability has also been noted in approximately 15% of sporadic endometrial cancers, but mutations in DNA repair genes have not yet been reported [33]. Finally, in a study of the putative tumor suppressor *E-cadherin* gene in 135 carcinomas of the endometrium and ovary only four mutations were detected [34]. The prognostic significance, if any, of this gene in ovarian and endometrial cancers remains to be determined.

10.2   Cervical cancer

Cancer of the uterine cervix is the second most common cancer in women worldwide and is particularly common in less-developed countries, where 80% of the world's cervical cancers occur. Current evidence indicates that the main cause of cervical cancer is cervical infection with certain types of human papillomavirus (HPV) that are transmitted sexually [35]. The overall incidence and mortality from cervical cancer has declined in Western countries and in most developing countries. In women under 40 years of age, however, mortality rates are leveling off or increasing in most countries [35]. The reasons for the overall decline in cervical cancer may be linked to improvements in the general standard of living and the increase in younger women may be due to the increasing prevalence of HPV infection. Screening for cervical cancer has undoubtedly led to a decline in incidence and mortality in many countries.

A number of prognostic factors have been linked to the progression of cervical cancers, including depth of stromal invasion, lesion depth and nodal involvement. In particular, HPV has been found to be integrated into malignant cells and the presence of the virus (in specific subtypes) in premalignant lesions indicates a poor prognosis. These anogenital-associated HPVs are further classified into low risk (e.g., HPV-6, HPV-11), associated with benign genital warts, and high risk (e.g., HPV-16, HPV-18), associated with lesions such as squamous intraepithelial neoplasia (SIN) which can progress to anogenital cancer. Approximately 85% of all cervical carcinomas are conservatively estimated to be high-risk HPV-positive.

A recent study set out to identify the determinants of high vs. low rates of cervical cancer in developing countries. Among behavioral measures, high fertility rates, early age at birth of the first child and high teenage birth rates were significantly associated with high cervical cancer rates. Countries with high cervical cancer rates were found to have fewer doctors per capita, less immunization coverage, more HIV infections and shorter life expectancies. Cervical cancer rates also tended to be higher in countries which had more spending on health but which had younger, less-educated populations [36].

Epidemiological studies show that infection with the high-risk HPVs 16, 18, 31, 33, 45 and minor types is the major risk factor for the subsequent development of cervical cancer. Experimental studies have shown that the *E6* and *E7* genes of the high-risk HPVs encode proteins that deregulate key controls in cell proliferation. E7 interacts with RB1, effectively blocking its transcriptional repressor function. E6 enhances the degradation of p53, thus compromising the ability of the cell to effect growth arrest in the face of DNA damage. As a consequence of these interactions, key cell cycle checkpoints are overridden, and mutations may be accumulated and transmitted to daughter cells. The HPV-infected cell may as a consequence acquire: defects in differentiation; the immortal phenotype; chromosomal aneuploidy and the increased probability of further mutations, which could lead to invasion [37,38].

Although it is well established that infection with high-risk types of HPV is central to the pathogenesis of cervical cancer many women are infected with high-risk types of HPV but only a subset will ever develop cervical cancer, suggesting that other cofactors must be present for the development of malignancy [39]. A pooled analysis of seven case–control studies involving 1263 patients with invasive cervical cancer and 1117 age-matched controls has clearly demonstrated that herpes simplex virus-2 (HSV-2) infection may act in conjunction with HPV infection to increase the risk of invasive cervical carcinoma [40].

HPV DNA is often integrated into the host genome and this frequently results in the loss of E2 protein which, since E2 can modulate HPV gene expression, results in the deregulated expression of E6 and E7. Also, since E2 is a potent inducer of apoptosis, the loss of E2 might be expected to result in increased cell proliferation [41]. However, HPV infection alone is not sufficient for malignant transformation. Estrogen and progesterone have both been shown to increase the levels of E2- and E7-induced apoptosis and one possibility is that in the presence of E2 these hormones might be protective against cervical cancer via their up-regulation of cell death. In the absence of E2 these two hormones might be a risk factor in cervical carcinogenesis via their effects on HPV or cellular gene expression, or via other yet to be determined pathways [41].

Although HPVs do not provoke strong systemic antibody or cell-mediated immune responses [42] the identification of HPV-specific immunity is providing the impetus for the development of vaccines both to prevent and to treat HPV-associated disease – the ultimate aim being to effect a worldwide eradication of high-risk HPV types thus preventing the high annual death rate from cancer of the cervix [43]. However, the relatively low immunization rates and the shortage of healthcare workers in countries with high cervical cancer rates present potential challenges to introducing prophylactic HPV vaccines [36]. Nevertheless, a recent proof-of-principle randomized, double-blind study, in which 2392 young women (16–23 years of age) received three doses of placebo or HPV-16 virus-like-particle vaccine resulted in the elimination of the risk of infection with HPV-16 in 100% of women who were not previously infected and all cases of HPV-16-related cervical intraepithelial cancer occurred in the placebo group [44]. This study clearly demonstrates the potential of a vaccine such as this in the prevention of cervical cancer if the challenges to vaccinating women in those countries with the highest cervical cancer rates can be overcome.

An alteration of microsatellite repeats is the result of slippage due to strand misalignment during DNA replication and is referred to as microsatellite instability (MSI). PCR-based microsatellite assay combined with tissue microdissection was recently used to search for MSI in 50 cervical cell carcinomas in Hong Kong women. The results suggested that MSI was present in a subgroup of cervical squamous cell carcinomas and that defects resulting in MSI may be related to tumor progression and possibly poor prognosis [45].

Sister chromatid exchanges (SCE) are reciprocal exchanges between sister chromatids. There have been inconsistent reports that the frequency of SCE in the peripheral blood lymphocytes of patients with cervical cancer is significantly higher than in normal individuals. An unmatched case-control study from Mexico concluded that SCE are associated with cervical cancer. The usefulness of this as a preclinical marker to identify those women at high-risk of developing cervical cancer will need to be ascertained in follow-up studies on precancerous lesions with high levels of SCE [46].

Work in the early 1990s showed that abnormal expression of several oncogenes also has prognostic value. Amplification and overexpression of *MYC* was found to be more frequent in advanced tumors of the uterine cervix when compared to early tumors. In tumor samples from 72 untreated patients with carcinoma of the cervix, overexpression was detected in 35% of cases. This overexpression was not a consequence of amplification which was seen in only 8% of cases. There was no relationship between *MYC* overexpression and stage, nodal status or age. However, there was an eightfold greater risk of relapse for patients with overexpression of *MYC*, which outweighed even nodal status as a prognostic factor [47]. In the same publication it was reported that elevated expression of *RAS* p21 was found in malignant as opposed to benign or premalignant lesions. In the small-cell type of squamous cell carcinomas, tumors with elevated expression of *RAS* p21 were shown to have a better prognosis than negative tumors. Expression of *RAS* p21 together with histological type may therefore be of prognostic significance in carcinomas of the cervix in specific histological types. Mutations at codon 12 of the *HRAS* gene were also found in cervical cancer and correlated with a poor prognosis [47]. However, confirmation of these findings has yet to be reported.

It has been suggested that *p53* regulates the radiosensitivity in human malignancies after irradiation but in cervical carcinoma its role is unclear due to the inactivation of functional p53 by infection with HPV. A study of 52 patients with carcinoma of the cervix who had received radiation therapy alone showed that *p53* mutations were significantly associated with local tumor recurrence and the authors concluded that *p53* may be of value as a predictive factor in radiation therapy for patients with stage IIIB squamous cell carcinoma of the cervix [48].

10.3 Testicular cancer

Testicular cancer is a rare disease and unlike most other solid tumors, the incidence rate does not increase as a function of age but reaches a maximum at about 25–34 years and then declines to become uncommon in men over the age of 55 years [49]. Although the incidence has been increasing over the years introduction of effective chemotherapy in the mid 1970s has resulted in reduced mortality rates.

Strong expression of p62 has been detected in differentiating areas of testicular tumors. In a flow-cytometric study of p62 expression in nuclei extracted from paraffin blocks, increasing levels of p62 correlated with increasing differentiation of teratomas. Patients who had no recurrence 3 years after diagnosis showed a significantly higher level of p62 than those who developed a recurrence in this time period. Expression of *MYC* may therefore prove to be of prognostic significance in this disease [50] although this has yet to be confirmed.

10.4 Renal cell cancer and bladder cancer

Although the incidence of bladder cancer is rising in most populations its mortality has begun to fall in a number of countries including the USA and Japan but not, on the whole, in continental Europe. Renal cell cancer (RCC) is common in societies with high rates of bladder cancer suggesting that these two functionally and anatomically related sites may be exposed to similar carcinogenic influences [51]. Both the incidence and mortality of RCC cancer have been steadily increasing in virtually every country and RCC accounts for 2–3% of adult malignancies.

The clinical behavior of renal cancer is unpredictable, although a few features, such as stage of tumor and ploidy, are partly correlated with prognosis. The disease is highly metastatic, with 30% of patients having metastases at the time of diagnosis and RCC is one of the most therapy-resistant cancers. Molecular and cytogenetic analyses have highlighted deletions, translocations, or loss of heterozygosity in 3p21-26, a putative RCC locus, as well as in 6q, 8p, 9pq and 14pq. Studies on phenotypic expression of human kidney tissue and on post-translational modifications in RCC have not yet provided a marker for early RCC diagnosis [52 and references therein].

A study of *MYCL* RFLPs in patients with renal cell cancer found no significant differences between the RFLP patterns of normal and tumor DNA. However, the presence of metastases was associated with lack of a particular allele suggesting that this RFLP may be a marker of genetic predisposition to the formation of metastases rather than the development of primary disease.

A recent study established a two-dimensional polyacrylamide gel electrophoresis (2-D PAGE) human kidney reference map in order to compare RCC to normal kidney tissue. Four proteins present in normal tissues were not expressed in ten out of ten RCC cases. The absence of two mitochondrial proteins suggested that mitochondrial dysfunction may play a major role in RCC genesis or evolution [52].

von Hippel-Lindau (VHL) disease is an autosomal dominant disorder and, eventually, the majority of patients with this disease will develop a renal cell carcinoma if they live long enough. Renal cell carcinoma is now the leading cause of death in patients with VHL disease [53]. The *VHL* disease gene has been mapped to the short arm of chromosome 3 and flanking markers have been identified [53,54 and references therein]. Statistical and molecular genetic studies suggest that the *VHL* gene functions as a tumor suppressor gene like *RB1*. Although chromosome 3p loci are involved in the pathogenesis of sporadic renal cell carcinomas, it is not yet known whether acquired mutations at the *VHL* locus are responsible for sporadic renal cell carcinomas [53,54]. VHL disease is the most frequent cause of inherited renal cell carcinoma but familial renal cell carcinomas, without additional features, also occur and are associated with a t(3;8) translocation. In a kindred study, renal cell carcinoma was only seen in translocation carriers, each of which had an 87% risk of developing the cancer by 60 years of age [55 and references therein]. Recent work has shown that the *VHL* gene product, pVHL, is a component of an SCF (Skp1-Cdc53-F-box)-like ubiquitin–ligase complex that targets the α-subunits of the hypoxia-inducible factor (HIF) heterodimeric transcription factor for polyubiquitylation and proteasomal degradation. It is postulated that overproduction of growth factors encoded by HIF target genes, such as vascular endothelial growth factor (VEGF), platelet-derived growth factor β chain (PDGFβ) and transforming growth factor-α (TGFα) contribute to tumor formation following pVHL inactivation [56,57].

The entire urinary tract is lined with transitional cell epithelium and a multistep model for the development of transitional cell carcinoma (TCC) of the bladder, based on experiments in rodents, has been proposed [58] (see *Figure 10.1*). EGF in rat urine has been found to be responsible for the potentiating effect of chemical carcinogens on the pathogenesis of TCC. Overexpression of EGFR protein is related to higher cytological grade and more aggressive disease [55]. In a prospective study of 101 new cases of TCC of the bladder, using immunohistochemical staining for EGFR, death was independently associated with advanced stage and EGFR positivity. In addition, those patients with EGFR-positive tumors were found to have a shorter time to recurrence and a higher recurrence rate [60].

Mutations and LOH of the *RB1* gene are also found in a significant proportion of bladder cancers and are associated with local muscular invasion and a high cytological grading, both of which are related to poor prognosis [55 and references therein]. Overexpression of p53 protein as detected by immunohistochemistry appears to identify those patients with

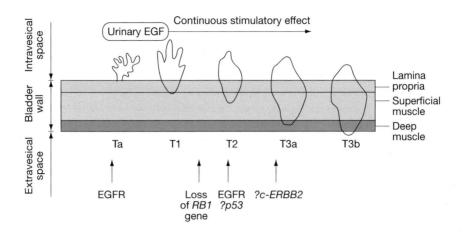

Figure 10.1

A sketch relating the progression of bladder TCC to known genetic results involved. Adapted from [58] with permission from Blackwell Science.

superficial bladder cancer at risk of the development of muscle invasive or metastatic disease. However, the role of p53 overexpression in patients with advanced or metastatic bladder cancer is not yet well established. In one study of 44 specimens from 44 patients with advanced bladder tumors (T2–T4) undergoing radical cystectomy, p53 overexpression was found to be a significant independent prognostic indicator. While this needs to be confirmed in prospective studies it provides a potential prognostic tool for decision making for defined subgroups of patients in the future [61].

Recent analyses have demonstrated that p53 accumulation may be also present in dysplasia and carcinoma *in situ*. However, the literature with regard to p53 accumulation and prognosis in patients with bladder cancer is conflicting and confusing with quite different results being reported by different research groups. This prompted a review of all papers using immunohistochemistry to assess p53 accumulation and correlate it with prognosis in bladder cancer and urothelial tumors of the upper urinary tract [62]. These authors analyzed 138 publications involving 43 trials comprising 3764 patients and found considerable differences in the results and conclusions of the studies, reflecting study design, patient selection, tissue samples, sample size, methods, evaluation of the immunostaining and endpoint variations between the studies. The authors concluded that standardization of the p53 immunohistochemistry technique is a priority and that prospective multicenter trials comparing p53 accumulation with disease outcome in patients with bladder cancer, stratified by tumor stage and grade, are essential.

In an attempt to overcome the problems with immunohistochemistry a number of alternative approaches have been tried. In a recent report expression of cyclin D1 and ki67, a proliferation marker, was evaluated by immunohistochemistry in human papillary superficial bladder cancers and was correlated with p27 (*KIP1*), p21 (*WAF1*) and *ERBB2* expression, *p53* gene status and protein expression, ploidy and cancer [63]. The study demonstrated that a high proliferation rate associated with a low nuclear expression of cyclin D1 and p27 could identify high-risk patients, especially those with T1 tumors and it was concluded that cell cycle-related proteins yield significant prognostic information in addition to tumor stage and grade. Using fluorescence *in situ* hybridization (FISH) it has also been suggested that amplification of *CCND1* or alterations in *MYC/CCND1* early in bladder cancer may have clinical relevance in promoting and predicting progression to detrusor muscle-invasive transitional cell carcinoma [64].

An alternative genetic approach was used to detect bladder cancer by microsatellite analysis of the *p16*, *RB1*, and *p53* tumor suppressor genes in 28 cytological urine samples. These markers reliably detected tumors of all stages and grades with 93% sensitivity and 100% specificity [65]. Although the sample size was small, if these results are confirmed in larger numbers of patients, microsatellite analysis of urine specimens could represent a novel, potentially useful, noninvasive clinical tool to detect bladder cancer.

A study designed to define the potential clinical relevance of identifying alterations affecting the p53 pathway in 140 patients with bladder cancer applied different methods, including oligonucleotide microchip array sequencing, conventional manual sequencing and immunohistochemistry to detect p53 status [66]. Combining the mutation-detection assays, 79 patients were found to have *p53* mutations. Direct sequencing identified 66 point mutations and five frameshift mutations. The microchip array detected 65 point mutations and four splice site mutations but missed all five frameshift mutations. The authors concluded that microarray-based sequencing technologies could be of value in routine molecular diagnostics, particularly for detecting *p53* base-exchange point mutations and that clinically oriented molecular assays need to incorporate multiple predictive markers.

Tumor-associated antigens (TAAs) of the melanoma antigen-encoding (MAGE) gene family have been identified in several human malignancies and are being evaluated as targets of specific immune responses. A monoclonal antibody to MAGE-A4 protein and tissue microarrays were used to evaluate MAGE-A4 expression in 2317 samples from 1849 patients with transitional cell carcinoma of the urinary bladder. MAGE-A4 positivity was found to be significantly correlated with the invasive phenotype, high tumor grade and decreased survival [67].

A candidate tumor suppressor gene in invasive (T3a/b) bladder TCC has been located to a 2.5-cM region at chromosome 10q23.3. This region harbors the *MMAC1/PTEN/TEP1* gene, which is frequently inactivated in a variety of cancers. Recent work indicates that *MMAC1* is not the primary target for inactivation in bladder TCC and that another gene, in close proximity to the *MMAC1* locus, within the region of frequent allelic losses, may be the target for inactivation [68 and references therein].

A recent report has summarized the multiple genetic alterations that have been identified in invasive human bladder cancers, which are present in different combinations and in different frequencies in the different types of bladder cancer. Approximately 67% of superficial papillary TCCs show early losses involving chromosome 9q, while very few show either a *p53* or a *CDKN2/p16* mutation. Loss of 9q may therefore be the earliest event in the initiation of papillary TCC. In contrast, bladder carcinoma *in situ* (CIS) and squamous cell carcinoma (SCC) show relatively low percentages of 9q loss. However, ~65% of bladder CIS contain a *p53* alteration and ~67% of bladder SCC contain a *CDKN2/p16* alteration. These authors propose a model in which the predominant genetic alteration may be the 'initiating event' in bladder cancer pathogenesis which may then play a role in determining the biological potential of the tumor [69].

The mitotic kinase-encoding gene *STK15/BTAK/AuroraA* is associated with aneuploidy and transformation in mammalian cells. *STK15* activates an unknown oncogenic pathway that involves centrosome amplification and results in missegration of chromosomes. A study employing immunohistochemistry, FISH and Southern blotting with 205 samples of bladder cancer found that *STK15* gene amplification and associated increased expression of the mitotic kinase it encodes are associated with aneuploidy and aggressive clinical behavior [70].

An immunohistochemical evaluation of the expression levels of the mismatch repair (MMR) proteins hMLH1 and hMSH2 in patients with TCC bladder cancer with a median follow-up of 5 years demonstrated that reduced expression of either MMR protein was associated with fewer recurrences of superficial tumors and fewer relapses in all tumors, compared to tumors which had normal expression. The results indicate that reduced expression of these MMR proteins may make an important contribution to the development of a subset of TCCs and supports a potential role for MMR expression as a prognostic indicator [71].

10.5 Prostate cancer

Prostate cancer is a common malignancy in men and the incidence rate has increased markedly during the last two decades, whereas mortality has remained relatively stable [72]. However, a recent report from the UK has shown a marked increase in mortality from 1971–1989 in men aged 45–64 years, indicating an increase in the incidence of fatal prostatic cancer in younger men [73]. The apparent increase in incidence may be a reflection of improved detection rates related to the use of prostate-specific antigen (PSA) [72] as well as to overdiagnosis due to the use of PSA. A computer simulation model of PSA testing and subsequent prostate cancer diagnosis and death from prostate cancer, among a hypothetical cohort of two million men who were 60–84 years old in 1988, showed that the observed trends in prostate cancer incidence were consistent with considerable overdiagnosis among PSA-detected cancers [74]. Populations of African descent appear to be more susceptible to developing prostate cancer and African American men have the highest incidence rate of prostate cancer of any ethnic group in the USA [75].

This disease also has an extremely variable natural history, ranging from a noninvasive to a rapidly metastatic disease which is fatal within a short time following diagnosis. It is frequently diagnosed incidentally at the time of surgery during transurethral resection for urinary obstruction. Although factors such as histological grade have been evaluated as prognostic factors, there is considerable uncertainty concerning their use.

A family history of prostate cancer is a consistent risk factor and can also be used to predict the presence of prostate cancer among asymptomatic men who undergo PSA screening. Approximately 10–15% of cases of prostatic cancer have a familial component and hereditary prostate cancer resembles nonhereditary prostate cancer in terms of age of onset, pathologic appearance and grade [76].

Previous studies have demonstrated multiple alterations of tumor suppressor genes in human prostate cancer, including *p53* on 17p, *PTEN1* and *Mxi-1* on 10q and/or *Kai-1* on 11p. Loss of heterozygosity at polymorphic loci on 8p, 10p, 16q and 18q has also been reported [77–83 and references therein]. Transfer of a normal human chromosomal fragment into cancer cells by microcell-mediated chromosome transfer (MMCT), provides an alternative approach to detecting the presence of tumor suppressor genes by testing the ability of normal genes to suppress the malignant phenotype. Using this approach 7q, 8q, 10q, 11q, 17q and 19p have all been demonstrated to have suppressor activity for metastasis or tumorigenicity when introduced into rat or human prostate cancer cell lines [83] and these authors have used MMCT to localize a prostate tumor suppressor gene to a segment of approximately 1.2 Mb on 10p15.1. In addition, the macrophage scavenger receptor 1 gene (*MRS1* also known as *SR-A*), located at 8p22, has been shown to be potentially important in determining susceptibility to prostate cancer in men of both African American and European descent [84].

Prostate-specific antigen (PSA) is highly overexpressed in prostate cancer. One important regulator of PSA expression is the androgen receptor (AR). Human AR is a ligand-dependent nuclear transcription factor that regulates the expression of genes necessary for the growth and development of normal and malignant prostate tissue. AR is able to up-regulate the expression of the *PSA* gene by directly binding and activating its promoter. This AR activity is repressed by p53. A p53 mutation found in metastatic prostate cancer severely disrupts the p53-negative activity on the AR, suggesting that the inability of p53 mutants to down-regulate AR is, in part, responsible for the metastatic phenotype [85].

Prostate stem cell antigen (PSCA) is a 123 amino acid glycoprotein [86] which is expressed most strongly in the prostate, where it is localized to the putative stem cell component of the prostate, the basal cells. PSCA has been demonstrated to be overexpressed by more than 80% of prostatic cancers and has been correlated with the aggressive features of high stage, high Gleason grade and anchorage independence [87]. PSCA has a limited normal tissue distribution and has recently been found to be also overexpressed in pancreatic adenocarcinoma [88].

The tumor suppressor gene *PTEN1* (*MMAC1/TEP1*) is lost frequently in advanced prostate cancer. However, the function of PTEN1 in tumorigenesis is not understood. A recent study has shown that expression of BCL2 in prostate tumors correlates with loss of the PTEN1 protein and loss of PTEN1 leads to up-regulation of the *BCL2* gene, thus contributing to survival and chemoresistance of prostate cancer cells [89].

Kruppel-like factor 6 (KLF6) is a zinc finger transcription factor of unknown function. The *KLF6* gene has been found to be mutated in a subset of human prostate cancer cases. LOH analysis revealed that one *KLF6* allele was deleted in 66% (17 of 22) of primary prostate tumors. Sequence analysis of the retained *KLF6* allele revealed mutations in 71% of cases. Functional studies confirmed that whereas wild-type KLF6 up-regulates p21 (*WAF1/CIP1*) in a p53-independent manner and significantly reduces cell proliferation, tumor-derived KLF6 mutants do not, thereby providing strong evidence that *KLF6* is a tumor suppressor gene involved in human prostate cancer [90].

Selective subtractive differential gene display has been recently used to identify a gene (tazarotene-induced gene 1 (*TIG1*)) which may be a tumor suppressor and whose diminished expression is involved in the malignant progression of prostate cancer [91].

Annexin I is one of the members of a recently described family of calcium-dependent binding proteins. Down-regulation of annexin I protein expression is a common finding in high-grade prostatic intraepithelial neoplasia and prostate cancer, suggesting that annexin I dysregulation may be an important early even in prostate cancer initiation. One likely mechanism for annexin I dysregulation is at the level of gene transcription [92].

A common therapy for non organ-confined prostate cancer involves androgen deprivation. An analysis of gene expression changes induced by dihydrotestosterone (DHT) in the androgen-responsive prostate cancer line, LNCaP, using serial analysis of gene expression (SAGE) and quantitative 2D-PAGE, indicated that 351 genes were significantly affected by DHT treatment, of which 147 were induced and 204 repressed by androgen, thus indicating the need for profiling at both the RNA and protein levels to obtain a comprehensive evaluation of the patterns of gene expression [93]. A novel gene on chromosome 2p24 that is differentially expressed between androgen-dependent and androgen-independent prostate cancer cells has also been identified recently [94].

A number of studies have implicated alleles at the CAG and GGC trinucleotide repeats of the *AR* gene with high-grade, aggressive prostate cancer. Individuals with CAG repeat lengths less than 20 and GGC repeats less than 16 have been associated with increased risk of developing prostate cancer [95 and references therein]. These authors have shown that populations of African descent possess significantly shorter alleles for these two loci than nonAfrican populations. Allelic diversity for both markers was higher among African Americans than any other population, including indigenous Africans from Sierra Leone and Nigeria.

E-cadherin and α-catenin are components of adherens junctions which mediate calcium-dependent cell–cell adhesion in a homotypic manner. A third component of adherens junctional complexes is β-catenin, which bridges E-cadherin with α-catenin

[96 and references therein]. Reduced expression of E-cadherin, has been reported in advanced-stage tumors. In a 3-year follow-up study, men whose tumors showed normal amounts of E-cadherin showed no disease progression while those whose tumors showed little expression of the protein had high PSA levels, and their disease had advanced [97]. The risk of prostate cancer associated with a C/A single-nucleotide polymorphism (SNP) at −160 bp relative to the transcription start site of the *E-cadherin* gene promoter has been shown to be substantial and may prove to be helpful in the future in identifying susceptible individuals [98]. Screening exon 3 of β-catenin from a panel of 81 primary prostate cancers, 22 lymph node metastases from untreated patients and 61 metastatic tissues from 19 patients who had died of hormone-refractory disease, using single-strand conformation polymorphism (SSCP) and DNA sequencing, has identified putative activating mutations (missense and deletion) in 5% of samples. One patient was found to have the same 72 base pair deletion in each of 19 separate metastases, indicating that this change was associated with a clonal population of metastatic cells. Complementary *in vitro* work showed increased β-catenin nuclear activity in prostate cancer cell lines. These data implicate the β-catenin signaling pathway in the development of a subset of prostate cancers [96].

cDNA microarray screening has also been used to compare gene expression patterns in benign prostate hyperplasia (BPH) and tumor samples. Screening of three individual cDNA arrays identified eight genes with three-fold greater expression in six tumor samples than in one nontumor sample and one BPH sample. Real-time PCR was used to confirm the overexpression of these eight genes, as well as 12 genes selected from the literature, against a panel of 17 tumors and 11 BPH samples. Two genes, δ-*catenin* (*delta-catenin/CTNND2*), an adhesion junction-associated protein that promotes cell scattering, and prostate-specific membrane antigen (*PSMA/FOLH1*) were found to be significantly overexpressed in the prostate cancer samples [99].

Estrogen-modulated transgenic mice, such as ER-knockouts (αERKO and βERKO), aromatase-knockout (ArKO) and aromatase-overexpressing (AROM+) mice have all contributed to an understanding of the roles of estrogens in male reproductive biology, including prostate growth and development. Varying pathological changes of the prostate have been identified as being the result of aberrant actions of estrogen, both directly through ER or indirectly by altering the endocrine status of the mice. Further characterization and manipulation of these estrogen-modulated mice should lead to a more complete understanding of the hormonal regulation of the prostate gland and potentially pave the way for improved therapies for prostate cancer [100].

Locally advanced, relapsed and metastatic prostate cancer has a dismal prognosis with conventional therapies offering no more than palliation. Gene expression profiling has shown that the polycomb group protein enhancer of zeste homolog 2 (EZH2) is overexpressed in hormone-refractory, metastatic prostate cancer. This study indicated that dysregulated expression of EZH2 may be involved in the progression of prostate cancer as well as being a marker that distinguishes indolent prostate cancer from those at risk of lethal progression [101].

A novel candidate susceptibility gene (*ELAC2*) has been identified recently on chromosome 17p. Carriers of mutations in *ELAC2* have been found to display a higher risk of developing prostate cancer. Overexpression of *ELAC2* in tumor cells causes a delay in G_2–M progression resulting in the accumulation of cyclin B levels [102].

Finally, locoregional administration of nonviral and viral vectors can yield impressive local gene expression and therapeutic effects in experimental models. As our understanding of these genes and the molecular biology of prostate cancer improves it is anticipated that gene therapy may make a significant contribution to the management of prostate cancer in the future [103].

References

1. **Mant, J.W.F. and Vessey, M.P.** (1994) In *Cancer Surveys, Volume 19: Trends in Cancer Incidence and Mortality*. (Doll, R., Fraumeni, J.F., Jr., and Muir, C.S., eds). Imperial Cancer Research Fund. Cold Spring Harbor Laboratory Press, p. 287.
2. **Banks, E.** (2000) In *Methods in Molecular Medicine, Vol. 39: Ovarian Cancer: Methods and Protocols* (Bartlett, J.M.S., ed.). Humana Press Inc., Totowa, N.J. Chapter 1, p. 3.
3. **Hedenfalk, I.A.** (2002) *J. Natl Cancer Inst.*, **94**, 960.
4. **Schink, J.C.** (1999) *Semin. Oncol.*, **26**, 2–7.
5. **Hough, C.D., Sherman-Baust, C.A., Pizer, E.S. et al.** (2000) *Cancer Res.*, **60**, 6281.
6. **Gallion, H.M., Pieretti, M., DePriest, P.D. and van Nagell, J.R., Jr.** (1995) *Cancer*, **76**, 1992.
7. **Gallion, H.H. and Smith, S.A.** (1994) *Semin. Surg. Oncol.*, **10**, 249.
8. **Berchuck, A.** (1995) *J. Cell. Biochem.*, **23** (Suppl.), 223.
9. **Jazaeri, A.A., Yee, C.J., Sotiriou, C. et al.** (2002) *J. Natl Cancer Inst.*, **94**, 990.
10. **Bertucci, F., Eisinger, F., Taget, R. et al.** (2002) *J. Natl Cancer Inst.*, **94**, 1506.
11. **Hilton, J.L., Geisler, J.P., Rathe, J.A. et al.** (2002) *J. Natl Cancer Inst.*, **94**, 1396.
12. **Schultz, D.C., Vanderveer, L., Berman, D.B. et al.** (1996) *Cancer Res.*, **56**, 1997.
13. **Eccles, D.M., Brett, L., Lessells, A. et al.** (1992) *Br. J. Cancer*, **65**, 40.
14. **Sheridan, E., Hancock, B.W. and Goyns, M.H.** (1993) *Cancer Lett.*, **68**, 83.
15. **Kihana, T., Tsuda, H., Teshima, S. et al.** (1992) *Jpn. J. Cancer Res.*, **83**, 978.
16. **Bosari, S., Viale, G., Radaelli, U. et al.** (1993) *Human Pathol.*, **24**, 1175.
17. **Righetti, S.C., Torre, G.D., Pilotti, S. et al.** (1996) *Cancer Res.*, **56**, 689.
18. **Santoso, J.T., Tang, D-C., Lane, S.B. et al.** (1995) *Gynecologic Oncol.*, **59**, 171.
19. **Lee, J-H., Kang, Y-S., Park, S-Y. et al.** (1995) *Cancer Genet. Cytogenet.*, **85**, 43.
20. **Russell, S.E.H.** (2000) In *Methods in Molecular Medicine, Vol. 39: Ovarian Cancer: Methods and Protocols* (Bartlett, J.M.S., ed.). Humana Press Inc., Totowa, N.J. Chapter 3, p. 25.
21. **Bast, R.C., Jr. and Mills, G.B.** (2002) In *Methods in Molecular Medicine, Vol. 39: Ovarian Cancer: Methods and Protocols* (Bartlett, J.M.S., ed.). Humana Press Inc., Totowa, N.J. Chapter 4, p. 37.
22. **Alaiya, A.A., Franzén, B., Fujioka, K. et al.** (1997) *Int. J. Cancer*, **73**, 678.
23. **Lal, A., Lash, A.E., Altschul, S.F. et al.** (1999) *Cancer Res.*, **59**, 5403.
24. **Baekelandt, M., Holm, R., Nesland, J.M. et al.** (2000) *J. Clin. Oncol.*, **18**, 3775.
25. **Petricoin, E.F., III., Ardekani, A.M., Hill, B.A. et al.** (2002) *Lancet*, **359**, 572.
26. **Riman, T., Dickman, P.W., Nilsson, S. et al.** (2002) *J. Natl Cancer Inst.*, **94**, 497.
27. **Huang, S., van Arsdall, M., Tedjarati, S. et al.** (2002) *J. Natl Cancer Inst.*, **94**, 1134.
28. **Obermair, A., Schmid, B.C., Packer, L.M. et al.** (2002) *Int. J. Cancer*, **100**, 166.
29. **Orsulic, S., Li, Y., Soslow, R.A. et al.** (2002) *Cancer Cell*, **1**, 53.
30. **Cramer, D., Harlow, B., Titus-Ernstoff, L. et al.** (1998) *Lancet*, **351**, 104.
31. **Sherman, M.E., Sturgeon, S., Brinton, L. and Kurman, R.J.** (1995) *J. Cell. Biochem.*, **23** (Suppl.), 160.
32. **Kanamori, Y., Kigawa, J., Itamochi, H. et al.** (2002) *Int. J. Cancer*, **100**, 686.
33. **Berchuck, A.** (1995) *J. Cell. Biochem.*, **23** (Suppl.), 174.
34. **Risinger, J.I., Berchuck, A., Kohler, M.F. and Boyd, J.** (1994) *Nature Genetics*, **7**, 98.
35. **Beral, V., Hermon, C., Munoz, N. and Devesa, S.S.** (1994) In *Cancer Surveys, Volume 19: Trends in Cancer Incidence and Mortality*. (Doll, R., Fraumeni, J.F., Jr., and Muir, C.S., eds). Imperial Cancer Research Fund. Cold Spring Harbor Laboratory Press, p. 265.
36. **Drain, P.K., Holmes, K.K., Hughes, J.P. and Koutsky, L.A.** (2002) *Int. J. Cancer*, **100**, 199.
37. **Munger, K.** (1995) *J.Cell. Biochem.*, **23** (Suppl.), 55.
38. **Stanley, M.A.** (2001) *Best Practice Res. Clin. Obstet. Gynaecol.*, **15**, 663.
39. **Hawes, S.E. and Kiviat, N.B.** (2002) *J. Natl Cancer Inst.*, **94**, 1592.
40. **Smith, J.S., Herrero, R., Bosetti, C. et al.** (2002) *J. Natl Cancer Inst.*, **94**, 1604.
41. **Webster, K., Taylor, A. and Gaston, K.** (2001) *J. Gen. Virol.*, **82**, 201.
42. **Man, S.** (2001) *Best Practice & Res. Clin. Obstet. & Gynaecol.*, **15**, 701.
43. **Stern, P.L., Faulkner, R., Veranes, E.C. and Davidson, E.J.** (2001) *Best Practice & Res. Clin. Obstet. & Gynaecol.*, **15**, 783.

44. **Koutsky, L.A., Ault, K.A., Wheeler, C.M.** *et al.* (2002) *New. Engl. J. Med.*, **347**, 1645.
45. **Chung, T.K.H., Cheung, T.H., Wang, V.W.** *et al.* (2001) *Gynecol. Obstet. Invest.*, **52**, 98.
46. **Cortés-Gutiérrez, E.I., Cerda-Flores, R.M. and Leal-Garza, C.H.** (2000) *Cancer Genet. Cytogenet.*, **122**, 121.
47. **Field, J.K. and Spandidos, D.A.** (1990) *Anticancer Res.*, **10**, 1.
48. **Ishikawa, H., Mitsuhashi, N., Maebayashi, K. and Niibe, H.** (2001) *Cancer*, **91**, 80.
49. **Forman, D. and Moller, H.** (1994) In *Cancer Surveys, Volume 19: Trends in Cancer Incidence and Mortality*. (Doll, R., Fraumeni, J.F., Jr., and Muir, C.S., eds). Imperial Cancer Research Fund. Cold Spring Harbor Laboratory Press, p. 323.
50. **Watson, J.V., Stewart, J., Evan, G.** *et al.* (1986) *Br. J. Cancer*, **53**, 331.
51. **McCredie, M.** (1994) In *Cancer Surveys, Volume 19: Trends in Cancer Incidence and Mortality*. (Doll, R., Fraumeni, J.F., Jr., and Muir, C.S., eds). Imperial Cancer Research Fund. Cold Spring Harbor Laboratory Press, p. 343.
52. **Sarto, C., Marocchi, A., Faraglia, B.** *et al.* (2002) *Int. J. Cancer*, **97**, 671.
53. **Hodgson, S.V. and Maher, E.R.** (1993) In: *A Practical Guide to Human Cancer Genetics*. Cambridge University Press. Part 3, p. 157.
54. **Gnarra, J.R., Duan, D.R., Weng, Y.** *et al.* (1996) *Biochim. Biophys. Acta*, **1242**, 201.
55. **Hodgson, S.V. and Maher, E.R.** (1993) In *A Practical Guide to Human Cancer Genetics*. Cambridge University Press. Part 2, p. 74.
56. **Kaelin, W.G.** (2002) *Nature Rev. Cancer*, **2**, 673.
57. **Kim, W. and Kaelin, G., Jr.** (2003) *Curr. Opin. Genet. Develop.*, **13**, 55.
58. **Leung, H.Y.** (1994) In *Cancer. A Molecular Approach*, Chapter 13, (Lemoine, N., Neoptolemos, J. and Cooke, T., eds). Blackwell Scientific Press. p. 223.
59. **Macdonald, F. and Ford, C.H.J.** (1997) *Molecular Biology of Cancer*. BIOS, Oxford.
60. **Neal, D.E., Sharples, L., Smith, K.** *et al.* (1990) *Cancer*, **65**, 1619.
61. **Kuczyk, M.A., Bokemeyer, C., Serth, J.** *et al.* (1995) *Eur. J. Cancer*, **31A**, 2243.
62. **Schmitz-Drager, B.J., Goebell, P.J., Ebert, T. and Fradet, Y.** (2000) *Eur. Urol.*, **38**, 691.
63. **Sgambato, A., Migaldi, M., Faraglia, B.** *et al.* (2002) *Int. J. Cancer*, **97**, 671.
64. **Watters, A.D., Latif, Z., Forsyth, A.** *et al.* (2002) *Br. J. Cancer*, **87**, 654.
65. **Sourvinos, G., Kazanis, I., Delakas, D.** *et al.* (2001) *J. Urol.*, **165**, 249.
66. **Lu, M-L., Wikman, F., Orntoft, T.F.** *et al.* (2002) *Clinical Cancer Res.*, **8**, 171.
67. **Kocher, T., Zheng, M., Bolli, M.** *et al.* (2002) *Int. J. Cancer*, **100**, 702.
68. **Liu, J., Babaian, D.C., Liebert, M.** *et al.* (2000) *Molec. Carcinogen.*, **29**, 143.
69. **Reznikoff, C.A., Sarkar, S., Julicher, K.** *et al.* (2000) *Urologic Oncol.*, **5**, 191.
70. **Sen, S., Zhou, H., Zhang, R-D.** *et al.* (2002) *J. Natl Cancer Inst.*, **94**, 1320.
71. **Catto, J.W.F., Xinarianos, G., Burton, J.L.** *et al.* (2003) *Int. J. Cancer*, **105**, 484.
72. **Jarup, L., Best, N., Toledano, M.B.** *et al.* (2002) *Int. J. Cancer*, **97**, 695.
73. **Post, P.N., Stockton, D., Davies, T.W. and Coebergh, J-W.W.** (1999) *Br. J. Cancer*, **79**, 13.
74. **Etzioni, R., Penson, D.F., Legler, J.M.** *et al.* (2002) *J. Natl Cancer Inst.*, **94**, 981.
75. **Brawley, O.W. and Kramer, B.S.** (1996) *Epidemiology of Prostate Cancer. In Comprehensive Textbook of Genitourinary Oncology*. (Volgelsang, N.J., Scardino, P.T., Shipley, W.U. and Coffey, D.S., eds). Williams and Wilkins, Baltimore.
76. **Narod, S.** (1998) *Biochim. Biophys. Acta*, **1423**, F1.
77. **Veronese, M.L., Bullrich, F., Negrini, M. and Croce, C.M.** (1996) *Cancer Res.*, **56**, 728.
78. **Takahashi, S., Shan, A.L., Ritland, S.R.** *et al.* (1995) *Cancer Res.*, **55**, 4114.
79. **Gao, X., Zacharek, A., Salkowski, A.** *et al.* (1995) *Cancer Res.*, **55**, 1002.
80. **Kubota, Y., Shuin, T., Uemura, H.** *et al.* (1995) *Prostate*, **27**, 18.
81. **Gao, X., Honn, K.V., Grignon, D.** *et al.* (1993) *Cancer Res.*, **53**, 2723.
82. **Dong, J-T., Lamb, P.W., Rinker-Schaeffer, C.W.** *et al.* (1995) *Science*, **268**, 884.
83. **Fukuhara, H., Maruyama, T., Nomura, S.** *et al.* (2001) *Oncogene*, **20**, 314.
84. **Xu, J., Zheng, S.L., Komiya, A.** *et al.* (2002) *Nature Genet.*, **32**, 321.
85. **Shenk, J.L., Fisher, C.J., Chen, S-Y.** *et al.* (2001) *J. Biol. Chem.*, **276**, 38472.
86. **Reiter, R.E., Gu, Z., Watabe, T.** *et al.* (1998) *Proc. Natl Acad. Sci. USA*, **95**, 1735.
87. **Gu, Z., Thomas, G., Yamashiro, J.** *et al.* (2000) *Oncogene*, **19**, 1288.
88. **Argani, P., Rosty, C., Reiter, R.** *et al.* (2001) *Cancer Res.*, **61**, 4320.
89. **Huang, H., Cheville, J.C., Pan, Y.** *et al.* (2001) *J. Biol. Chem.*, **276**, 38830.

90. **Narla, G., Heath, K.E., Reeves, H.L. *et al.*** (2001) *Science*, **294**, 2563.
91. **Jing, C., El-Ghany, A., Beesley, C. *et al.*** (2002) *J. Natl Cancer Inst.*, **94**, 482.
92. **Kang, J.S., Calvo, B.F., Maygarden, S.J. *et al.*** (2002) *Clin. Cancer Res.*, **8**, 117.
93. **Waghray, A., Feroze, F., Schober, M.S. *et al.*** (2001) *Proteomics*, **1**, 1327.
94. **Chang, G.T.G., Steenbeek, M., Schippers, E. *et al.*** (2001) *Eur. J. Cancer*, **37**, 2129.
95. **Kittles, R.A., Young, D., Weinrich, S. *et al.*** (2001) *Hum. Genet.*, **109**, 253.
96. **Cheshire, D.R., Ewing, C.M., Sauvageot, J. *et al.*** (2000) *The Prostate*, **45**, 323.
97. **Nelson, N.J.** (1995). *J. Natl Cancer Inst.*, **87**, 1281.
98. **Verhage, B.A.J., van Houwelingen, K., Ruijter, T.E.G. *et al.*** (2002) *Int. J. Cancer*, **100**, 683.
99. **Burger, M.J., Tebay, M.A., Keith, P.A. *et al.*** (2002) *Int. J. Cancer*, **100**, 228.
100. **Jarred, R.A., McPherson, J., Bianco, J.J. *et al.*** (2002) *Trends Endocrin. Metab.*, **13**, 163.
101. **Varambally, S., Dhanasekaran, S.M., Zhou, M. *et al.*** (2002) *Nature*, **419**, 624.
102. **Korver, W., Guevara, C., Chen, Y. *et al.*** (2003) *Int. J. Cancer*, **104**, 283.
103. **Harrington, K.J., Spitzweg, C., Bateman, A.R. *et al.*** (2001) *J. Urol.*, **166**, 1220.

Leukemia and lymphoma

11

11.1 Introduction

The age-adjusted death rates for leukemia in the major industrialized countries are shown in *Table 11.1*.

Chromosome abnormalities are a common feature in leukemias and lymphomas. For example, in acute myelogenous leukemia (AML) over 160 structural chromosomal abnormalities have been observed to date [2]. These genomic aberrations have sometimes, as in chronic lymphocytic leukemia (CLL), been found to be important independent predictors of disease progression and survival [3]. Many oncogenes have been identified in leukemias and lymphomas because of the presence of consistent chromosomal translocations which can be identified cytogenetically [4]. While there was relatively little evidence that the wide variety of genes known to be associated with chromosome translocations were actually capable of causing tumors *in vivo*, it was assumed from their consistent association with translocations that they were oncogenes although functional evidence was often lacking [5]. However, considerable evidence has accumulated over the last decade to implicate them in the oncogenic process and the breakpoint in one of the chromosomes has frequently been shown to occur in, or close to, an oncogene, some of which are indicated in *Table 11.2*.

The numerous chromosomal alterations and chromosomal translocations that occur in leukemias and lymphomas are thought to cause oncogenic activation through one of two mechanisms. The first is gene activation after translocations involving the T cell receptor (TCR) or immunoglobulin gene loci and the second is gene fusion, which occurs when the TCR or immunoglobulin genes are not involved in the translocation process and results in the fusion of the coding regions of two genes which in turn leads to a fusion or chimeric protein with altered functional properties [5,6 and references therein]. Transcription factors regulate cell growth, differentiation, and apoptosis by binding to

Table 11.1 Age-adjusted death rates for leukemia in the major industrialized countries of the world (2000)

Country	Age-adjusted death rates per 100 000 population	
	Male	Female
USA	6.6	4.2
Canada	6.2	3.9
France	6.1	3.9
Germany	5.7	3.9
Russian Federation	5.0	3.4
United Kingdom	4.9	3.3
Japan	4.1	2.6

Abstracted from [1].

Table 11.2 Chromosomal translocations in leukemia and lymphoma associated with oncogenes

Disease	Translocation	Oncogene
Burkitt's lymphoma	t(8;14)(q24.1;q32.3) t(8;22)(q24.1;q11) t(2;8)(p11;q24.1)	MYC
B-CLL	t(11;14)(q13;q32.3)	BCL1 (CCND1/Cyclin D1)
	t(14;19)(q32;q13)	BCL3
Follicular lymphoma	t(14;18)(q32.3;q21.3)	BCL2
CML (AML and ALL)	t(9;22)(q34;q11)	ABL/BCR
AML	t(8;21)(q22;q22)	AML1/CBFβ/ETO
AML/CML	t(16;21)(p11.2;q22.2)	TLS/ERG
APL	t(15;17)	PML/RARα
ALL	t(4;11)(q21;q23)	ALL-1
T-ALL	t(11;14)(p15;q11)	RBTN1 (TTG1)
	t(11;14)(p13;q11) t(2;8)(q24;q24)	RBTN2 (TTG2) MYC
	t(8;14)(q24;q11)	MYC

specific DNA sequences within the promoter regions of growth-regulatory genes and modulating expression of these genes. Oncogenic activity can result from activation of existing transcription factors (including lineage-specific transcription factors which are involved in normal hematopoietic differentiation), abnormal expression of transcription factors, abnormal expression of transcription factor genes, or the formation of chimeric transcription factor genes [7,8]. Several fusion proteins generated by chromosomal translocations are chimeric transcription factors, i.e. new transcription factors formed by the fusion of portions of two transcription factors. The activity of these fusion proteins differs considerably from their normal counterparts and they are thought to deregulate the expression of target genes controlled by the wild-type transcription factors. An example of this is the BCR–ABL fusion protein (see Section 11.5).

Oncogene products, which can include growth factors, kinases and GTP-binding proteins, in addition to transcription factors, participate in the signaling pathways from the cell surface to the nucleus. Stem cell self-renewal, commitment to differentiation and progenitor cell expansion or terminal differentiation are controlled by the activity of growth factors and nuclear regulators. Alterations in any of the signaling pathways, such as those induced by chromosomal abnormalities, have the potential to induce transformation [7].

This is complicated by the fact that mammalian transcription factors are often members of large protein families that bind to similar DNA sequences, which raises the question of whether members of a particular family regulate expression of overlapping or unique sets of genes. In general, *in vitro* DNA-binding assays indicate that factors having similar DNA-binding motifs bind to similar DNA sequences. It is this commonality of DNA-binding motifs that raises the problem that if a group of different transcription factors can all bind to the same sequence, how does one determine which member of the family regulates a particular target gene? The corollary of this is how can increased levels of activity of one member of a family of transcription factors cause neoplastic transformation whereas high levels of other, very similar, family members do not lead to loss of cell growth control? [9].

A 'master gene' model has been postulated to explain the consistent observation in acute leukemias that genes involved in transcriptional regulation are activated by association with TCR or immunoglobulin genes after chromosomal translocation. These activation processes cause the transcription of the master genes in cells that would not normally express them. In cancers, the master gene products bind to their specific activation targets and cause responder target genes to be transcribed. These in turn may bind to specific targets in the genome and cause responder target genes to be transcribed [5,10].

A detailed discussion of all the transcription factors that have been identified in association with translocations in leukemias and lymphomas is beyond the scope of this chapter. Only some of these are discussed below and for a more detailed treatment the reader is referred to the excellent reviews that exist on this topic [7,11,12]. In addition, other oncogenes commonly identified in human tumors, such as the *RAS* family, have also been found to be associated with leukemias and lymphomas and are discussed below.

11.2   *TEL/JAK2*

The ETS (external transcribed spacer) family of transcription factors plays important roles in B cell development, hematopoiesis and signal transduction. TEL (translocation, ETS, leukemia) is an ETS family member that is a common target of gene rearrangements in leukemia. TEL is a site-specific DNA-binding transcription factor. *TEL* is rearranged in AML, chronic myelomonocytic leukemia (CMML), myelodysplastic syndrome (MDS) and acute lymphoblastic leukemia (ALL). There are more than 20 different chromosomal translocations in human leukemia involving the *TEL* gene [9]. Different regions of the *TEL* gene are retained in different leukemias. For example, *TEL/MNI* fusion results in the disruption of *ETS* target genes. The *TEL/AML1* fusion is frequent in ALL and is the most common gene rearrangement in pediatric malignancy. AML1 normally binds to a DNA sequence in the transcriptional regulatory region of the TCR and in a variety of cytokines and their receptors. The TEL/AML1 fusion protein is therefore thought to deregulate transcriptional activity of *AML1* target genes. It has been proposed that the TEL/AML1 fusion protein is a dominant interfering protein that can efficiently down-regulate all AML-regulated target genes and that it is the loss of AML-activated gene expression that causes leukemia.

TEL/JAK2 (Janus kinase 2) fusion protein is found in both lymphoid and myeloid leukemias. The JAKs are receptor-associated tyrosine kinases involved in the intracellular signaling pathways of several cytokines and growth factor receptors [7].

11.3   *MYC*

The involvement of the *MYC* gene in Burkitt's lymphoma (BL) is probably one of the most studied of all events involving oncogenes. The translocation between chromosomes 8 and 14 is a highly consistent rearrangement found in approximately 80% of BL patients. In the remaining patients the translocation occurs between chromosomes 8 and either 2 (~15%) or 22 (~5%). This translocation places the *MYC* gene from chromosome 8 next to the immunoglobulin (Ig) heavy chain locus on chromosome 14 or the κ and λ constant region exons, on chromosomes 2 and 22, 3' of the *MYC* gene on chromosome 8 (*Figure 11.1*). In the commonest translocation the breakpoint on chromosome 8 occurs on the 5' side of the second exon of the *MYC* gene. This means that the two coding exons of the gene – exons 2 and 3 – are always translocated and placed close to the constant region of the immunoglobulin heavy chain locus on chromosome 14. In the two less common translocations the breakpoint is distal to the *MYC* gene. In all cases, the coding exons of *MYC* do not undergo any structural rearrangement. Instead the translocation results in deregulation of *MYC* with the timing and level of expression of MYC protein being changed dramatically [4]. Following the general rule that the development of

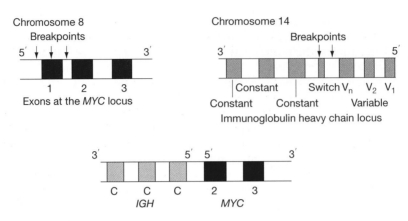

Figure 11.1

Molecular basis of the translocation between the *MYC* gene on chromosome 8 and the immunoglobulin heavy chain on chromosome 14 which occurs in the majority of cases of Burkitt's lymphoma. Adapted from [13].

malignancy is a multistage event, other changes are necessary for the development of BL. In one model the first change is chronic stimulation and activation of the B cells by infectious agents, primarily Epstein–Barr virus, leading to the emergence of immortalized clones. The second stage is the translocation involving *MYC*. A second model proposes that these two events occur in the reverse order. Additional changes in other oncogenes are likely to follow these two steps [14], for example *p53* mutations have been detected in approximately one-third of fresh BL tumor specimens [15].

Although a sensitive marker for BL, *MYC* rearrangements are not specific as they are occasionally seen in larger B cell lymphomas, lymphoblastic lymphomas, in an unusual subset of aggressive transformed follicular lymphomas and in a significant fraction of late-stage multiple myelomas [15]. Flow-cytometric detection of p62 has demonstrated a correlation between MYC protein levels and the aggressiveness of malignant non-Hodgkin's lymphomas of B-cell origin (B-NHLs).

The position of the chromosomal breakpoints in *MYC* and the various Ig genes is widely dispersed, which has made it difficult to design polymerase chain reaction-based tests for *MYC/Ig* fusion genes [15]. However, there is some evidence that there is variation in an RFLP associated with the *MYCL* locus, which plays a role in both survival and susceptibility to nonHodgkin's lymphoma and acute lymphoblastic lymphoma.

Little is known of the mechanisms that cause chromosomal breaks in *MYC*. Experiments with transgenic mice have shown that juxtaposition of Ig enhancer sequences and *MYC* creates an oncogene [15]. The recent use of cDNA microarrays to identify gene expression signatures specific for a variety of B cell lymphomas [16] should permit a much more detailed description of gene and protein expression in primary BL cells in the near future (see Section 11.16).

11.4 *BCL1* and *BCL2*

In some B cell lymphoproliferative diseases, including 30–39% of mantle cell lymphomas (MCLs), a t(11;14)(q13;q32) translocation has been observed [17,18]. This translocation involves the immunoglobulin heavy chain locus at chromosome 14q32 and the chromosome 11q13 region and is analogous to the t(8;14) translocation described earlier where *MYC* is placed adjacent to the immunoglobulin heavy chain enhancer. The translocated

11q13 region is known as the *BCL1* locus The *CCND1* gene has been implicated as the relevant proto-oncogene within the *BCL1* locus and its protein product is cyclin D1. *CCND1* is involved in the G_1 to S transition in the cell cycle [18 and references therein] (see Chapter 4).

A translocation between chromosomes 14 and 18 is found in approximately 85% of follicular lymphomas and involves the oncogene *BCL2* on chromosome 18, placing it next to the immunoglobulin heavy chain locus. Progression of follicular lymphomas to a more aggressive form is frequently associated with overgrowth of subclones containing chromosomal changes, in addition to the characteristic 14;18 translocation. In some cases this is associated with an 8;14 translocation involving the *MYC* gene, as seen in Burkitt's lymphoma. The *BCL2* translocation is not entirely specific for follicular lymphomas; it has also been reported in occasional cases of CLL, ALL, and small noncleaved lymphoma, as well as in diffuse large-cell lymphomas and a variable number of Hodgkin's disease cases. The presence of the translocation may indicate a derivation from follicular lymphoma in these last two situations; in other neoplasms, however, it may represent a common first step to tumorigenesis, which may be followed by unique additional events [17 and references therein]. BCL2 protein expression similarly lacks diagnostic specificity because of its ubiquity. This protein is normally expressed in hematopoietic precursors and basal cells and in long-lived cells, such as medullary thymocytes and peripheral T cells and the B cells of the mantle zone of the follicles. The protein is not expressed in reactive germinal centers, a feature of diagnostic importance when compared with the strong immunostaining of neoplastic follicles. In addition to the follicular lymphomas, overexpression of BCL2 protein has been reported in diffuse lymphomas of both T and B cell lineages and both small and large cell type, CLL, ALL, plasma cell neoplasms, myeloid dysplasia and Hodgkin's disease. There must, therefore, be other mechanisms, besides the t(14;18) translocation, that can lead to *BCL2* overexpression, either by deregulating the oncogene or by acting at the translational or post-translational level. One of these mechanisms is the up-regulation of BCL2 protein expression by EBV [19].

It is now known that BCL2 and related cytoplasmic proteins are key regulators of apoptosis, the cell suicide program critical for development, tissue homeostasis, and protection against pathogens. Those most similar to BCL2 promote cell survival by inhibiting adapters needed for activation of the proteases (caspases) that dismantle the cell. More distant relatives promote apoptosis instead, apparently through mechanisms that include displacing the adapters from the pro-survival proteins. Thus for many, but not all apoptotic signals, the balance between these competing activities determines cell fate. BCL2 family members are essential for maintenance of major organ systems, and mutations affecting them are implicated in a variety of cancers [20,21 and references therein]. At least 20 BCL2 family members have been identified in mammalian cells which are divided into three subfamilies; BCL2 (promotes cell survival – BCL2, BCLXL, BCLW, MCL1, A1); BAX (pro-apoptosis – BAX, BAK, BOK) and BH3 (pro-apoptosis – BIK, BLK, HRK, BNIP3, BIMl, BAD, BID) [21]. All members possess at least one of four conserved motifs known as BCL2 homology domains (BH1 to BH4). Most pro-survival members, which can inhibit apoptosis in the face of a wide variety of cytotoxic insults, contain at least BH1 and BH2, and those most similar to BCL2 have all four BH domains. The two pro-apoptotic subfamilies differ markedly in their relatedness to BCL2. BAX, BAK and BOK (also called MTD), which contain BH1, BH2 and BH3, resemble BCL2 fairly closely. In contrast the seven other known mammalian 'killers' possess only the central short BH3 domain; they are otherwise unrelated to any known protein, and only BIK and BLK are similar to each other [20,21].

Apoptosis normally eliminates cells with damaged DNA or an aberrant cell cycle, i.e. those most likely to give rise to a neoplastic clone. With the discovery of the anti-apoptotic function of the *BCL2* oncogene the concept emerged that a raised threshold for apoptosis represents a central step in tumorigenesis [17]. The oncogenic impetus of *BCL2* translocation, seen in most follicular lymphomas and some cases of diffuse large cell lymphoma and CLL has been verified in transgenic mice. All pro-survival *BCL2*-like genes are potentially oncogenic, and some mutations probably increase their expression indirectly. In hematopoietic cells, oncoproteins such as MYB, RAS, and AML1-ETO induce *BCL2* expression which is often elevated in myeloid leukemia. The risk of B lymphoid tumors is enhanced probably because BCL2 can countermand the apoptotic action of other oncoproteins such as MYC. Expression of *BCL2* may also constitute a major barrier to the success of genotoxic cancer therapy [22 and references therein]. It should be noted here that although the evidence from human tumors that cancer generally requires impaired apoptosis is not overwhelming, the hypothesis is strongly supported by experimental models [21].

Despite the above, the exact mechanism(s) of BLC2's anti-apoptotic function is not yet clear and, as indicated earlier, BCL2 expression does not correlate consistently with a worse outcome or with resistance to anticancer therapy as for example in pediatric ALL patients, where response and survival following chemotherapy is independent of BCL2 expression. These findings indicate that expression of BCL2 alone may not be enough to protect cells from apoptosis and the discovery that BCL2 is post-translationally modified by phosphorylation supports another level of regulation of function for this family of proteins [23].

Diffuse B cell nonHodgkin's lymphomas with t(14;18)(q32.3;q21.3) have a worse prognosis than those without this translocation and the development of a sensitive PCR assay has enabled one NHL cell in 10^5 normal cells to be detected, making the screening of bone marrow for residual malignant cells after therapy feasible [24].

11.5 *ABL/BCR*

The Philadelphia chromosome (Ph) is the result of a reciprocal recombination between chromosomes 9 and 22 which generates a shortened chromosome 22 (Ph) and an elongated chromosome 9. The discovery of the Ph chromosome and its consistent involvement in CML was the first time that a relationship between a cytogenetic abnormality and malignancy was demonstrated [22]. A 9;22 translocation is found in approximately 90% of all cases of CML. It is not diagnostic for this disease, as it is also present in 25% of adult ALLs, in 2–10% of childhood ALLs and in occasional cases of AML. The breakpoints on chromosome 9 are located in the first and second introns of the *ABL* oncogene. The *ABL* gene comprises two alternative first exons, 1a and 1b, and the common exons 2–11. An exceptionally long first intron spaces the first exon (1b) 200 kb 5′ from exon 1a (*Figure 11.2*). Two transcripts can be formed by alternative splicing, 1a (6 kb) and 1b (7 kb). The two mRNA molecules encode two proteins of about 145 kDa in size which belong to the family of nonreceptor tyrosine kinases. The b form is myrostilated at an amino terminal glycine residue which directs the protein to the plasma membrane. The ABL protein contains the aminoterminal part encoded by exon 1a, lacks the myrostilation signal and is predominantly present in the nucleus [25]. Overexpression of *ABL* results in cell cycle arrest, similar to tumor suppressor genes such as *p53* and *RB*.

On chromosome 22, the breakpoint occurs within the breakpoint cluster region (*BCR*) gene. It spans a region of 135 kb and comprises 23 exons (*Figure 11.2*). The *BCR* gene is a member of a gene family of unclear function which also includes the *BCR2*, *BCR3* and *BCR4* genes. Gene expression gives rise to two mRNA types of 4.5 and 6.7 kb which are both translated into a 160 kDa protein but several different BCR proteins can be identified ranging in size from 83–190 kDa [25]. Several different structural regions have been identified in the BCR protein (*Figure 11.3*). A role for BCR protein in cell cycle regulation has been suggested.

In the majority of patients with CML breakpoints in intron 1 or 2 of the *ABL* gene and in the major breakpoint cluster (MBCR) of the *BCR* gene (*Figure 11.2*) result in the formation

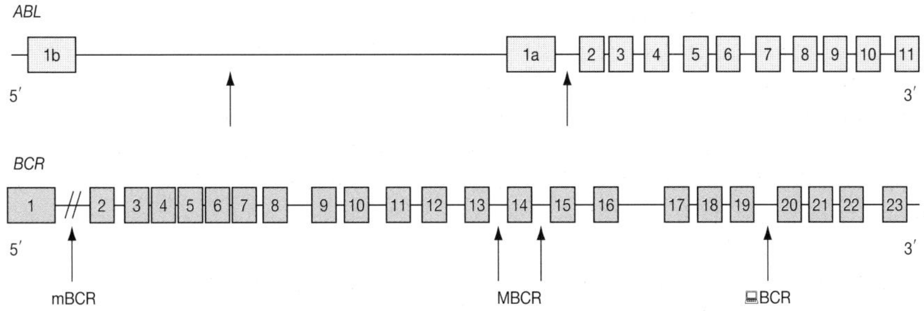

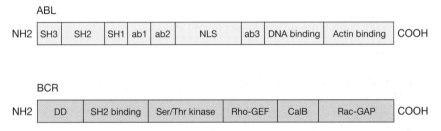

Figure 11.2

The *ABL* and *BCR* genes. Boxes indicate exons and spaces between boxes are introns. Breakpoint regions are indicated with arrows. Three breakpoint clusters have been identified in the *BCR* gene. The minor breakpoint cluster region (mBCR) is located in the first large intron of the *BCR* gene and is associated with the Philadelphia chromosome-positive ALLs. The major BCR (MBCR) region is present on the 3′ side of the gene and breakpoints in this cluster are between exons 13 and 14 or 14 and 15. Philadelphia chromosome-positive leukemias with breakpoints in this region are most frequently CMLs. The ⊟BCR region is located between exons 19 and 20. Adapted from [25].

Figure 11.3

ABL and BCR protein functional domains. SH3 and SH2 domains at the amino terminal part of ABL are involved in protein–protein interactions. The SH1/tyrosine kinase domain plays a crucial role in the phosphorylation of targets such as CRKL and 62 [Kdadok] in CML. Between residues 525 and 717 there are three proline-rich stretches, known as abl binding sites 1, 2 and 3, which can bind to other SH3-containing proteins. A nuclear localization signal (NLS) is located between ab2 and ab3 which can direct the protein to the nucleus. Near to the carboxyterminal end a stretch of acidic residues are involved in the binding of ABL to DNA. At the carboxyterminal end an actin-binding site is present which is crucial for the localization of the cytoplasmic BCR–ABL fusion protein thus determining the availability of substrates for tyrosine kinase. At the amino terminal end of BCR there is a dimerization domain (DD) which is involved in coiled-coil oligomerization of BCR and also mediates dimerization of the BCR–ABL protein which is probably essential for the transforming potential of BCR–ABL. This is followed by a domain which can bind at the SH2 domain of ABL, followed by a region with a unique serine/threonine kinase activity, a region of homology with Rho guanine-nucleotide exchange factors (Rho-GEF), a region which might be involved with calcium-dependent lipid binding, and a region which is implicated with Rac-GTPase activating protein (Rac-GAP) function. The exact role of these domains and how they work in concert has to be elucidated. Adapted from [25].

of a *BCR–ABL* gene expressing either the b2a2 or b3a2 mRNA variant (*Figure 11.4*). Both types of mRNA are translated into a fusion protein of 210 kDa with elevated constitutive protein-tyrosine kinase activity. Since a large number of in-frame *BCR–ABL* fusions are possible, the number of variant *BCR–ABL* genes leading to functional fusion proteins are in

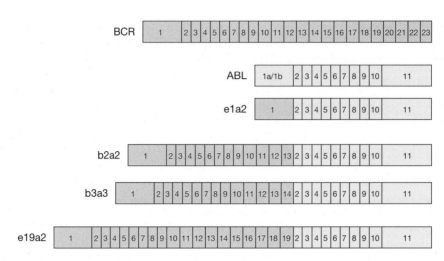

Figure 11.4

BCR–ABL fusion proteins. Breakpoints in the mBCR region of the *BCR* gene give rise to the fusion of *BCR* exon 1 (e1) with *ABL* exon 2, generating a fusion protein of 185 kDa. Breakpoints in the MBCR region give rise to the fusion of either *BCR* exon 13 (b2) with *ABL* exon 2 or to the fusion of *BCR* exon 14 (b3) with *ABL* exon 2. Both fusions generate a protein of about 210 kDa. Breakpoints in the μBCR give rise to the fusion of *BCR* exon 19 (c3) with *ABL* exon 2 generating a fusion protein of about 230 kDa. Adapted from [25].

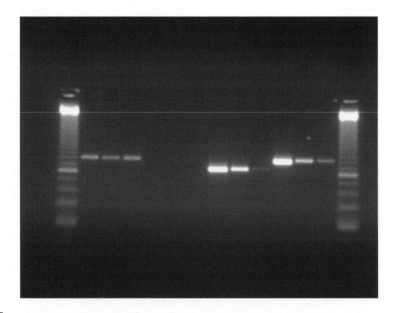

Figure 11.5

RT-PCR for *BCR–ABL* in CML patients. Outside lanes contain size markers. All tests are in triplicate. Lanes 1–3 are samples from a CML patient with a low level of minimal residual disease with a b3a2 transcript. Lanes 4–6 are from a patient with no evidence of MRD. Lanes 7–9 are b2a2 fusion controls in 10^{-2}, 10^{-3} and 10^{-4} copies. Lanes 10–12 are b3a2 fusion controls at the same dilutions. Figure courtesy of Jo Mason and Rachel Doak, West Midlands Regional Genetics Laboratory.

theory vast [25]. Recent work suggests that BCR/ABL inhibits the expression of a key cell cycle inhibitor, p27 (Kip1), which is a widely expressed inhibitor of CDK2 [26].

A much greater heterogeneity in breakpoints has been found for *BCR* compared to *ABL* (*Figure 11.4*). As a result of the reciprocal recombination an *ABL–BCR* gene is also formed but no ABL–BCR protein has been detected in samples from patients so far so its biological role and/or clinical relevance remains unknown at present [25].

STAT (signal transduction and activation of transcription) proteins are latent cytoplasmic transcription factors that affect gene transcription via the activation of the Janus kinases (JAK) and the subsequent recruitment of STAT to an activated receptor complex. STAT5 is constitutively activated in *BCR/ABL* expressing cell lines. STAT5 activation is involved in antiapoptotic activity and cell cycle progression induced by *BCR/ABL* and has been shown to be important for *BCR/ABL*-mediated leukemogenesis [7].

The ability to detect the presence of a *BCR/ABL* fusion gene by PCR has also allowed the development of a very sensitive diagnostic test for CML (*Figure 11.5*). In addition, it has been used successfully in the detection of the presence of residual disease following bone marrow transplantation. This might permit early therapy for patients with residual disease and thereby help to sustain remission. The major problem with this technique is also the reason for its use, that is, its high sensitivity. The persistence of low levels of Ph-positive cells has been a consistent finding and it is not yet clear how many of these cells are necessary to re-establish disease. In the near future long-term follow-up of these patients will hopefully clarify this point and allow post-transplant therapy to be initiated where necessary.

11.6 *AML1/CBFβ/ETO(MTG8)*

A number of cases of AML are not associated with a consistent chromosome translocation but rather with small mutations in the coding region of lineage-specific transcription factors. In many of the cases of AML that are associated with a common chromosome translocation, the resulting translocation product disrupts the expression of the same lineage-specific factors [8]. In fact, the most common genetic mechanism associated with AML as well as ALL is the aberrant expression of a transcription factor or the production of an abnormal transcription factor, supporting the view that disruption of normal differentiation is a key component in the development of these malignancies [8].

The AML1 and CBFβ subunits of core-binding factor (CBF) are involved in several chromosomal abnormalities frequently associated with acute leukemias. As a result, the CBFβ-SMMHC, AML1-ETO and AML1-MDS1/EVI1 fusion proteins are expressed in subsets of AML and TEL-AML1 is expressed in B-lineage ALL [7,11]. CBFα2 (AML1) is a DNA-binding subunit in the family of CBFs which are heterodimeric transcriptional factors that contain a common CBFβ subunit bound to one of the CBFα subunits. Expressed in hematopoietic cells, it is a key regulator of early hematopoiesis.

The *AML1* gene α unit has a DNA-binding domain homologous to the product of *Drosophila* segmentation gene, *RUNT*. The *ETO(MTG8)* gene is localized on chromosome 8q22 and encodes the mammalian homolog of the *Drosophila* protein NERVY. Identification of other *ETO* family members involved in translocations with *AML1* suggest that the *ETO* sequences are critical for the transforming activity of these fusion proteins.

The first genetic rearrangement to be discovered was t(8;21) in patients with a form of AML. The *AML1* gene is located at the translocation breakpoint on chromosome 21 in the t(8;21)(q22;q22) translocation (*AML1/ETO*) and the fusion protein is present in 12% of AML cases. The encoded fusion protein consists of the N-terminal 177 amino acids of AML1 fused in-frame to almost the complete ETO protein. It is still able to heterodimerize with CBFβ and interact with CBF DNA-binding sites, however, the fusion protein binds through its ETO sequences to both ETO/MTG family members and a corepressor complex which results in repression of genes that are normally activated by AML1/CBFβ, thereby inhibiting transcription. In addition, AML1/ETO may indirectly repress transcription through interactions with other transcription factors [27]. AML1/ETO also activates the promoters of genes encoding BCL2 and the M-CSF receptor through AML1-binding sites and is expressed in myeloid progenitor cells [7]. For a review of the expression and function of the *ETO* gene family readers are referred to [28].

Although p14/19 (ARF) is a tumor suppressor protein that blocks MDM2-mediated p53 degradation [29] it has been shown that the *p14* gene is a direct target for repression by AML1/ETO, which may begin to explain why p53 is not mutated frequently in acute leukemias [30].

11.7 *TLS/ERG*

The t(16;21)(p11.2;q22.2) translocation is found in myeloid leukemia and is associated with a poor prognosis in AML, secondary AML associated with MDS, and CML in blast crisis. It juxtaposes the *TLS/FUS (TLS)* gene on chromosome 16 and the *ERG* gene (*ETS-related gene*) on chromosome 21 and forms the *TLS/FUS* fusion gene.

The *TLS* gene was originally identified in liposarcomas and encodes an RNA-binding protein. *ERG* encodes a transcriptional activator of the external transcribed spacer (ETS) proto-oncogene family. The chimeric protein that is produced by the fusion gene contains an altered transcriptional activating and DNA-binding domain [7]. The exact mechanisms by which the TLS fusion proteins lead to transformation are not yet established.

11.8 *PML/RARα*

Cells typically contain 10–30 nuclear bodies/nucleus. Multiple different biological functions have been attributed to the various nuclear body components – many seem to be involved in transcriptional regulation.

The *PML* gene was originally cloned as the t(15;17) chromosomal partner of the retinoic acid receptor *RARα* (a member of a superfamily of nuclear hormone receptors) in acute promyelocytic leukemia (APL) in which fusion genes encoding PML-RARα and RARα-PML fusion proteins are generated [12]. In APL, PML, together with the other nuclear body components, is delocalized from the nuclear body into abnormal nuclear structures. Research on PML in the past few years has revealed an important role for this protein in essential cellular functions such as cell proliferation, differentiation and tumor suppression [7]. PML and its modification by SUMO-1 (an ubiquitin-like protein) are essential for the proper formation of the nuclear body. PML and the nuclear body can control gene expression through regulation of the transcriptional activity of nuclear receptors and other transcriptional factors.

11.9 *ALL-1* and 11q23 translocations

A gene, variously designated as *ALL-1*, *MLL* (myeloid-lymphoid or mixed lineage leukemia), *HRX* or *HTRX*, has been shown to have a critical role in multiple leukemic groups, as well as in occasional lymphomas, primarily by forming a 'fusion gene' with many partners from different chromosomes resulting in the synthesis of chimeric RNAs and most likely of chimeric proteins. Alterations in the *ALL-1* gene are found in 5–10% of all human leukemias and the *ALL-1* gene is rearranged in the vast majority of 11q23 abnormalities. The various acute leukemias, both lymphoid and myeloid, with cytogenetic and molecular evidence for alterations in *ALL-1*, uniformly carry a poor prognosis

[31]. The t(4;11) translocation is the most common genetic alteration in early childhood leukemia and this, along with other rearrangements of the *ALL-1* gene, sometimes submicroscopic, apparently contributes to at least 70% of the acute leukemias occurring in children under the age of 1 year. Secondary leukemias with 11q23 rearrangements typically have a myeloid phenotype and, as with the *de novo* leukemias, these are clinically aggressive and respond poorly to therapy.

ALL-1 is involved in translocations with an astonishing number of chromosomes and thus genes on other chromosomes. Over 40 different translocations have been identified and in 16 of them the breakpoints have been cloned, all leading to in-frame fusions and, presumably, to functional fusion proteins [4]. Critical studies of *ALL-1* function, as have been carried out for *AML1* for example, will not be possible until the target gene or genes of *ALL-1* have been identified [4].

11.10   *RBTN* T cell oncogenes

RBTN1 (*TTG1*) is located on chromosome 11p15 and is activated by the t(11;14)(p15;q11) translocation (see *Table 11.2*). *RBTN2* is located on 11p13 and is activated by the translocation t(11;14)(p13;q11). *RBTN3* is located on chromosome 12p12-13 and has not yet been found to be associated with a chromosomal translocation. The *RBTN* family of genes encode proteins with the cysteine-rich LIM domain that is similar to the electron transport protein ferredoxin and to zinc finger proteins. The LIM domain has been found in a number of proteins that are thought to be involved in developmental regulation [5]. While it is assumed that the *RBTN* family of genes are T cell oncogenes that are activated by chromosomal translocations, the only direct evidence that *RBTN1* and *RBTN2* can participate in tumorigenesis has been obtained from studies in transgenic mice [5].

11.11   Cyclin-dependent kinase 4 inhibitor, *CDKN2* (*p16/MTS1/INK4a*)

Approximately 10% of patients with ALL have cytogenetic abnormalities involving 9p21-22, the locus encoding the putative tumor suppressor gene *CDKN2* (see Chapter 4). Homozygous deletion seems to be a major mechanism for activating *CDKN2* in ALL, CLL, malignant lymphoma, lymphoid transformation of chronic myelogenous leukemia and adult T cell leukemia. Less frequently, mutation of one allele and concomitant loss of the remaining allele is responsible for *CDKN2* inactivation, and this occurs preferentially in T phenotype ALL [32 and references therein]. In addition, recent studies have shown that inappropriate methylation of *p16*'s promoter is a major mechanism of gene silencing in leukemia [33]. Whatever the mechanism, it appears that the disruption of 9p21 is an early event in the development of leukemia [34].

Deletions at the *INK4a/ARF* locus often encompass and inactivate at least two other nearby potential tumor suppressor genes, *p15* (*INK4b/MTS2*) and *p14* (*ARF*) indicating that one event can result in the silencing of two or more genes [33,35].

11.12   *RAS*

RAS is one of the most frequently deregulated proteins in leukemias, estimated to be mutated in at least 30% of cases [36]. In reality, RAS deregulation is probably significantly higher, as deregulation of guanine-nucleotide exchange factors (GEFs) or GTPase-activating proteins (GAPs), overexpression of growth factor receptors, autonomous cytokine production, mutation within the *RAS* promoter or activation by cooperating oncogenes, such as *BCR–ABL*, may all lead to constitutive signaling of RAS without mutations in RAS itself [37 and references therein].

The elucidation of the mitogen-activated protein kinase (MAPK) pathway has been one of the major breakthroughs in signal transduction research in recent years [25]. Deregulation of members of this cascade, such as RAS, have been investigated and are of great importance in, for example, AML. Although RAS mutations are very rare in chronic-phase CML cells, the MAPK pathway is probably very important for the transforming activity of BCR–ABL [25 and references therein]. Research on the role of RAS is ongoing but not all signaling by RAS goes through the MAP kinase pathway.

Mutations in *RAS* genes, particularly *NRAS*, occur frequently in AML and less commonly in CML [38]. In MDS and AML, a mutation has been found in the *RAS* gene in 20–30% of cases: this is usually in *NRAS*, occasionally in *KRAS* and infrequently in *HRAS*. Several studies have suggested that patients with MDS and a *RAS* mutation are more likely to progress to AML and hence have the worst prognosis. However, as there are examples of MDS patients with a *RAS* mutation in their multipotent stem cells who have stable disease, this marker is unlikely to be of use on its own as a prognostic indicator. Secondly, AML patients with or without *RAS* mutations do not have a significantly different prognosis and no relationship with degree of differentiation can be seen [38]. Alternative roles for the RAS mutation may be as a marker for monitoring the effects of chemotherapy or for detecting minimal residual disease [36].

In a large study of childhood ALL, point mutations were found in 6% of cases at codons 12 or 13 of *NRAS*. These cases had a significantly higher risk of hematological relapse and a trend towards a lower rate of complete remission than those without *RAS* mutations. Presence of *RAS* mutations was independent of other high-risk factors for ALL and may therefore be a useful prognostic indicator. Patients with AML and *RAS* mutations have been shown to have a poor response to chemotherapy and low remission rates [37].

Despite extensive debate on its role, RAS deregulation in leukemia is accepted as contributing to disease aggression, immune evasion and reduced patient survival [37].

11.13   *p53*

p53 mutations occur moderately often in hematological malignancies and are particularly associated with progression of disease in both lymphoid and myeloid leukemias and lymphomas. In addition, *p53* mutations occur very frequently in Burkitt's and other high-grade B cell lymphomas. Also, the Reed–Sternberg cells in Hodgkin's lymphoma express high levels of mutant p53, suggesting a major contribution of these cells to the disease [39]. A high incidence of *p53* gene mutation has also been found in mucosa-associated lymphoid tissue (MALT) lymphomas suggesting that *p53* plays an important role in the development of these lymphomas, particularly in high-grade transformations [40]. There is also a close association between *p53* abnormality and the replication error (RER+) phenotype, suggesting that the two affect each other [40]. Elevated *BCL2* expression may be related to loss of *p53* expression. Functional p53 not only acts as a repressor of *BCL2* gene expression but also promotes transactivation of BAX. Thus loss of p53 not only disables a cell's intrinsic mechanism to survey DNA damage, but may also dramatically alter the anti- to pro-apoptotic ratio of BCL2 family proteins [37]. It has been demonstrated recently that the *BCL2* promoter is the target for repression by p53 [41]. The prognostic significance of these associations has yet to be determined.

11.14   *MDM2*

The murine double minute-2 (*MDM2*) gene has been studied in patients with B cell CLL or B cell NHL. *MDM2* gene expression was low in normal B cells, whereas 28% of the patients with B-CLL or NHL had more than 10-fold higher levels of *MDM2* gene expression.

This overexpression was found more frequently in patients with the low-grade type of lymphoma (56%) than in those with intermediate-/high-grade types (11%) and was found significantly more frequently in patients at advanced clinical stages. These results suggest that *MDM2* gene overexpression may play an important role in the tumorigenicity and/or disease progression of CLL and low-grade lymphomas of B cell origin [42].

11.15 Detection of minimal residual disease (MRD)

The detection of residual cells derived from the malignant clones which remain after treatment has been the focus of many studies the purpose of which has been to determine whether cure or relapse can be predicted from the number of these MRD cells. Many approaches have been developed for the detection of MRD including cytogenetic, western blotting, Southern blotting, FISH, phenotypical characterization and functional analysis, and PCR (standard, competitive quantitative and real time) [25,43] (*Figure 11.6*).

The degree of tumor load reduction after therapy is an important prognostic factor for patients with CML. Conventional metaphase analysis has been considered to be the 'gold

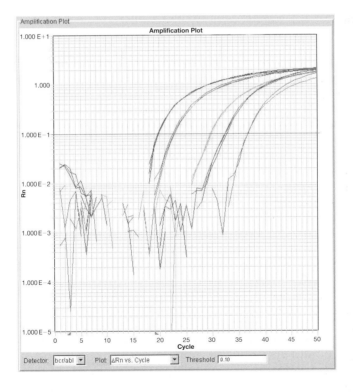

Figure 11.6

Real-time amplification curves of serially diluted plasmid (*lps-bcr-abl*) containing a *BCR–ABL* insert representing 1×10^{-6}, 1×10^{-5}, 1×10^{-3}, 1×10^{-2} and ten copies respectively. PCR cycle number has been plotted on the x axis and the change in fluorescent reporter dye on the y axis (DRn). The cycle threshold (Ct) is defined as the cycle number at which the DRn crosses a defined threshold above baseline. Each sample has been run in triplicate and the plots shift to the right with each dilution therefore the Ct value increases as the amount of RNA decreases. Figure courtesy of Suzanna Akiki, Mike Griffiths and Jo Mason, West Midlands Regional Genetics Laboratory.

standard' for evaluating patient response to treatment but this technique normally requires bone marrow aspiration and is therefore invasive. Molecular methods can be performed on peripheral blood specimens with obvious advantages [43].

Detection of MRD by cytogenetic analysis and the measurement of surface or immuno-logical markers using flow cytometry (phenotyping) has improved the sensitivity over histochemical techniques. However, PCR analysis has revolutionized the sensitivity of detection as chromosomal rearrangements can now be detected with a theoretical sensi-tivity of one malignant cell in a population of 10^5–10^6 cells [6 and references therein]. The Ig and/or TCR clonal rearrangements present at diagnosis in lymphoid leukemias and lymphomas provide a useful target which marks the abnormal cell progeny. Following PCR amplification of one or more of the clonal antigen receptor gene rearrangements present at diagnosis, a clone-specific probe can be produced to aid in the detection of residual cells in post-therapy samples.

However, the extreme sensitivity of the PCR has brought its own problems including the risk of contamination resulting in false-positive results. Questions must also be addressed concerning the nature of the residual cells harboring the disease-specific marker. Many patients are PCR positive after achieving clinical remission status but progress to become PCR negative. Are PCR-positive cells present in a biopsy from a patient during clinical remission, clonogenic? It may not be ethical to intervene with additional therapy in patients who are judged PCR positive but who have no other clini-cal manifestations of disease [44]. Some CML patients, for example, can remain PCR pos-itive for *BCR/ABL* without hematological relapse for years. Efforts are being directed at establishing more quantitative assays in order to determine not only whether residual or disseminated disease is present, but also at what level [45]. Quantitative determination of residual disease levels after treatment for patients with CML may be achieved by a variety of methods including real time quantitative PCR, which may shortly become a routine basis for clinical decision making, not only for CML (*Figure 11.7*) but also in other leukemias with specific molecular markers [43].

The potential value of PCR analysis can be illustrated by the following examples. In CML and ALL there is evidence that transient *BCR–ABL* positivity is not usually followed by hematological relapse, while patients who have serial samples which are positive have a high risk of relapse [46]. In Philadelphia-positive ALL an increase in *BCR–ABL* transcript numbers on sequential analysis after transplantation permits the early identification of individual patients who are likely to progress to cytogenetic and hematological relapse [47]. The use of quantitative RT-PCR to monitor the kinetics of residual *BCR–ABL* tran-scripts over time has shown that those patients destined to relapse are characterized by the reappearance and/or rising levels of *BCR–ABL* transcripts [43] (*Figure 11.6*).

However, while molecular analysis of remission specimens can predict a significant fraction of clinical relapses, the clinical value of MRD detection at this level of sensitivity still remains unclear. The hope remains that the results of molecular analysis will enable the implementation of new treatment strategies for patients whose remission marrows remain positive. The statement that in order to clarify the role of molecular analysis in the detection of MRD, prospective studies are essential is as valid today as it was when it was first made [6].

11.16   Refining the classification of leukemias and lymphomas

Expression profiling using microarray technology (see Chapter 14) is being used to refine the classification of cancer with the greatest advances being seen in leukemias, lym-phomas and breast cancers [48]. By comparing the gene expression profile patterns of pri-mary diffuse large B cell lymphomas (DLBCL) with those of normal lymphoid cells at

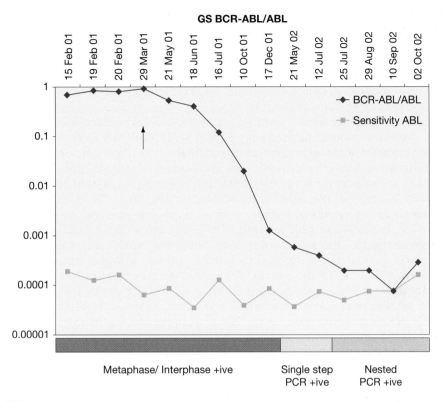

Figure 11.7

Taqman quantitative assessment of MRD in a CML patient following treatment with Glivec. The number of copies of *BCR-ABL* normalized against a housekeeping gene is plotted on the y axis against time on the x axis. An assessment of assay sensitivity determined by the detectable number of copies of the housekeeping gene at each time point is also illustrated. Cytogenetic and end point PCR positivity is indicated below the x axis. Figure courtesy of Suzanna Akiki, Mike Griffiths and Jo Mason, West Midlands Regional Genetics Laboratory.

different stages of differentiation it was found that DLBCL could be separated into at least two major expression subclasses: one that resembled activated normal B cells, with a poor prognosis, and the other similar to germinal center B cells, with a good prognosis [16,48]. In collaboration with the Lymphoma/Leukemia Molecular Profiling Project [49] the analysis was expanded to 240 patients with DLBCL which confirmed the presence of an activated B cell-like subgroup with a poor prognosis (median survival ~2 years) and a germinal center-like subgroup with a good prognosis (median survival ~7 years). A third subgroup was also identified with a survival profile (~2 years) which resembled the B cell-like subgroup [48,49]. The analysis of gene clusters revealed five patterns of expression signatures: the germinal center signature, MHC class II signature, and the lymph node signature were associated with a good prognosis and genes in the other categories were associated with a poor prognosis [48].

With survival as the major endpoint a set of 13 genes was identified that clearly separated the clinical outcomes of 58 patients with DLBCL into a good outcome group (5-year overall survival of 70%) and a bad outcome group (5-year overall survival of 12%) [50]. Some of these genes (*NORI, PDE4B, PKC-β*) overlapped with the prognostic set identified by Alizadeh and colleagues [16] and showed utility in defining prognosis individually [48].

Some cancers with overlapping light-microscopic features have distinct genetic etiologies, different responses to therapeutic intervention and varying overall clinical outcomes hence accurate diagnosis is crucial for effective treatment and follow-up of patients, e.g. acute leukemias, nonHodgkin's lymphomas. Tumor subtypes in ALL are usually readily diagnosed by a combination of histopathology, histochemistry, immuno-phenotyping and cytogenetics, but there are examples of cases that are difficult to classify and some that show mixed lineage [51].

Analysis of 6500 genes with a 'training set' of 38 samples enabled the identification of a subset of genes expressed differentially between ALL and AML [52]. Using a weighted correlation method these investigators found that 50 marker genes could be used to classify an unknown leukemia sample with an accuracy of >85%. They were also able to separate ALL samples into two groups, reflecting either B or T cell differentiation. It was demonstrated subsequently that mixed-lineage leukemias (MLL) (characterized by translocations involving *ALL-1*) were distinct from classical ALL and AML [53].

A more recent analysis of gene expression in 360 pediatric leukemias demonstrated the extent to which gene expression profiles can readily distinguish leukemias with different phenotypes (e.g. B cell ALL or T cell ALL), or with different oncogenic translocation products (e.g. E2A-PBX1, BCR–ABL, TEL-AML1 or ALL-1). The application of classification algorithms to these signatures resulted in an almost 100% predictive accuracy among leukemias of different molecular subclasses. Importantly these algorithms were able to predict the presence of alterations in the *TEL* gene in the absence of a positive RT-PCR test for *TEL-AML1* rearrangements, suggesting that these new molecular tools might significantly augment current diagnostic methods [51].

However, a major limitation of the studies carried out to date is the small number of samples that have been used. Significantly greater numbers of samples are required to rigorously validate tumor class prediction, and to confirm the discovery of new, clinically relevant molecular subclasses of cancer. This problem is clearly illustrated in lymphomas by the differences in the prognostic value of genes identified in DLBCL, and in general by the fact that few studies of particular cancers have identified substantially overlapping gene sets that best distinguish subgroups of those cancers [51].

The use of meta-analysis on pooled data may help in this regard. However, despite the current limitations, tumor expression profiling is providing new insights into the biology and therapeutic responsiveness of several cancers, including leukemias and lymphomas and this can only expand in the years to come [48].

References

1. **American Cancer Society.** (2003) *Cancer Around the World, 2000, Death Rates per 100,000 Population for 45 Countries.* This can be found under Professionals, Facts and Figures, Cancer Facts & Figures 2003 at http://www.cancer.org/
2. **Mrüzek, K., Heinonen, K. and Bloomfield, C.D.** (2001) *Best Prac. & Res. Clin. Haematol.*, **14**, 19.
3. **Döhner, H., Stilgenbauer, S., Benner, A. *et al.*** (2000) *New Engl. J. Med.*, **343**, 1910.
4. **Rowley, J.D.** (1998) *Annu. Rev. Genet.*, **32**, 495.
5. **Rabbitts, T.H.** (1993) *Biochem. Soc. Trans.*, **21**, 809.
6. **Boxer, L.M.** (1994) *Annu. Rev. Med.*, **45**, 1.
7. **Crans, H.N. and Sakamoto, K.M.** (2001) *Leukemia*, **15**, 313.
8. **Tenen, D.G.** (2003) *Nature Rev. Cancer*, **3**, 89.
9. **Boyd, K.E. and Farnham, P.J.** (1999) *Proc. Soc. Exp. Biol. Med.*, **222**, 9.
10. **Look, A.T.** (1995) *Adv. Cancer Res.*, **67**, 25.
11. **Friedman, A.D.** (1999) *Leukemia*, **13**, 1932.

12. **Zhong, S., Slomoni, P. and Pandolfi, P.P.** (2000) *Nature Cell Biology*, **2**, E85.
13. **Macdonald, F. and Ford, C.H.J.** (1997) *Molecular Biology of Cancer*. BIOS Scientific Publishers, Oxford.
14. **Klein, G. and Klein, E.** (1986) *Cancer Res.*, **46**, 3211.
15. **Hecht, J.L. and Aster, J.C.** (2000) *J. Clin. Oncol.*, **18**, 3707.
16. **Alizadeh, A.A., Eisen, M.B., Davis, R.E.** *et al.* (2000) *Nature*, **403**, 503.
17. **Lim, L-C., Segal, G.H. and Wittwer, C.T.** (1995) *Am. J. Clin. Pathol.*, **104**, 689.
18. **Schuuring, E.** (1995) *Gene*, **159**, 83.
19. **Inghirami, G. and Frizzera, G.** (1994) *Am. J. Clin. Pathol.*, **101**, 681.
20. **Adams, J.M. and Cory, C.** (1998) *Science*, **281**, 1322.
21. **Cory, S. and Adams, J.M.** (2002) *Nature Rev. Cancer*, **2**, 647.
22. **Cory, S.** (1995) *Annu. Rev. Immunol.*, **13**, 513.
23. **Ruvolo, P.P., Deng, X. and May, W.S.** (2001) *Leukemia*, **15**, 515.
24. **Dyer, M.J.S.** (1995) In *Oncology. A Multidisciplinary Textbook,* Chapter 19 (Horwich, A., ed.) Chapman & Hall Medical, London, p. 585.
25. **Thijsen, S.F.T., Schuurhuis, G.J., van Oostveen, J.W. and Ossenkoppele, G.J.** (1999) *Leukemia*, **13**, 1646.
26. **Gesbert, F., Sellers, W.R., Signoretti, S.** *et al.* (2000) *J. Biol. Chem.*, **275**, 39223.
27. **Licht, J.D.** (2001) *Oncogene*, **20**, 5660.
28. **Davis, J.N., McGhee, L. and Meyers, S.** (2003) *Gene*, **303**, 1.
29. **Michael, D. and Oren, M.** (2002) *Curr. Opin. Genet. Develop.*, **12**, 53.
30. **Linggi, B., Muller-Didow, C,. Van de Locht, L.** *et al.* (2002) *Nature Med.*, **8**, 743.
31. **Canaani, E., Nowell, P.C. and Croce, C.M.** (1995) *Adv. Cancer Res.*, **66**, 213.
32. **Nakao, M., Yokota, S., Kaneko, H.** *et al.* (1996) *Leukemia*, **10**, 249.
33. **Rocco, J.W. and Sidransky, D.** (2001) *Exp. Cell Res.*, **264**, 42.
34. **Ragione, F.D., Mercurio, C. and Iolascon, A.** (1995) *Haematologica*, **80**, 557.
35. **Pines, J.** (1995) *Adv. Cancer Res.*, **66**, 181.
36. **Bos, J.L.** (1989) *Cancer Res.*, **49**, 4682.
37. **O'Gorman, D.M. and Cotter, T.G.** (2001) *Leukemia*, **15**, 21.
38. **Sweetenham, J.W.** (1994) *Exp. Hem.*, **22**, 5.
39. **Immamura, J., Miyoshi, I. and Koeffler, H.P.** (1994) *Blood*, **84**, 2412.
40. **Peng, H., Chen, G., Du, M.** *et al.* (1996) *Am. J. Pathol.*, **148**, 643.
41. **Wu, Y.L., Mehew, J.W., Heckman, C.A.** *et al.* (2001) *Oncogene*, **20**, 240.
42. **Watanabe, T., Hotta, T., Ichikawa, A.** *et al.* (1994) *Blood*, **84**, 3158.
43. **Hochhaus, A., Weisser, A., La Rosée, P.** *et al.* (2000) *Leukemia*, **14**, 998.
44. **McCarthy, K.P. and Wiedman, L.M.** (1995) In *PCR Applications in Pathology. Principles and Practice* (Latchman, D.S., ed.). Chapter 10, p. 216. Oxford University Press, Oxford.
45. **Amabile, M., Gannini, B., Testoni, N.** *et al.* (2001) *Haematologica*, **86**, 252.
46. **Gaiger, A., Lion, T., Kahls, P.** *et al.* (1993) *Leukemia*, **1**, 1766.
47. **van Rhee, F., Marks, D.I., Lin, F.** *et al.* (1995) *Leukemia*, **9**, 329.
48. **Liu, E.T.** (2003) *Curr. Opin. Genet. Develop.*, **13**, 97.
49. **Rosenwald, A., Alizadeh, A.A., Widhopf, G.** *et al.* (2001) *J. Exp. Med.*, **194**, 1639.
50. **Shipp, M.A., Ross, K.N., Tamayo, P.** *et al.* (2002) *Nature Med.*, **8**, 68.
51. **Hampton, G.M. and Frierson, H.F., Jr.** (2003) *Trends Molec. Med.*, **9**, 5.
52. **Golub, T.R., Slonim, D.K., Tamayo, P.** *et al.* (1999) *Science*, **286**, 531.
53. **Armstrong, S.A., Staunton, J.E., Silverman, L.B.** *et al.* (2002) *Nature Genet.*, **30**, 41.

Childhood solid tumors 12

12.1 Introduction

Cancer in children and young adults is relatively rare but is still the major cause of death by disease in children aged one to 14. Mortality rates have however declined by about 47% since 1975. Details of childhood cancers can be found at www.cancerindex.org/ccw/.

12.2 Retinoblastoma

Retinoblastoma is a relatively uncommon tumor in childhood which arises in the retina and accounts for 11% of cancers developing in the first year of life and about 3% of cancers in children under the age of 15 [1]. The incidence is approximately 1 in 25 000 and most cases are diagnosed under the age of 5 years and rarely presents after 7 years of age. The tumors are generally confined to the eye and as a result children have a 5-year survival rate of over 90%. However if the tumor has extended to the soft tissues surrounding the eye, to the optic nerve or into the brain and meninges, the 5-year survival is less than 10%.

As described in Chapter 3, retinoblastoma has been the paradigm for studies into tumor suppressor genes and formed the basis of Knudson's two hit hypothesis because of the presentation of the disease. Approximately 60% of cases are sporadic and 40% are hereditary although the majority of cases with germ-line mutations are due to *de novo* mutations rather than familial transmission. Seventy-five percent of cases of retinoblastoma tumors are unilateral and the remainder are bilateral, all of which have germ-line mutations [2]. Tumors in patients with germ-line mutations tend to occur at an earlier age than sporadic cases. Penetrance of familial retinoblastoma is 85–95% [3]. Patients with germ-line mutations have a high incidence of second tumors, primarily osteosarcomas, soft tissue sarcomas and melanomas [4]. Treatment of retinoblastoma depends on the extent of the disease within the eye and whether it has spread beyond the eye into the brain or other parts of the body but includes enucleation, external beam radiation, cryotherapy and brachytherapy. Systemic chemotherapy is under investigation for cases to reduce tumor volume prior to local management. Genetic counseling of families plays an important role in the management of families with both unilateral and bilateral tumors.

Once the *RB1* gene had been mapped to chromosome 13q, it became possible to identify those individuals at high risk of retinoblastomas in familial cases by the use of linked polymorphic markers [5]. However this is not an option for the majority of cases as they are either sporadic or *de novo* tumors. Once the gene had been identified, direct mutation analysis therefore became the method of choice for all cases [6]. Approximately 10% of cases have a cytogenetically visible deletion [5]. Small insertions, deletions, point mutations and promoter methylation in the *RB1* gene are found in up to 90% of cases with germ-line *RB1* [7]. The *RB1* gene has 27 exons and mutations are found spread throughout them all (see www. d-lohmann.de/Rb/mutations.html.) [8]. Most mutations create a premature termination codon. Recurrent mutations, accounting for 15–18% of cases, occur at C to T transitions found at 13 CGA codons within the gene [8]. There is some evidence of genotype–phenotype correlations. Nonsense mutations usually result in a more severe phenotype. Milder disease has been associated with germ-line promoter mutations [9], missense mutations [10] and splicing mutations [11] probably because these mutations compromise rather than completely abolish the function of RB. Molecular diagnosis enables earlier treatment and better health outcomes of RB patients and has also been shown to be more cost effective than conventional surveillance [7].

12.3 Wilms' tumor

Wilms' tumor is an embryonal renal tumor and is one of the most common solid tumors of childhood affecting approximately 1 in 10 000 children usually under the age of 5 years [12]. The disease is sporadic in the majority of cases but some cases are caused by germ-line mutations. Approximately 10% of cases are bilateral and of these 1–2% show familial transmission with the remainder of the germ-line mutations occurring as the result of a *de novo* event. Wilms' tumor has a high cure rate of over 85% achieved by a combination of surgery, chemotherapy and radiotherapy. The tumors are believed to arise because of failure of the metanephric blastema to undergo its normal developmental pathway to form nephrons and connective tissue. The tumor has a premalignant stage termed the nephrogenic rests [13].

Wilms' tumors are often found in association with five predisposing conditions each individually accounting for around 0.5–2% of cases [12,14]. The most important condition is WAGR syndrome (Wilms' tumor, aniridia, genitourinary abnormalities and mental retardation) in which approximately 30% of patients will develop Wilms' tumors. Denys Drash syndrome (DDS) is characterized by renal nephropathy and ambiguous genitalia and over 90% of cases will develop Wilms' tumor. Beckwith–Wiedemann syndrome (BWS) is an overgrowth syndrome defined by growth abnormalities such as hemihypertrophy, macroglossia and exomphalos which predisposes to several pediatric malignancies including Wilms' tumor. The risk of tumor development in this condition is approximately 8%. Perlman syndrome, which comprises nephromegaly with renal dysplasia, macrosomia, cryptorchidism and multiple facial anomalies has also been found to predispose to Wilms' tumor [15]. Finally Wilms' tumor can be inherited in an autosomally dominant manner and linkage has been shown to both 17q and 19q.

Wilms' tumor, as might be expected from the above, displays genetic heterogeneity with multiple genes involved in familial and sporadic forms as described in Section 3.7 [14]. The genes which have been identified are described below but it seems clear that others remain to be isolated [16].

Early studies of LOH in sporadic Wilms' tumor together with the identification of deletions at 11p13 in WAGR suggested the existence of a tumor suppressor gene in this region and led to isolation of the *WT1* gene in 1990 [17,18]. *WT1* has ten exons and encodes a zinc finger protein which functions as a transcription factor as described in Section 3.7. Mutations have been described in tumors, including large and small deletions as well as point mutations [19,20]. Mutations in the zinc finger domain result in loss of DNA-binding activity whereas those in the transactivation domain convert the protein into a transcriptional activator rather than a repressor. There is considerable variation in the phenotype associated with *WT1* mutations with different classes of mutations confering different degrees of genitourinary malformations. Patients with WAGR syndrome usually have a deletion of one allele of *WT1* [21]. DDS is associated primarily with missense mutations in the zinc finger-encoding region of the gene which affect amino acids involved in DNA binding and zinc finger structure. One mutation, a C to T transition at codon 394 of exon 9 has been described in approximately half of the cases [22]. Other mutations are frequently found in exon 9 suggesting that this is a hotspot for mutations in DDS. There is marked phenotypic variation seen between DDS patients with the same common mutation. The severity of the DDS phenotype

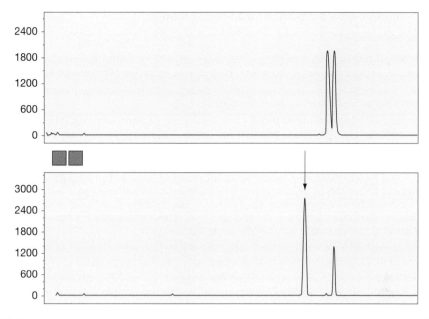

Figure 12.1

Mosaic uniparental disomy identified with the microsatellite marker D11S1984 in a sporadic case of BWS. The top panel shows the two alleles at this locus in the mother and the lower panel shows the alleles in the child with BWS. No sample was available from the father but the arrowed allele must be paternal in origin. There is a clear dosage effect seen for the paternal allele over the maternal allele indicating mosaic paternal disomy. Courtesy of Anita Luharia, West Midlands Regional Genetics Laboratory.

is because mutant WT1 forms dimers with wild-type protein and therefore acts in a dominant negative fashion, resulting in loss or reduction of DNA binding [23].

BWS maps to chromosome 11p15, a region which contains a cluster of imprinted genes including the growth factor, *IGF2*, which is paternally expressed and the maternally expressed growth-inhibitory genes *H19* and *p57^{KIP2}* (*CDKN1C*) [24]. In Wilms' tumors showing loss of heterozygosity for 11p15 markers, over 90% show preferential loss of the maternal allele. In Wilms' tumors without LOH of 11p15, there is biallelic expression of *IGF2* and loss of expression of *H19* indicating the loss of imprinting [25]. In both scenarios there is therefore overexpression of the growth-promoting gene, *IGF2*, and loss of expression of growth-inhibitory genes *H19* and *p57^{KIP2}*. In BWS, 2–3% of cases have a chromosome abnormality which is either a paternal duplication of 11p15 or a maternally derived inversion or balanced translocation [24]. About 20% of sporadic cases of BWS show paternal mosaic disomy (*Figure 12.1*) and up to 50% of cases show loss of methylation of the differentially methylated region (DMR) of the *KCNQ10T1* gene known as *KvDMR1*. There are recognized genotype–phenotype correlations in BWS. In particular embryonal tumors such as Wilms' tumor are associated with UPD or *H19* methylation but not with aberrant methylation of *KvDMR1* [26,27].

In rare cases where Wilms' tumor is inherited in an autosomal dominant fashion, two other genes have been mapped by linkage, *FWT1* at chromosome 17q21 [28] and *FWT2* at 19q13.3–13.4 [29]. Finally LOH has been identified in Wilms' tumors on chromosomes 16q, 1p and 11q implying the presence of further genes associated with tumor development. Wilms' tumor development is therefore complex and looks to involve several different pathways.

Table 12.1   International Neuroblastoma Staging System

Stage	Disease features
1	Localized tumor with complete gross excision with or without microscopic residual disease; representative ipsilateral lymph nodes negative for tumor microscopically
2A	Localized tumor with incomplete gross excision; representative ipsilateral lymph nodes negative for tumor microscopically
2B	Localized tumor with or without complete gross excision with ipsilateral nonadherent lymph nodes positive for tumor. Enlarged lymph nodes must be negative microscopically
3	Unresectable unilateral tumor infiltrating across the midline, with or without regional lymph node involvement; or localized unilateral tumor with contralateral regional lymph node involvement; or midline tumor with bilateral extension by infiltration or by lymph node involvement. The midline is defined as the vertebral column. Tumors originating on one side and crossing the midline must infiltrate to or beyond the opposite side of the vertebral column
4	Any primary tumor with dissemination to distant lymph nodes, bone, bone marrow, liver, skin, and/or other organs (except as defined for stage 4S)
4S	Localized primary tumor (as defined for stages 1, 2A or 2B) with dissemination limited to skin, liver and/or bone marrow (limited to infants <1 year of age). Marrow involvement should be minimal (<10% of total nucleated cells identified as malignant by bone biopsy or by bone marrow aspirate). More extensive bone marrow involvement would be considered to be stage 4 disease. The results of the metaiodobenzylguanidine (MIBG) scan (if done) should be negative for disease in the bone marrow

12.4   Neuroblastoma

Neuroblastoma is the most common solid tumor in early childhood [30]. The mean age at diagnosis in sporadic cases is 30 months and drops to 9 months for the much rarer familial cases. Neuroblastomas orginate in the adrenal medulla or the paraspinal sites where sympathetic nervous system tissue is present. Approximately 70% of patients will have metastatic disease at diagnosis [31]. Prognosis is related to the age at diagnosis, clinical stage of disease and regional lymph node involvement. Staging of tumors is now carried out according to the International Neuroblastoma Staging System as described in *Table 12.1*. Children with localized neuroblastoma or those under one year of age at diagnosis, even with metastatic disease, have a high likelihood of long-term survival with minimal therapy. Older children with advanced-stage disease have a significantly decreased chance of cure and diagnosis in the adolescent or adult results in a worse prognosis regardless of stage or site. Advances in the biology and genetics of neuroblastomas have allowed classification into low-, intermediate- and high-risk groups which means that the most appropriate intensity of therapy can be selected. The most important biological variables in neuroblastomas are Shimada histology [32], aneuploidy of the tumor and amplification of the *MYCN* oncogene [30,33].

Neuroblastomas show complex acquired genetic abnormalities including deletions of chromosomes 1p and 11q, amplification of *MYCN* and gains of 17q (*Figure 12.2*) [34]. Amplification of *MYCN* is one of the best-established prognostic indicators and is increasingly being used for treatment decisions. *MYCN* amplification is most frequently found in children aged over 1 year with advanced stage of disease and almost always correlates with rapid tumor progression and poor outcome irrespective of tumor stage [33,35,36]. Patients with stage 4S neuroblastomas are those who frequently show spontaneous remission and this is strongly associated with lack of *MYCN* amplification [36]. In localized

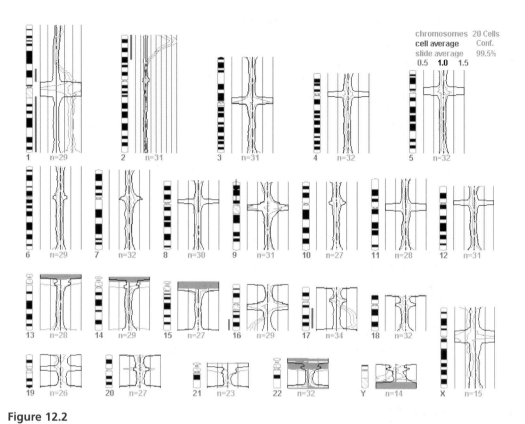

Figure 12.2

Comparative genomic hybridization of a neuroblastoma showing *MYCN* amplification, and gain of 17q. Courtesy of Sara Dyer, West Midlands Regional Genetics Laboratory.

neuroblastomas, lack of *MYCN* amplification is the major factor identifying a group of patients who do not require aggressive therapy [37]. Measurement of *MYCN* amplification is therefore an essential component of the routine diagnostic evaluation of new neuroblastoma patients. Amplification can be detected by fluorescence *in situ* hybridization (FISH) (*Figure 12.3*) [38] and PCR [39] and more recently by real time PCR [40] allowing accurate quantification.

Gain of chromosome 17q is the most frequent chromosomal abnormality in neuroblastomas and correlates closely with *MYCN* amplification. It has also been shown to be an important prognostic factor and is strongly associated with an unfavorable prognosis [41]. Detection of this abnormality is therefore important at the time of diagnosis. The position of the 17q breakpoints has been shown to identify subgroups with different outcomes, with those with breakpoints proximal to the *ERBB2* oncogene having a significantly better survival than those with distal breakpoints [42]. Dosage alteration of genes within the region are the likely cause of this effect. The *NM23* tumor suppressor gene has been shown to be at the edge of the common region of 17q again suggesting that increased expression of this gene plays an important role in the poor prognosis [43].

11q deletions are found in 15–20% of cases of neuroblastomas and have been primarily associated with a subgroup of patients with stage 4 neuroblastomas without *MYCN* amplification but with 3p loss. This subgroup also shows a poor prognosis similar to those with

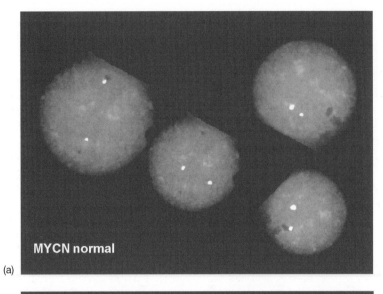

(a)

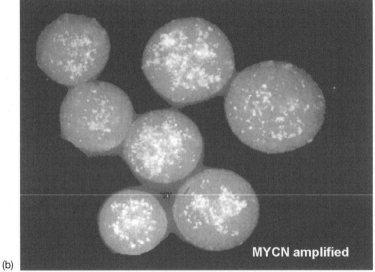

(b)

Figure 12.3

Fluorescence *in situ* hybridization (FISH) of the *MYCN* oncogene showing (a) a normal cell and (b) a cell with *MYCN* amplification. Courtesy of Sara Dyer, West Midlands Regional Genetics Laboratory.

MYCN amplification [44]. LOH has mapped the smallest region of deletion to 11q14-23, which is therefore the most likely location of an 11q-associated tumor suppressor gene [45].

Deletion of 1p is a common finding in neuroblastomas and LOH of 1p is found in 25–35% of tumors [46]. In general, deletion of 1p is large and is associated with *MYCN* amplification and with a decreased event-free survival though not with overall survival [46]. This region is believed to contain a tumor suppressor gene and the smallest region of overlap has been mapped to 1p36.2-3 [46]. Although no confirmed gene has been isolated, the *UBE4B/UFD2* gene stands out as a possible candidate [47].

Finally, expression of the cyclin-dependent kinase inhibitor p27 is also a prognostic indicator, independent of *MYCN* amplification [48].

12.5   Rhabdomyosarcoma

Rhabdomyosarcomas are a heterogeneous group of malignant tumors and are the most common cause of soft tissue sarcomas in childhood, accounting for 7–8% of all childhood cancers and for two-thirds of sarcomas diagnosed under the age of 14 years [49]. The incidence is highest in children under 4 years. They are divided into two histological groups, the alveolar and embryonal subtypes which are present in different age groups and at different sites in the body [50]. The more common type is the embryonal group which includes the spindle cell and botryoid variants according to the newest classification of these tumors, The International Classification. Embryonal rhabdomyosarcomas account for approximately 60% of cases, usually occurring in children under 4 years of age and are found located in the head and neck, retroperitoneum and genitourinary tract. The rarer alveolar form is characterized by small round cells with dark hyperchromatic nuclei held together by strands of intracellular collagen. Alveolar rhabdomyosarcomas are more often found in adolescence and are likely to occur on the limbs. They are associated with higher-stage disease and a poor prognosis. Rhabdomyosarcomas are highly malignant tumors arising in the mesenchymal cells which develop into striated muscle and are characterized immunohistochemically by the presence of structural proteins of muscle such as alpha actin, fast myosin and myosin heavy chain. These markers are useful for the differential diagnosis of rhabdomyosarcomas from other tumors such as Ewing's sarcomas and neuroblastomas.

Most rhabdomyosarcomas are sporadic but some cases are associated with specific familial cancer predisposition [51]. An increase in CNS abnormalities and GU abnormalities similar to those seen in Wilms' tumor has been described and a number of cases of BWS have also been associated with rhabdomyosarcomas. Up to 6% of neurofibromatosis type 1 cases also have rhabdomyosarcomas and several cases have been found in families with Gorlin syndrome. Finally, Li and Fraumeni identified a high incidence of breast cancer, sarcomas and acute leukemias in family members of children with rhabdomyosarcomas leading to the identification of the Li–Fraumeni syndrome (see Section 5.6).

Alveolar rhabdomyosarcomas are associated with specific chromosomal abnormalities. The most consistent are the t(2;13)(q35;q14) translocation seen in 55% of cases and the t(1;13)(p36;q14) translocation occurring in a further 22% [52]. Physical mapping and cloning studies revealed that the two loci disrupted by these translocations were two members of the *PAX* gene family, *PAX3* on chromosome 2 and *PAX7* on chromosome 1 (see Section 2.2.2). These genes encode highly related members of the paired box transcription factor family and contain an N-terminal DNA-binding domain with a paired box and a homeobox motif and a C terminal transcriptional activation domain (*Figure 12.4*). These two genes are thought to be important in muscle development during embryogenesis when condensation of the mesoderm, known as somites, are formed from which skeletal muscle develops [53]. The gene on chromosome 13 involved in both translocations is the *FKHR* gene which encodes a novel member of the forkhead transcription factor family, the product of which has an N terminal forkhead DNA-binding domain and a C terminal transcriptional activation domain. The translocations disrupt the *PAX* genes in intron 7 and *FKHR* gene in intron 1 resulting in the fusion of the DNA-binding domains of the *PAX* genes with the transactivation domain of the *FKHR* gene. The fusion genes produced by the translocations, *PAX3-FKHR* and *PAX7-FKHR* are expressed as chimeric transcripts which encode chimeric proteins and activate transcription of PAX-binding sites and are 10–100-fold more potent than wild-type protein [51]. The presence of these fusion genes is a useful marker to aid diagnosis of alveolar rhabdomyosarcoma as an adjunct to

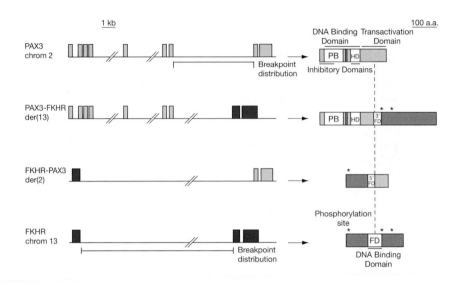

Figure 12.4

Chimeric genes produced by the t(2;13) translocation in alveolar rhabdomyosarcoma.
Adapted from [51].

histological diagnosis and can be detected by both FISH and RT-PCR [53]. Data suggest
that the *PAX3-FKHR* fusion gene is associated with a poorer prognosis than *PAX7-FKHR*
[52,54]. Amplification of other chromosomal regions such as 12q13-15, 1q21 and 2p24
has also been identified in alveolar rhabdomyosarcomas [51,54]. 2p24 contains the
MYCN gene, and this has been shown to be amplifed in alveolar but not in embryonal subtypes. There is some evidence to suggest that amplification correlates with poorer survival
rate. Embryonal rhabdomyosarcomas do not show a consistent chromosomal translocation but do show LOH at 11p15 as well as loss of imprinting of genes within this region as
described for Wilms' tumor and BWS [54,55], an observation which links the rhabdomyosarcomas with these other embryonal neoplasms.

12.6   Ewing's sarcoma

Ewing's sarcoma is the second most common pediatric bone tumor after osteosarcoma
and accounts for 10–15% of all primary bone tumors usually occurring between 10 and
20 years of age [56,57]. It is a small, round cell tumor of neuroectodermal origin but with
limited neural differentiation. The tumor can occur in any bone but is most commonly
found in bones of the lower extremities, then the pelvis, upper extremities, axial skeleton
and ribs, and finally the face. The disease does not appear to be associated with any familial cancer syndromes. Tumors of similar histology, peripheral primitive neuroectodermal
tumors (pPNET), and Askin tumors, also arise in soft tissues and are the second most
common childhood soft tissue malignancies. These tumors are found in the chest wall,
paraspinal tissues, abdominal wall, head and neck and extremities [57]. Ewing's and
pPNET have been shown to represent a common entity following the identification of a
nonrandom chromosome rearrangement in both tumor types and are now recognized as
the Ewing's sarcoma family of tumors (ESFT).

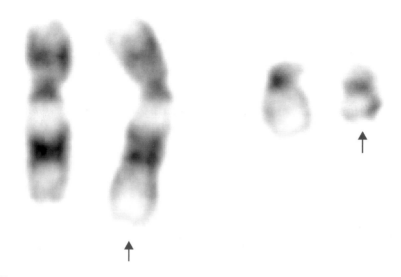

Figure 12.5

A partial karyotype of chromsomes 11 and 22 showing the derived chromosomes in a case of Ewings sarcoma. Courtesy of Dom McMullan, West Midlands Regional Genetics Laboratory.

The most common rearrangement in these tumors is the t(11;22)(q24;q12) translocation found in 85% of cases [58,59] (*Figure 12.5*). The gene on chromosome 22 is the *EWS* gene, a gene with a nuclear localization domain at the N terminus and an RNA-binding, glutamine-rich domain at the C terminus. The exact function of this gene however remains unknown. The gene on chromosome 11 is the *FLI1* gene, a member of the ETS transcription family. The position of the breakpoints in both genes is quite variable and can be between introns 7 to 10 of EWS and from introns 3 to 9 of *FLI1*. The result of the translocation is the production of a fusion gene, *EWS-FLI1*, the product of which has increased transcriptional activity compared to wild-type FLI1 [56,57]. This chimeric gene has been shown to act as a dominant oncogene, capable of transforming the NIH3T3 cell line which neither *EWS* or *FLI1* alone can do. The targets of the *EWS-FLI1* gene are just beginning to be identified. A recent study has shown that the regulators of the G_1 phase of the cell cycle are possible targets. The EWS-FLI1 fusion protein has been shown to interact with the promotor of the cyclin-dependent kinase inhibitor, p21, to negatively regulate its activity [60] and this finding may define a target for therapeutic intervention in the future. In cases of Ewing's without the t(11;22), other rearrangements have been found, all involving the *EWS* gene on chromosome 22 but partnering a number of other ETS family members on other chromosomes. The most common of these is the t(21;22)(q22;q12) which results in the fusion of *EWS* with the *ERG* gene on chromosome 21.

Identification of *EWS-ETS* rearrangements is important for the differential diagnosis of Ewings sarcoma so that appropriate clinical management can be undertaken. The translocations can be identified cytogenetically in many cases (*Figure 12.5*). However accuracy can be increased by the use of RT-PCR or FISH. PCR-based approaches also enable the exact breakpoints within *EWS* and *FLI1* to be determined. This has important prognostic significance as different translocations have different outcomes. Those cases, termed type 1, in which exons 1–7 of *EWS* are joined to exons 6–9 of *FLI1*, are found in 60% of cases of Ewings and have been suggested to indicate those patients with a favorable outcome, at least in patients with localized disease [57]. Patients with other sized transcripts appear to have a poor prognosis. Further studies to confirm these preliminary findings are underway.

References

1. http://seer.cancer.gov/publications/childhood/retinoblastoma.pdf.
2. **Knudson, A.G.** (1971) *Proc. Natl Acad. Sci.*, **68**, 820.
3. **Naumova, A. and Dapienza, C.** (1994) *Am. J. Hum. Genet.*, **54**, 264.
4. **Gallie, B.L., Dunn, J.M., Chan, H.S. et al.** (1991) *Ped. Clin. North Am.*, **38**, 299.
5. **Wiggs, J., Nordenskjold, M., Yandell, D. et al.** (1988) *N. Engl. J. Med.*, **318**, 151.
6. **Dunn, J.M., Phillips, R.A., Becker, A.J. and Gallie, B.L.** (1988) *Science*, **241**, 1797.
7. **Richter, S., Vandezande, K., Chen, N. et al.** (2003) *Am. J. Hum. Genet.*, **72**, 253.
8. **Lohmann, D.R.** (1999) *Hum. Mut.*, **14**, 283.
9. **Cowell, J.K., Bia, B. and Akoulitchev, A.** (1996) *Oncogene*, **12**, 431.
10. **Cowell, J.K. and Bia, B.** (1998) *Oncogene*, **16**, 3211.
11. **Schubert, E.L., Strong, L.C. and Hansen, M.F.** (1997) *Hum. Genet.*, **100**, 557.
12. **Coppes, M.J. and Egeler, R.M.** (1999) *Sem. Urol. Oncol.*, **17**, 2.
13. **Beckwith, J.B., Kiviat, N.B. and Bonadio, J.F.** (1990) *Ped. Pathol.*, **10**, 1.
14. **Brown, J.W. and Malik, K.T.A.** (2001) *Exp. Rev. Mol. Med.* www.ermm.cbcu.cam.ac.uk/01003026h.htm.
15. **Henneveld, H.T., van Lingen, R.A., Hamel, B.C. et al.** (1999) *Am. J. Med. Genet.*, **86**, 439.
16. **Rapley, E.A., Barfoot, R., Bonaiti-Pellie, C. et al.** (2000) *Br. J. Cancer*, **83**, 177.
17. **Gessler, M., Poustka, A., Cavenee, W.K. et al.** (1990) *Nature*, **343**, 774.
18. **Call, K.M., Glaser, T., Ito, C.Y. et al.** (1990) *Cell*, **60**, 509.
19. **Huff, V. and Saunders, G.F.** (1993) *Biochim. Biophys. Acta*, **1155**, 295.
20. **Little, M. and Wells, C.** (1997) *Hum. Mut.*, **9**, 209.
21. **Baird, P.N., Groves, N., Haber, D.A. et al.** (1992) *Oncogene*, **7**, 2141.
22. **Jeanpierre, C., Denamur, E., Henry, I. et al.** (1998) *Am. J. Hum. Genet.*, **62**, 824.
23. **Hastie, N.D.** (1992) *Hum. Mol. Genet.*, **1**, 293.
24. **Maher, E.R. and Reik, W.** (2000) *J. Clin. Invest.*, **105**, 247.
25. **Feinberg, A.P.** (1999) *Cancer Res.*, **59**, 1743s.
26. **Engel, J.R., Smallwood, A., Harper, A. et al.** (2000) *J. Med. Genet.*, **37**, 921.
27. **DeBaun, M.R., Niemitz, E.L., McNeil, D.E. et al.** (2002) *Am. J. Hum. Genet.*, **70**, 604.
28. **Rahman, N., Abidi, F., Ford, D. et al.** (1998) *Hum. Genet.*, **103**, 547.
29. **McDonald, J.M., Douglass, E.C., Fisher, R. et al.** (1998) *Cancer Res.*, **58**, 1387.
30. **Brodeur, G.M.** (2003) *Nat. Rev. Cancer*, **3**, 203.
31. **Cotterill, S.J., Pearson, A.D., Pritchard, J. et al.** (2000) *Eur. J. Cancer*, **36**, 901.
32. **Shimada, H., Ambros, I.M., Dehner, L.P. et al.** (1999) *Cancer*, **86**, 364.
33. **George, R.E., Variend, S., Cullinane, C. et al.** (2001) *Med. Ped. Oncol.*, **36**, 169.
34. **Bown, N.** (2001) *J. Clin. Pathol.*, **54**, 897.
35. **Schmidt, M.L., Lukens, J.N., Seeger, R.C. et al.** (2000) *J. Clin Oncol.*, **18**, 1260.
36. **Ikeda N., Iehara T., Tsuchida Y. et al.** (2002) *Br. J. Cancer*, **86**, 1110.
37. **Rubie, H., Hartmann, O., Michon, J. et al.** (1997) *J. Clin. Oncol.*, **15**, 1171.
38. **Mathew, P., Valentine, M.B., Bowman, L.C. et al.** (2001) *Neoplasia*, **3**, 105.
39. **De Cremoux, P., Thioux, M., Peter, M. et al.** (1997) *Int. J. Cancer*, **72**, 518.
40. **Raggi, C.C., Bagnoni, M.L., Tonini, G.P. et al.** (1999) *Clin. Chem.*, **45**, 1918.
41. **Bown, N., Lastowska, M., Cotterill, S. et al.** (2001) *Med. Ped. Oncol.*, **36**, 14.
42. **Lastowska, M., Cotterill, S., Bown, N. et al.** (2002) *Genes, Chrom. Cancer*, **34**, 428.
43. **Godfried, M.B., Veenstra, M., v Sluis, P. et al.** (2002) *Oncogene*, **21**, 2097.
44. **Plantaz, D., Vandesompele, J., van Roy, N. et al.** (2001) *Int. J. Cancer*, **91**, 680.
45. **Maris, J.M., Guo, C., White, P.S. et al.** (2001) *Med. Ped. Oncol.*, **36**, 124.
46. **Maris, J.M., Weiss, M.J., Guo, C. et al.** (2000) *J. Clin. Oncol.*, **18**, 1888.
47. **Krona, C., Ejeskar, K., Abel, F. et al.** (2003) *Oncogene*, **22**, 2343.
48. **Bergmann, E., Wanzel, M., Weber, A. et al.** (2001) *Int. J. Cancer*, **95**, 176.
49. **McDowell, H.P.** (2002) *Arch. Dis. Child.* **88**, 354.
50. **Newton, W.A., Gehan, E.A., Webber, B.L. et al.** (1995) *Cancer*, **76**, 1073.
51. **Pressey, J.G. and Barr, F.G.** (2001) In *Molecular Genetics of Cancer* (Cowell, J.K., ed.). BIOS Scientific Publishers, Oxford.
52. **Sorensen, P.H.B., Lynch, J.C., Qualman, S.J. et al.** (2002) *J. Clin. Oncol.*, **20**, 2672.

53. **Anderson, J., Gordon, A., Pritchard-Jones, K. and Shipley, J.** (1999) *Genes, Chrom. Cancer,* **26**, 275.
54. **Kelly, K.M., Womer, R.B., Sprensen, P.H.** *et al.* (1997) *J. Clin. Oncol.,* **15**, 1831.
55. **Gordon, T., McManus, A., Andersen, J.** *et al.* (2001) *Med. Ped. Oncol.,* **36**, 259.
56. **Sandberg, A.A. and Bridge, J.A.** (2000) *Cancer Genet. Cytogenet.,* **123**, 1.
57. **Burchill, S.A.** (2003) *J. Clin. Pathol.,* **56**, 96.
58. **Aurias, A., Rimbaut, C., Buffe, D.** *et al.* (1984) *Cancer Genet. Cytogenet.,* **12**, 21.
59. **Turc-Carel, C., Philip, I., Berger, M.P.** *et al.* (1984) *Cancer Genet. Cytogenet.,* **12**, 1.
60. **Nakatani, F., Tanaka, K., Sakimura, R.** *et al.* (2003) *J. Biol. Chem.,* **278**, 15105.

Therapeutic applications

13

13.1 Introduction

The medical need for better cancer treatment is clear. In the developed world, roughly one in three people contracts cancer and around one in four of those die from the disease. The worldwide incidence of cancer is set to double from 10 to 20 million over the next two decades and the death rate will increase from 6 to 10 million. Advances in treatment with surgery, radiotherapy and chemotherapy have had a limited impact on mortality. Cures can be achieved in childhood cancers, testicular cancer and lymphoma, and improvements in survival rates have been made as a result of adjuvant drug treatment of breast and colorectal cancer. However, the majority of human cancers are difficult to treat, especially in their advanced, metastatic forms. There is thus a pressing need for new and effective forms of systemic therapy and the discovery of novel, mechanism-based agents directed against the molecular pathology of cancer offers huge potential [1].

In earlier chapters we presented compelling evidence for the importance of oncogenes and tumor suppressor genes in the etiology of many human tumors. These genes, their RNA transcripts, and their protein products are therefore potential targets for attack by specific therapeutic agents. In this chapter we will discuss the various ways in which this is being realized.

13.2 Receptors as targets

13.2.1 Receptor tyrosine kinases – ERBB

Receptor tyrosine kinases (RTKs) are important regulators of intercellular communication controlling cell growth, proliferation, differentiation, survival and metabolism. Deregulation of protein tyrosine kinase (PTK) activity usually results in RTKs with constitutive or greatly enhanced signaling capacity leading to malignant transformation [2]. PTKs are prime targets for the development of anticancer agents because in several cancers their activity is up-regulated by gain-of-function mutations or overexpression. PTK activity can be up-regulated by several mechanisms: genomic rearrangements, e.g. *BCR-ABL* in CML; point mutations, e.g. Flt-3 in AML and c-kit (the receptor for stem cell factor) in gastrointestinal stromal tumors; overexpression, e.g. epidermal growth factor receptor (EGFR) in various cancers; and ectopic or inappropriate expression of growth factors, such as vascular endothelial growth factor (VEGF) and its receptors on endothelial cells which are involved in angiogenesis [3]. ERBB receptors belong to the epidermal growth factor (EGF) family of structurally related RTKs and four ERBB members have been identified so far: ERB1 (EGFR, HER1); ERBB2 (HER2, Neu); ERBB3 (HER3); and ERBB4 (HER4) [4].

A variety of approaches towards the prevention or inhibition of RTK signaling has been developed, including selective targeting of the extracellular ligand-binding domain, the intracellular tyrosine kinase or the substrate-binding region. Pharmacological agents such as monoclonal antibodies (Mabs), antibody conjugates (see Section 13.3.2), antisense oligonucleotides (see Section 13.5) and small chemical compounds have been developed for these purposes, for example Imatinib (Gleevec/Glivec) which is being used for the treatment of CML and gastrointestinal tumors [2–4] (see also Section 13.2.2).

Small molecule tyrosine kinase inhibitors that have been developed for the treatment of ERBB1- and ERBB2-expressing tumors include ZD1839 and OSI-774 (ERBB1) and tryphostins, 4,5-dianilinophthalamide and emodin (ERBB2). These agents have shown considerable promise in *in vitro* and preclinical animal models. Both ZD1839 and OSI-774 have shown activity in phase I and II clinical trials and further clinical trials in a variety of tumor types are being currently undertaken. Second-generation inhibitors are under development by a number of pharmaceutical companies [4].

Additional strategies for the inhibition of RTKs include the use of immunotoxins. One promising immunotoxin is the EGF fusion protein DAB389EGF, which contains the enzymatically active and membrane translocation domains of diphtheria toxin and sequences for human EGF. A variety of EGFR-expressing tumors, such as breast cancer and nonsmall-cell lung cancer, have been shown to be sensitive to DAB389EGF in preclinical studies and this recombinant toxin is now under evaluation in phase II clinical trials [2].

13.2.2   BCR-ABL

The tyrosine kinase activity of the BCR-ABL oncoprotein results in reduced apoptosis and thus prolongs survival of chronic myelogenous leukemia (CML) cells. The tyrosine kinase inhibitor Imatinib (STI571, Glivec) was reported to selectively suppress the proliferation of *BCR-ABL*-positive cells and is an example of a rationally designed, molecular-targeted drug for the treatment of a specific cancer (CML) [5,6]. Three large multinational studies in patients with late chronic-phase CML, in whom previous interferon treatment had failed, have shown that achievement of a hematological and cytogenetic response increased the earlier the treatment was started with Imatinib in the course of the disease and that these responses were associated with improved survival and progression-free survival [6]. Preclinical studies have shown that the combination of Imatinib with various anticancer agents might have synergistic effects and several phase I/II studies are evaluating the feasibility of combining Imatinib with interferon, polyethylene glycol (PEG)ylated interferon, cytarbine and other single-agent or combination chemotherapy regimens, in patients with either chronic-phase or advanced CML [6]. Assuming that Imatinib could be included in pretransplantation conditioning therapies, it has also been tested whether combinations of Imatinib and γ-irradiation or alkylating agents such as busulfan or treosulfan would display synergistic activity in *BCR-ABL*-positive CML cell lines. The combination of Imatinib + γ-irradiation proved to be significantly synergistic over a broad range of cell growth inhibition levels in *BCR-ABL*-positive cell lines and produced the strongest reduction in primary CML colony-forming progenitor cells. Combinations of Imatinib + busulfan and Imatinib + treosulfan showed only additive to antagonistic effects. Imatinib did not potentiate the effects of irradiation or cytotoxic agents in *BCR-ABL*-negative cells. Such data provide the basis to further develop Imatinib-containing conditioning therapies for stem cell transplantation in CML [5].

13.2.3   Estrogen receptor (ER)

Estrogen (estradiol) is a steroid hormone that affects growth, differentiation and function of the female reproductive organs, including the breast, uterus and ovaries and also plays several other important physiological roles, e.g. in maintaining bone density and protecting against osteoporosis. Importantly, estrogen also promotes cancer cell growth in the breast and the uterus. All of these effects are mediated by estrogen binding to ERs and the ER regulates gene transcription both directly, by binding to an estrogen-responsive element (ERE) in gene promoters, and indirectly, by binding through other transcription factors [7].

Estrogen has been a major target in the treatment of breast cancer since the end of the 19th century and tamoxifen was the first selective estrogen receptor modulator (SERM) to be developed. It has estrogen-like actions in maintaining bone density and in lowering circulating cholesterol, but antiestrogenic actions in the breast. It has proved to be valuable in the treatment of ER-positive breast cancer. The finding that tamoxifen could inhibit the growth of breast cancer, but at the same time stimulate the growth of endometrial cancer in the nude mouse model, indicated that its mode of action is specific to a target tissue. The overall conclusion from clinical trial data is that there is a 2–3-fold increase in the risk of endometrial cancer in tamoxifen-treated postmenopausal patients. Another SERM, raloxifene, binds to ERs to competitively block estrogen-induced DNA transcription in both the breast and the endometrium. Originally developed for the treatment of breast cancer its poor bioavailability and its short biological half-life mean it is not as effective an antitumor agent as tamoxifen [8].

The role of tamoxifen in chemoprevention (i.e. breast cancer prevention) in high-risk pre- and postmenopausal women is more controversial with conflicting results being reported from four studies that have addressed this question [8]. Clinical studies have shown that the use of raloxifene in postmenopausal women with osteoporosis decreased the risk of vertebral fractures, increased bone mineral density in the spine and reduced the risk of invasive breast cancer by 72% and the risk of ER-positive breast cancer by 84% [9]. A phase III, double-blind trial of tamoxifen and raloxifene in which postmenopausal women are randomized to tamoxifen or raloxifene orally for 5 years hopes to recruit 22 000 volunteers and will compare the relative merits of raloxifene and tamoxifen for the prevention of invasive breast cancer, as well as effects on the cardiovascular system and bones [8].

The molecular determinants for the tissue specificity of SERMs are currently under investigation and it is known that tissue-specific co-regulator expression levels determine tamoxifen's different effects on breast and endometrial tissue. This improved understanding of the mechanism of action of SERMs will hopefully lead to better SERMs without carcinogenic side effects [10].

13.2.4   Retinoic acid receptor (RAR) and retinoid X receptor (RXR)

Retinoids are natural derivatives of vitamin A or retinol. The retinoid signal is mediated through RARs and RXRs on target cells, each of which comprise three isotypes – α, β, γ – as well as several isoforms. RARs and RXRs are transcription factors that act predominantly as RAR-RXR heterodimers, positively or negatively modulating gene transcription. Natural and synthetic retinoids are effective inhibitors of tumor cell growth *in vitro* and *in vivo* but the natural derivatives have limited therapeutic use due to their toxicity. Synthetic compounds selective for the different retinoid receptor isotypes are currently undergoing clinical evaluation. In addition, the combination of retinoids with other chemotherapeutic agents may also be of value in cancer therapy [11] (see also Section 11.8 on APL).

13.2.5   Peroxisome proliferator-activated receptor γ (PPARγ)

PPARγ is a nuclear receptor and transcription factor that regulates the expression of many genes relevant to carcinogenesis. Deficient expression of PPARγ can be a significant risk factor for the development of cancer but, paradoxically, in some cases overexpression can enhance carcinogenesis. In experimental models ligands for PPARγ have been shown to suppress breast carcinogenesis and to induce differentiation of human liposarcoma cells. By analogy to the SERM concept, it has been suggested that PPARγ modulators (SPARMs), designed to have desired effects on specific genes and target tissue without undesirable effects on others, will be clinically important in the future for chemoprevention and chemotherapy of cancer [12].

13.3 Immunologic intervention

13.3.1 Vaccines

A detailed treatment of the extensive evaluation and use of cancer vaccines in experimental animals and in humans that has occurred over the last decade is beyond the space limitations in this section. A good starting point for those wishing to obtain details are the recent reviews of this topic and the references therein [13–19].

Tumor immunology has been advancing on two major fronts: first, the elucidation of the basic mechanisms of tumor recognition and rejection, and protection from recurrence; and, second, the application of that knowledge to cancer as a clinical problem. The basic science front has advanced rapidly and provided tools with which to explore two different approaches to disease – immunotherapy and cancer vaccines. Immunotherapy relies on administration of preformed effector mechanisms and is more suited for application to the treatment of established tumors [16].

Approaches that specifically activate the immune system to control cancer growth in vivo have been a longstanding goal in cancer immunology and medical oncology. The objective for cancer vaccines is prevention. Based on the successes that vaccines have had in preventing infectious diseases, efforts are being made to define tumor targets and immunization protocols that would be effective in priming the immune system to eliminate cancer before it manifests itself clinically. Although the basic studies can be well designed and preclinical animal models continue to be developed and used to test the newly discovered scientific principles, there has been hesitation in translating these findings to cancer prevention in humans. With the exception of efforts to vaccinate against microorganisms known to cause human cancers, such as HPV, and in that way prevent not only infection but also (indirectly) cancer (see Section 10.2), hundreds of clinical trials of cancer vaccines are currently being carried out in patients who already have cancer. These studies encompass 'therapeutic vaccines' which are driven more by the practicality of the approach than by strong supporting evidence from basic and preclinical research [16].

A large number of different approaches to cancer vaccination have been evaluated during the last decade. Initial studies focused on first-generation vaccines based on whole cell preparations or tumor lysates derived from autologous or allogeneic tumors. The aims of these trials were the evaluation of toxicity, immunological effects and clinical improvement in patients with advanced disease. These studies established the feasibility of immunizing cancer patients against their own tumors. Significant clinical benefits have been reported, both in terms of long-term survival and recurrence rate, in some of these trials.

More recently, however, cancer vaccines targeting well-characterized tumor-associated antigens (TAAs), i.e. molecules selectively or preferentially expressed by cancer cells but not by key normal cells, have been designed and used in clinical trials, mostly in patients with advanced disease. The use of overexpressed or mutated proteins, or mutated oncogenic growth factor receptors, as tumor-associated antigens yields rational targets for specific immunoprevention. Results to date with these second-generation vaccines suggest that they are safe and that they can elicit humoral and cellular responses against tumor antigens, without inducing unacceptable clinical signs of autoimmunity [13]. Analyses of clinical immunization trials have demonstrated the advantage of vaccination strategies using defined antigens over whole tumor cells. However, a correlation between T cell responses and tumor regression has not yet been established and some patients have been found to exhibit tumor regression with an unexpectedly low frequency of antivaccine T cells [20].

Much effort has been expended in molecularly characterizing many of these TAAs and this research has yielded a large number of TAA peptide epitopes. The advantages of using peptide-based vaccines include: (1) easy and relatively inexpensive production of synthetic peptides; (2) the simplicity of peptide administration in a clinical setting; (3) the possibility of treating only those patients whose tumors express the relevant epitopes, thus avoiding the useless immunization of patients whose tumors are TAA-negative; and (4) the availability of *in vitro* or *ex vivo* assays that can assess patients' immune response to vaccine epitopes. When bound to major histocompatibility complex molecules, TAA peptides are recognized by T cells. Clinical studies have therefore been initiated to assess the therapeutic potential of active immunization or vaccination with TAA peptides in patients with metastatic cancer. So far, only a limited number of TAA peptides, mostly those recognized by CD8+ T cells in melanoma patients, have been clinically tested. In some clinical trials, partial or complete tumor regression was observed in approximately 10–30% of patients. No serious side effects have been reported. The clinical responses, however, were often not associated with a detectable T-cell-specific antitumor immune response when patients' T cells were evaluated in *ex vivo* assays [18].

Strategies involving various forms of peptides have been used either alone or combined with different cytokines, adjuvants or dendritic cells to enhance specific immune responses. Although individual patients have benefited, no strategy has emerged as universally applicable; neither has any route of administration. Increasingly sensitive methods have correlated clinical responses with measurable immune responses to vaccination in some patients [17].

Peptide-based vaccination against neoplastic disease has made enormous progress and remains an active and crucial area of investigation [21], holding promise for improving the clinical outcome in cancer patients. However, several hurdles need to be overcome. One of these is the ability of tumor cells to evade a strong tumor-specific immune response.

Most antigen-discovery methods have employed tumors as sources of antigen. Many TAAs have been found to be expressed very early in the process of transformation and are frequently expressed in premalignant dysplasia and metaplasia. Several human tumors, such as colon, breast, prostate and lung, yield themselves to early detection and patients with familial history or other risk factors for these malignancies are under regular monitoring. Surgery is an option for removal of a small tumor but, with the exception of polyp removal for colon cancer prevention, there are no general strategies for dealing with premalignant changes. The accurate evaluation of vaccine-induced immunological effects has become the major end-point of clinical vaccine trials and monitoring of specific immune response in relation to clinical developments is only possible if the vaccine antigen is known. Using well-defined immunogens such as antigenic peptides – either alone or in conjunction with a cytokine or adjuvant – the kinetics, intensity and distribution of specific immune responses have been characterized.

It is hoped that the use of highly sensitive methods for the assessment of antigen-specific immune responses will help to identify the most promising vaccination strategies in the next generation of cancer vaccine trials. The consideration of individual tumor features predictive of the susceptibility to T-cell targeting (i.e. expression of antigen and MHC class I/II, and the presence of cellular infiltrates) as prognostic factors will contribute to a better selection of patients for antigen-specific immunotherapy. The analysis and correlation of spontaneous and vaccine-induced immunity and clinical improvements will help to identify the most promising antigens and the most effective strategies of immunization, for example by enhancing immunogenicity.

The conclusions of a workshop that addressed the immunologic monitoring of cancer vaccine therapy were that current data and opinion support the use of a functional assay like enzyme-linked immunospot assay or cytokine flow cytometry in combination with a quantitative assay such as a tetramer-based assay for immune monitoring. At present assays appear to be most useful as measures of vaccine potency. Careful immune monitoring in association with larger-scale clinical trials may ultimately enable the correlation of monitoring results with clinical benefit [22].

DNA vaccination is currently being explored as a potential strategy for combating cancer. However, tumor antigens are often weak and the immune system of patients may be compromised. For B-cell tumors, immunoglobulin idiotypic antigens provide defined targets but are poorly immunogenic. Antigenic differences between normal and malignant cells form the basis of clinical immunotherapy protocols. Because the antigenic phenotype varies widely among different cells within the same tumor mass, immunization with a vaccine that stimulates immunity to a broad array of tumor antigens expressed by the entire population of malignant cells is likely to be more effective than immunization with a vaccine for a single antigen. One strategy is to prepare a vaccine by transfer of DNA from the patient's tumor into a highly immunogenic cell line. Weak tumor antigens, characteristic of malignant cells, become strongly antigenic if they are expressed by immunogenic cells. In animal models of melanoma and breast cancer, immunization with a DNA-based vaccine is sufficient to inhibit tumor growth and to prolong the lives of tumor-bearing mice.

Delivery of target antigens in a DNA vaccine format allows flexibility of design so that antibody and/or cellular responses can be induced. The ease of manipulation and relatively low cost of production have facilitated the rapid evaluation of various formulations. One critical problem for cancer vaccines is that the patient's immune system has apparently failed to respond to the developing tumor and the immune capacity may be compromised by chemotherapy. DNA fusion vaccines incorporating various immune-enhancing molecules can be used to promote immunity against attached tumor antigens. Strategic use of these molecules should circumvent tolerance, and different designs can be used to activate specific pathways of attack. Testing in pilot clinical trials shows insignificant toxicity, opening the way for assessment of efficacy [15].

DNA fusion vaccine strategy has proved to be effective in DNA vaccines against a variety of antigens using different immuno-enhancing sequences. Because the fusion proteins are expressed *in vivo* in transfected host cells, some molecules that are difficult to produce *in vitro* may now be assessed. It is possible to design DNA fusion vaccines containing various molecules with different immunological properties or functions, thus opening the avenue of manipulating the immune system to achieve the optimal immunological responses. For example, fusion of a sequence derived from tetanus toxin to the genes encoding idiotypic determinants has proved highly effective in activating protective antitumor immunity. DNA fusion vaccines containing immuno-enhancing sequences can augment and direct immune attack on a range of target antigens. Gene-based fusion vaccines offer ease of manipulation and flexible design to activate an effective attack on cancer [15].

Another strategy to increase the potency of DNA vaccines is to improve antigen presentation by professional antigen-presenting cells (APCs). Cytotoxic T lymphocyte antigen 4 (CTLA-4) is one of the critical co-stimulatory surface molecules that determine the early outcome of stimulation through the T cell receptor (TCR). It has recently been suggested that manipulation of CTLA-4 has considerable promise as a strategy for the immunotherapy of cancer [23]. Fusion of CTLA-4 molecules was shown to increase the potency of DNA vaccines and provide protection against viral challenge. It was thought that the enhanced immune responses were due to more efficient antigen presentation through the interaction between CTLA-4 and its surface receptors B7-1 and B7-2 expressed by professional

APCs. Improved vaccine potency was also observed when cytokines and chemokines were used as fusion partners to target professional APCs for antigen presentation [15].

Dendritic cells (DCs) are potent APCs that are specialized in the initiation of T cell immunity. DCs are being investigated as adjuvants for active immunization in humans and several studies have been published using DC-based immunotherapy in cancer patients with melanoma, myeloma, lymphoma, prostate, renal cell, ovarian, breast or colon cancer. Evidence of clinical improvement (e.g. regression of metastases) and enhanced T cell immunity (antigen-specific proliferative responses and delayed-type hypersensitivity reactions) was obtained in some cases, even in late-stage disease. Because of the differences in protocol design, effects of concurrent treatments, DC source, routes and frequency of administration, and lack of generally accepted measures of immunity, the results are difficult to compare [24–26]. Nevertheless, it seems likely that within the next several years the *ex vivo* DC vaccination approach will be improved and standardized to an extent that there will be consensus on its potential and the need for more vigorous research programs [26].

Another co-stimulatory molecule is 4-1BB. Tumor cells can turn off the effects of some co-stimulatory molecules, thus ensuring their own survival. Recent experiments in mice have shown that monoclonal antibodies against the T cell activation molecule 4-1BB have been effective in the treatment of established mouse tumors. One method of constructing a DNA vaccine that will stimulate the immune system, in the same manner as a Mab, is to transfect genes that encode single-chain Fv fragments (scFv) for expression at the cell surface. To create a vaccine that stimulates the immune system similarly to the efficacious Mab anti-4-1BB, 1D8, a vector was constructed encoding cell-bound single-chain Fv fragments from 1D8. The vector was transfected into the K1735 melanoma, selected because of its low immunogenicity and very low expression of major histocompatibility complex class I molecules. The transfected cells induced a strong type I T-helper response, for which CD4+ but not CD8+ T lymphocytes were necessary and that involved natural killer cells. Vaccinated mice rejected established wild-type K1735 tumors growing as subcutaneous nodules or in the lung. An analogous approach may be effective against micrometastases in patients, including tumors whose expression of major histocompatibility class I is very low [27].

Like other forms of cancer therapy, tumors in mice treated solely with the vaccines tended to recur. The heterogeneity of the tumor-cell population made it likely that a cellular variant, resistant to the immunity engendered by the vaccine, escaped destruction. Several different tumor escape mechanisms have been described that enable tumor cells to proliferate in the presence of active antitumor immunity. Nevertheless, the strong antitumor immune responses that follow immunization with DNA-based vaccines raises the possibility that an analogous approach can be devised to prepare a vaccine that, along with other established forms of therapy, can be used in the treatment of cancer patients.

There are several potential ways to avoid the escape of tumor cells: (1) patients with early-stage disease could be vaccinated to cope with immune tolerance or immunosuppression caused by factors released by tumor cells; (2) multi-epitope vaccines could be used to bypass the heterogeneity in TAA expression; (3) cytokine adjuvants, such as granulocyte-macrophage colony-stimulating factor (GM-CSF), could be used to recruit DCs at the vaccination site and improve TAA presentation. IL-2 and/or IL-12 given systemically could be used to help to expand anti-tumor T cells. IL-2 could be used to restore the functions of patients' T cells; (4) the expression of peptide–MHC complexes on target cells could be increased by the systemic administration of IFN-α or IFN-γ. Finally, as indicated by many studies in animal models and in humans, class II-restricted HLA epitopes should be provided, even in the form of promiscuous determinants, to augment the strength and duration of the immune response.

Cancer vaccination is not a therapeutic modality of generally broad applicability. Based on recent experiences, individualized approaches tailored according to defined patient and disease parameters may lead to the successful application of antigen-specific vaccination to prevent or treat human cancer [17]. There is currently an 'embarrassment of riches' with regard to multiple vaccine strategies in the clinic with no single approach proving to be superior [19]. Immune prevention could also be considered for individuals whose tumors have been eradicated by standard therapy. Clinical data on patients treated with monoclonal antibodies, and experimental data in mice immunized after tumor surgery, show that prevention of recurrence in conditions of minimal residual disease is a permissible immunological challenge. However, since the immune system has been imprinted by a previous encounter with a progressing tumor, tolerance to TAAs may need to be overcome in order to elicit an efficient response [16].

Advances in tumor biology and tumor immunity have helped to better understand the mechanisms displayed by a number of tumors to escape host immunity. This bulk of new knowledge will be used to design future cancer vaccines, which will likely target multiple TAAs, presented by different antigen-presentation formats, in association with synthetic adjuvants and/or immunostimulatory cytokines. However, there are several outstanding issues which need to be addressed, for example vaccine therapy often leads to partial tumor regression in both experimental animals and patients and tumors often recur. What additional steps should be taken to prevent recurrence? What role(s) do tumor antigens play in the pathogenesis of the neoplasm, and how can this knowledge be used to generate more effective vaccines? Since DNA vaccines include the expression of normal cellular constituents, is there a danger of development of autoimmune disease and, if so, how can it be prevented? How can DNA-based tumor vaccines be used to identify new classes of tumor-associated antigens? [14].

Lastly, specific assays for measuring immune responses qualitatively and quantitatively are critically needed in order to establish a meaningful correlation between clinical outcome and immune responses in patients receiving experimental vaccines [13].

13.3.2 Antibodies

The idea of targeting drugs to cancer cells using antibodies as carriers is not a new concept: it was proposed early in the twentieth century by Paul Ehrlich. However, it is only in the last 30 years that this approach has begun to be tested experimentally and a major impetus to its evaluation was the development of monoclonal antibodies (Mabs) by Kohler and Milstein in the mid 1970s [28]. As they are derived from the clonal expansion of a single antibody-producing cell, monoclonal antibodies are homogeneous and react with only a single antigenic determinant. The production of such antibodies was clearly a major advance over the polyclonal antisera available previously which contained mixtures of antibodies of varying specificities. In the clinical as well as the experimental setting, this lack of definition made polyclonal antisera less useful for purposes where high specificity was required.

Antibodies may be effective via a variety of mechanisms. They can block essential cellular growth factors or receptors, directly induce apoptosis, bind to target cells and recruit 'effector functions', such as antibody-dependent cellular cytotoxicity (ADCC) or complement-dependent cytotoxicity, and act as delivery systems/carriers for the site-specific delivery of molecules such as radioisotopes, chemotherapeutic drugs and toxins [29].

Several Mabs, human-mouse chimeric and fully humanized anti-ERBB1 and ERBB2 antibodies have been generated for ERBB targeting. Herceptin (Trastuzumab), a Mab against ERBB2, was the first genomics-based therapeutic agent to be applied selectively on the basis of the genetic characteristics of the tumor, i.e. ERBB2-overexpression in breast cancers [2]. Trastuzumab shows high affinity for the extracellular domain of ERBB2 and is a potent inhibitor of ERBB2-expressing breast cancer cells [4,30]. Trastuzumab promotes accelerated ERBB2 internalization and degradation, and induces antibody-dependent cytotoxic responses depending on the level of ERBB2 expression. It inhibits the growth of ERBB2-positive human tumor xenografts and can eradicate established tumors and enhance the anticancer activity of a range of anticancer agents. As a single agent disease stabilization was seen in 14% of evaluable patients and when combined with anthracyclines or paclitaxel the overall response rate improved to 45% and survival time was increased by 25% [31]. In a phase II study in patients with breast cancer with high levels of expression of ERBB2, 23% of patients had tumor shrinkage of greater than 50% [32]. Recently, a multicenter phase III study in patients with metastatic ERBB2-positive breast cancer showed a longer median survival for those treated with Trastuzumab [31].

Cetuximab is a chimeric Mab directed at the extracellular domain of the ERBB1 receptor which blocks ERBB1 signaling, resulting in arrest of ERBB1-expressing tumor cells in G_1 of the cell cycle. It has been shown to inhibit a wide range of human tumor ERBB-expressing xenografts in nude mice and several phase I and II clinical trials with Cetuximab as a single agent, or in combination with chemotherapeutic drugs, have been undertaken. Partial responses to therapy have been reported in patients with colorectal cancer. Cetuximab-based combination regimens are now being evaluated in randomized trials [4]. A fully humanized Mab has entered a phase I clinical trial and early clinical studies are ongoing with a bispecific antibody recognizing ERBB1 and the IgG receptor (CD64).

Mabs linked to radioisotopes have been used for radioimmunolocalization of tumors in patients as well as for radioimmunotherapy. A recent phase I study set out to determine the maximum tolerated dose (MTD) of iodine-131 (^{131}I) labeled 81C6 anti-tenascin Mab that could be administered into surgically created resection cavities in patients with malignant gliomas and to identify any objective responses. Newly diagnosed patients with malignant gliomas with no prior external-beam therapy or chemotherapy were treated with a single injection of ^{131}I-labeled 81C6 into the resection cavity. The initial dose was 20 mCi and escalation was in 20-mCi increments up to 180 mCi. Patients were observed for toxicity and response until death or for a minimum of 1 year after treatment. Dose-limiting toxicity was observed at doses greater than 120 mCi and consisted of delayed neurotoxicity. None of the patients developed major hematological toxicity. Median survival for patients with glioblastoma multiforme and for all patients was 69 and 79 weeks, respectively [33].

Encouraging results with a three-step avidin-biotin pretargeting approach to target ^{90}Y-biotin to the tumor in patients with recurrent high-grade glioma led to the application of the same strategy in an adjuvant setting, to evaluate: (i) time to relapse; and (ii) overall survival. Thirty-seven high-grade glioma patients, 17 with grade III glioma and 20 with glioblastoma, were enrolled in a controlled open nonrandomized study. All patients received surgery and radiotherapy and were disease-free by neuroradiological examinations. Nineteen patients (treated) received adjuvant treatment with radioimmunotherapy. In the treated glioblastoma patients, the median disease-free interval was 28 months (range 9–59 months); median survival was 33.5 months and one patient was still without evidence of disease. All 12 control glioblastoma patients died after a median survival from diagnosis of 8 months. In the treated grade III glioma patients median disease-free interval was 56 months (range 15–60 months) and only two patients within this group died. Three-step radioimmunotherapy promises to have an important role as adjuvant treatment in high-grade gliomas, particularly in glioblastoma where it impedes progression, prolonging time to relapse and overall survival. The authors concluded that a further randomized trial was justified [34].

Recombinant immunotoxins are composed of a very potent protein toxin fused to a targeting moiety such as a recombinant antibody fragment or growth factor. These molecules bind to surface antigens on cancer cells and kill the target cells by catalytic inhibition of protein synthesis. Recombinant immunotoxins have been developed for solid tumors and hematological malignancies and have been characterized intensively for their biological activity *in vitro* on cultured tumor cell lines as well as *in vivo* in animal models of human tumor xenografts. The excellent *in vitro* and *in vivo* activities of recombinant immunotoxins have led to the initiation of clinical trial protocols. Recent trial results have demonstrated potent clinical efficacy in patients with malignant diseases that are refractory to the traditional modalities of cancer treatment. The results demonstrate that such strategies can be developed into a separate modality of cancer treatment with the basic rationale of specifically targeting cancer cells on the basis of their unique surface markers. Efforts are now being made to improve the current molecules and to develop new agents with better clinical efficacy. This can be achieved by development of novel targeting moieties with improved specificity that will reduce toxicity to normal tissues [35].

Although there are extensive data on the efficacy of immunoconjugates of Mabs linked to chemotherapeutic drugs *in vitro* and in preclinical *in vivo* animal models, trials of these agents in patients has been limited. Antibody-directed enzyme prodrug therapy (ADEPT), first described by Bagshawe [36], is a strategy for selective treatment of cancer, in which a nontoxic prodrug is enzymatically converted into a cytotoxic compound selectively at the surface of malignant cells by employing an enzyme-immunoconjugate.

Because of its high cytotoxicity the antibiotic CC-1065, isolated from *Streptomyces zelensis*, is a good candidate for ADEPT. Simplistically, the active molecule is a ring of three carbon atoms which are under strain and when the ring bursts, cause damage to DNA. The prodrug is like the natural antibiotic but without the strained ring and with a sugar 'safety-catch'. Once the sugar is removed, the molecule rearranges itself into a three-atom ring and can cause its toxic effect. An antibody linked to an enzyme is used to remove the sugar. Preclinical testing of this prodrug in mice has shown very promising results and it remains to be seen whether these results can be reproduced in humans [37].

In another example of this approach, a fusion protein incorporating the scFv of an anti-carcinoembryonic antigen (CEA) antibody and a plant-derived β-galactosidase (linamarase), which is capable of splitting the sugar linamirin to release cyanide has been developed. This system – known as antibody-guided enzyme nitrile therapy (AGENT) – has shown efficacy *in vitro*. However, *in vivo* uptake studies in mice indicated that AGENT could not be activated by normal mammalian cell metabolism and work is now progressing with a human enzyme from the same family so that preclinical and then clinical studies can be undertaken with this system [38].

Antibody engineering has made it possible to design antibodies with optimal characteristics for delivery of radionuclides for tumor imaging and therapy. A humanized divalent-Fab' cross-linked with a bis-maleimide linker referred to as humanized divalent-Fab' maleimide was produced as a result of this design process. It is a humanized divalent antibody with no Fc, which can be produced in bacteria and has enhanced stability compared with F(ab')2. Results of a clinical study in patients with colorectal cancer using humanized divalent-Fab' maleimide, generated from the anti-CEA antibody A5B7, radiolabeled with iodine-131 have been reported recently. Ten patients received an i.v. injection of ^{131}I-labeled A5B7 humanized divalent-Fab' maleimide, positive tumor images were obtained by gamma camera imaging in eight patients with known lesions and one previously undetected lesion was identified. True negative results were obtained in two patients without tumor. Analysis of serial blood gamma counting and gamma camera images showed a higher tumor to blood ratio compared to A5B7 mF(ab')2 used previously in the clinic, implying this new molecule may be superior for radioimmunotherapy. Dose calculations showed a relatively high radiation dose to the kidney, which may limit the amount of activity that could be administered in radioimmunotherapy. However the reduction in immunogenicity was also a major advantage for A5B7 humanized divalent-Fab' maleimide over murine versions of this antibody suggesting that humanized divalent-Fab' maleimide should be a useful vehicle for repeated therapies [39].

Malignant hematopoietic cells express lineage-restricted antigens that serve as suitable targets for antibody-directed therapy. Although several highly specific, potent and relatively nontoxic, engineered antibodies, immunoglobulin fragments and conjugates have been developed, only three have gained approval for clinical use. Of these, a chimeric mouse-human anti-CD20 antibody has yielded the most impressive clinical results. Encouraging data with the other approved antibodies, and with agents in clinical trials, suggest that rational antibody design will generate effective products for several different hematological malignancies. Despite these advances, significant challenges remain in the identification of optimal cellular targets, antibody forms and treatment schedules for therapeutic applications [40].

Questions which still need to be addressed to optimize antibody-directed therapies include: (1) Can antibodies and immunoconjugates be better designed? (2) Can tumor-specific cytotoxicity be enhanced? (3) Are antibody-based treatments most effective as monotherapy or in combination with other modalities? The pace of discovery and the clinical applications of antibody-directed therapies are rapidly accelerating. Antibody-based approaches currently offer a versatile, potent and, in many cases, less-toxic alternative to conventional treatments. Although effective as single agents, the greatest benefit of antibodies and immunoconjugates might be as components of combination or adjuvant therapies [40].

13.3.3 Immune cells

The identification of many human tumor antigens as potential targets for immunotherapy has led to clinical trials to augment the tumor immune response through the use of adoptively transferred immune cells. A comprehensive treatment of the literature on this topic from *in vitro*, to preclinical, to clinical studies is beyond the scope of this section and those interested in more details should consult these references [41–43].

The major rationale for the use of T cells is that these cells have the capacity to specifically kill tumor cells, to proliferate, and to persist after transfer and thereby be available to eliminate residual tumor cells or newly emerging tumor cells. Most studies involve the *ex vivo* expansion of autologous T cells and generate either a polyclonal proliferation by mimicking antigen signaling by using anti-TCR or anti-CD3 antibodies and the cytokine IL-2, or an antigen-specific proliferation by the addition of autologous dendritic cells that have been loaded with the relevant antigen [42].

Adoptive immunotherapy with EBV-specific cytotoxic T cells (CTLs) has proved to be a feasible strategy for reconstructing immunity and preventing or treating EBV–lymphoproliferative disease (LPD) in patients at high risk for this complication after hematopoietic stem-cell transplantation. This approach is being extended to Hodgkin's lymphoma and could be applied to other EBV-associated malignancies such as Burkitt's lymphoma and nasopharyngeal carcinoma in the future. Treatment with EBV-specific CTLs has been undertaken mainly in patients with recurrent Hodgkin's disease and although the results are promising they do not include any cures [43].

Treatment with autologous T cell transfer and high-dose IL-2 therapy after non-myeloblative lympho-depleting chemotherapy in patients with metastatic melanoma reportedly resulted in the rapid growth *in vivo* of clonal populations of T cells specific for the MART-1 melanocyte differentiation antigen, and resulted in the destruction of metastatic tumors. However, autoimmune attack on normal tissues that expressed MART-1 was also observed [44].

Infusions of natural killer (NK) cells, lymphokine-activated killer (LAK) cells from peripheral blood, and polyclonal cell populations isolated from tumor-infiltrating lymphocytes (TILs) and nonspecifically expanded *in vitro* with IL-2 have provided some evidence of efficacy in randomized controlled trials in patients with melanoma and renal cell carcinoma. However, the low response rates and the toxicity due to the high doses of IL-2 that have to be injected to maintain cell survival have limited this approach. In two randomized trials in lung cancer patients, adoptive immunotherapy with TILs and IL-2 infusions, or LAK and IL-2 therapy, have shown beneficial effects but it is uncertain in both cases whether the adoptively transferred cells contributed to the effect or whether it was mainly due to the IL-2. To date there are no published randomized controlled trials that have convincingly demonstrated clinical benefit from adoptively transferred cells [42].

Recent advances in immunological monitoring now enable more-direct detection of individual antigen-specific T cells on the basis of structural or functional properties. These advanced assays provide a degree of specificity and sensitivity that was not previously possible with classical cytolytic or proliferation assays. Careful use of these advanced assays to monitor the immune response in clinical trials may be used to address questions related to the magnitude, homing property, function and avidity of antitumor T cells that are required for effective therapy. An important corollary to an evaluation of the T-cell response will be an understanding of the tumor response to immune manipulation, particularly an evaluation of potential mechanisms of immune escape. Preclinical and clinical studies have shown that tumor cells might circumvent the immune response through defective antigen expression and presentation, inhibition of T-cell effector function and induction of antiapoptotic mechanisms [45].

Experiments in mouse model systems indicate that the transfer of TCRs, rather than the T cells themselves, into recipient cells (by gene transfer) can be used as a strategy for the transfer of tumor-specific T-cell immunity. This work is still in the preclinical phase of development and has not been tested yet in clinical trials [46].

An alternative approach to adoptively transferring immune cells involves the stimulation of specific cell populations *in vivo*. Natural killer T (NKT) cells are innate lymphocytes that share receptor structures and functions with conventional T cells and NK cells. NKT cells are specific for glycolipid antigens bound by the major histocompatibility complex class I-like protein CD1d. NKT cells rapidly produce large amounts of cytokines in response to T-cell receptor engagement, suggesting that activated NKT cells can modulate adaptive immune responses. Recent preclinical studies have revealed significant efficacy of NKT-cell ligands, such as the glycolipid α-galactosylceramide, for the treatment of metastatic cancers suggesting that appropriate stimulation of NKT cells could be exploited for prevention or treatment of human diseases including cancer [47].

Alpha-galactosylceramide (α-GalCer) is a glycosphingolipid that was originally isolated from the sea sponge *Agelas mauritianus* as an agent that clears metastatic tumors from mice. It was subsequently reported that α-GalCer is a potent stimulator of murine NKT cells. α-GalCer and its synthetic equivalent, KRN7OOO, bind CD1d molecules from mouse and human and stimulate NKT cells from either species. The striking conservation of this immune recognition system makes NKT cells particularly attractive as targets for the development of novel immune therapies.

α-GalCer has shown significant efficacy in the treatment of a variety of experimental tumors. Although α-GalCer-activated NKT cells can kill a wide range of tumor cell targets *in vitro,* accumulating evidence indicates that NK cells are the major effector cells responsible for the antimetastatic activities of α-GalCer *in vivo*. There is also evidence that IFN-γ induced by α-GalCer potentiates the generation of tumor-specific CD8+ cytotoxic T cells which might contribute to the antimetastatic activities of α-GalCer as well. Thus α-GalCer enhances components of both innate and adaptive antitumor cytotoxicity.

Clinical trials of α-GalCer for treatment of human cancers have been initiated but efficacy data are not yet available. One potential problem with applying the results of the animal studies to the treatment of human disease is that humans have fewer NKT cells than mice. Thus, in humans it might be more challenging to elicit strong NKT-cell responses with drugs such as α-GalCer. Another concern is that α-GalCer therapy in mice can have potent side effects although none have so far been noted in the clinical studies. It seems likely that α-GalCer may be evaluated as a vaccine adjuvant to enhance protective immune responses against pathogens and tumors [47].

13.3.4 Cytokines

In addition to the targets represented by the protein products of oncogenes, there is evidence that the expression of mutant genes is closely linked to aberrant cytokine gene expression. Early work demonstrated that expression of a mutant *RAS* oncogene in at least two different human cell types is associated with significant alterations in the regulation of genes encoding several cytokines, including IL-1A, IL-1B, CSF2, CSF3 and IL-6 [48]. In a mouse P815 mastocytoma model, tumor cells transfected with a human IL-13 cDNA in a plasmid expression vector, when injected into mice, were rejected and the mice developed systemic specific antitumor immunity leading to long-lasting specific antitumor protection [49]. In another study in mice, the vascular delivery of retroviral producer cell lines encoding the cytokines IL-4 and IL-2, directed tumoricidal inflammatory responses to established metastases. Cytokine gene targeting inhibited metastasis formation and caused significant overall reduction in tumor burden [50], indicating the potency of this approach.

Some growth factors are therapeutically useful partly because restricted expression of their receptors limits their action to particular cell types. However, no unique stimulating factor is known for many clinically relevant cell types. It has been shown that it is possible to target the soluble α receptor (Rα) component for IL-6 and ciliary neurotrophic factor (CNTF) to cells expressing particular surface markers, thereby rendering these cells responsive to IL-6 or to CNTF. This approach may open the way for 'designer cytokines' that are tailored to stimulate cells of choice, especially those for which a unique cytokine may not exist [51].

In addition, as discussed earlier, cytokines are also an integral part of vaccine (see Section 13.3.1) and adoptive immunotherapy (see Section 13.3.3) protocols.

Chemokines belong to a superfamily of cytokine-like, secreted proteins (8–11 kDa) that regulate leukocyte transport by mediating the adhesion of leukocytes to endothelial cells, the initiation of transendothelial migration and tissue invasion. The majority fall into two broad groups – CC chemokines with two adjacent cysteines, and CXC chemokines, in which the equivalent two cysteine residues are separated by another amino acid. To date 24 human CC chemokines (CCLl–CCL28), 15 human CXC chemokines (CXCLl–16) and one each of the CX_3C and C chemokine subclasses, which are represented by fractalkine (CX3CL1) and lymphotactin (XCLl), respectively, are known [52].

The role of chemokines in tumor biology is complex. Besides their action on hematopoietic cells, recent studies have shown that chemokines also induce distinct effects in non-hematopoietic cells such as stromal and solid tumor cells. Chemokines can act as autocrine or paracrine growth factors, induce angiogenesis or angiostasis, regulate metastasis and have a role in the host's immune response against tumor cells.

Findings in experimental tumor models have shown that the introduction of chemokines such as CCLl (I-309), CCL2 (monocyte chemoattractant protein-1; MCP-1), CCL3 (macrophage inflammatory protein-1a; MIP-1a), CCL5 (regulated upon activation normal T cell expressed and secreted, RANTES), CCLl6 (human β CC chemokine 4; HCC-4),

CCL19 (MIP-3b), CCL2O (MIP-3a), CCL21 (6Ckine), CXCL1O (interferon-γ inducible protein-10, IP-10) and XCL1 (lymphotactin) alone can induce tumor regression and immunity to subsequent tumor challenge [52].

However, the antitumor efficacy of chemokines alone is limited and new approaches are being developed that combine a chemoattractant (e.g. CCL19, CCL21, CXCL9, CXCL10, CXCL12 and XCL1) together with cytokines (IL-2, IL-12, GM-CSF), which are known for their stimulating properties on T cells, NK cells or tumor antigen-pulsed dendritic cells (DCs). It is hoped that chemokines might act as potent natural adjuvants for experimental antitumor immunotherapy and their combination with tumor peptide-pulsed DCs and direct coupling to tumor antigen or immunostimulatory cytokines has been shown to result in synergistic antitumor activity [52].

13.4   Other targets

13.4.1   Proteasomes

Protein degradation is fundamental to cell viability and the primary component of the protein degradation pathway in the cell is the 26S proteasome which is a large multiprotein complex present in the cytoplasm and the nucleus of all eukaryotic cells. The central role of the proteasome in controlling the expression of regulators of cell proliferation and survival has led to interest in developing proteasome inhibitors as novel anticancer agents. Studies *in vitro* and *in vivo* have shown that proteasome inhibitors have activity against a variety of tumors and one of these agents, PS-341 (bortezomid, VELCADE™), has been tested in clinical trials. These phase I trials showed that the treatment was well tolerated as a single agent and preliminary evidence of biological activity was seen in some patients, thereby providing the rationale for phase II and III trials in multiple myeloma. Phase II trials in several hematological malignancies and solid tumor types are also in progress and additional trials of bortezomib, in combination with other cytotoxic regimens, will focus on its activity in solid tumors [53]. Drugs that affect protein degradation by the proteasome are a potentially promising class of agents that are just beginning to be explored.

13.4.2   *p53*

Analysis of the gene encoding *p53* could serve to evaluate the effectiveness of a cancer treatment. Mutations in this gene occur in half of all human cancers and regulation of the protein is defective in a variety of others. Novel strategies that exploit our knowledge of the function and regulation of *p53* are being actively investigated. Strategies directed at treating tumors that have *p53* mutations include gene therapy, viruses that only replicate in *p53*-deficient cells, and the search for small molecules that reactivate mutant *p53*. Potentiating the function of *p53* in a nongenotoxic way in tumors that express wild-type protein can be achieved by inhibiting the expression and function of *MDM2* or viral protein [54].

Over 6000 papers have described *p53* alterations in human tumors – 15 121 somatic and 196 germ-line mutations in *p53* are catalogued in the International Association of Cancer Registries (IARC) database [55]. Over 1700 different mutations in p53 have been reported. The mutations are found throughout the open reading frame (ORF) as well as at splice junctions, and although the most common site for mutations is in exons 5–8, which encode the DNA binding domain of the protein, over 13% of mutations lie outside this region [54].

In the past decade the genetic and biochemical analysis of the *p53* pathway that leads from cellular stress (through *p53* activation) to growth arrest and apoptosis, has identified many targets for therapeutic development. It has also led to the growing realization that the toxicity and efficacy of many of the current treatments are also profoundly affected by the activity of the *p53* pathway. Most cytotoxic drugs induce the *p53* response in normal tissues, hence contributing to their toxicity, whereas tumors that retain the normal *p53* gene function are in many cases more responsive to treatment [54,56,57].

The various therapeutic approaches based around the *p53* pathway have been recently summarized as follows: (1) treatments for tumors in which the *p53* gene is mutant – including gene therapy (see Section 13.7) with wild-type *p53*, exploiting the absence of *p53* to enable selective drive of therapeutic gene expression, exploiting the absence of *p53* to enable selective viral replication, exploiting small-molecule inhibitors of the *p53* response, mimicking the function of downstream genes, reactivating mutant *p53*; (2) treatments for tumors in which the *p53* gene is wild-type (activating the function of the endogenous *p53* gene in the tumor) – including inhibiting *MDM2* (a negative regulator of *p53*), blocking the *p53–MDM2* interaction, inhibiting nuclear export and mimicking p14^{ARF} which is a small protein-activator of the *p53* response [54].

Mutation of *p53* is associated with poor prognosis in many types of human cancer. Because of the central role of *p53* in cell-cycle arrest and apoptosis, it has been proposed that lack of functional *p53* in tumors, either through mutation or by other mechanisms, such as overexpression of *MDM2*, could affect the efficacy of standard radiation and chemotherapy. The relationship between *p53* status and sensitivity to chemotherapy has been extensively studied in breast and ovarian cancers. The majority of findings from these studies indicate that mutation or alteration in *p53* can lead to decreased sensitivity and resistance to cytotoxic drugs. In terms of effects on tumor response to radiation therapy, numerous *in vitro* and *in vivo* studies have shown that loss of *p53* function increases post-irradiation clonogenic cell survival. This correlates with an abrogated G_1 checkpoint control and changes in apoptosis [57].

Collectively, the evidence indicates an association between lack of functional *p53* and inability of tumor cells to undergo apoptosis in response to chemotherapy and/or radiotherapy. Restoration of normal *p53* function in tumors might restore the apoptotic pathway and therefore lead to an increased response to conventional therapeutics [57].

It has been recently reported that a low-molecular-weight compound (PRIMA-1) has been identified which is capable of inducing apoptosis in human tumor cells through restoration of the transcriptional transactivation function to mutant p53. This molecule restored sequence-specific DNA binding and the active conformation to mutant p53 proteins *in vitro,* and *in vivo* in mice it showed an antitumor effect without apparent toxicity. The authors suggest that this molecule may serve as a lead compound for the development of anticancer drugs targeting mutant *p53* [58].

Molecular studies on p53 protein conformation and on the regulation of its DNA-binding activity have shown that the protein is oligomeric, it is in an equilibrium between different conformational states, that its regulatory domains control its core DNA-binding domain conformation, and that cancer-prone mutations can alter the normal conformational equilibrium within the tetramer. Numerous low-molecular-weight agents have been identified that are capable of reactivating both wild-type and mutant p53 *in vivo,* and these hold great promise for treatment in the future [59].

Death receptors – members of the tumor necrosis factor receptor (TNFR) superfamily – signal apoptosis independently of p53. Decoy receptors, in contrast, are a nonsignaling subset of the TNFR superfamily that attenuates death receptor function. Agents that are designed to activate death receptors (or block decoy receptors) might therefore be used to kill tumor cells that are resistant to conventional cancer therapy. Concomitant with the evaluation of the safety and efficacy of such agents in preclinical models is the identification of suitable candidates for clinical investigation. The identification of more TNF and

TNFR superfamily members through the Human Genome Project has yielded novel apoptosis-based approaches that have the potential to expand cancer therapy in a new direction [60].

13.4.3 Raf kinases

Raf kinases are proto-oncogenes that work at the entry point of the mitogen-activated protein kinase/extracellular-signal-regulated kinase (MAPK/ERK) pathway, a signaling module that connects cell surface receptors and RAS proteins to nuclear transcription factors. The pathway is hyperactivated in 30% of human tumors and impinges on all the functional hallmarks of cancer – immortalization, growth factor-independent proliferation, insensitivity to growth-inhibiting signals, ability to invade and metastasize, ability to attract blood vessels, and evasion of apoptosis. Raf is an attractive target for therapy as a single inhibitor could block several cancer-promoting elements at once, and thereby cause the collapse of synergistic processes that sustain the cancer cells, without the necessity to inhibit any single element completely [61].

Although Raf activation is still incompletely understood, three approaches are currently under investigation to inhibit the Raf-MEK (MAPK/ERK kinase) pathway. The first is the use of antisense RNA (see Section 13.5) to down-regulate Raf-1 protein levels. ISIS 5132 is an antisense oligonucleotide which has been shown to reduce Raf-1 mRNA levels in the blood cells from treated patients in phase 1 clinical trials. The results of phase II trials are awaited. The second is the use of chemical Raf inhibitors such as BAY 43-9006 which has entered phase I trials after encouraging preclinical results. The third approach is inactivation of MEK by Raf and PD184322 is a drug that does this effectively in preclinical studies with colon cancer xenografts in nude mice and which is now proceeding to clinical trials [61].

13.4.4 Cyclin-dependent kinases (CDKs)

With the recent understanding of the role of CDKs in cell cycle regulation (see Chapter 4) and the discovery that approximately 90% of all neoplasia is the result of 'CDK hyperactivation' leading to the abrogation of the *RB* pathway, novel CDK modulators are being developed. Most CDK inhibitors have antiproliferative properties associated with apoptosis-inducing activity and display antitumor activity. However, their cellular targets remain to be identified [62]. The first two CDK modulators tested in clinical trials, flavopiridol and UCN-01, demonstrated significant preclinical activity in hematopoietic models. Both compounds have also demonstrated activity in some patients with non-Hodgkin's lymphoma. The best schedule to be administered, combination with standard chemotherapeutic agents and demonstration of CDK modulation in tumor samples from patients in these trials are important issues that need to be answered in order to ensure the best possible use of these agents [63].

13.4.5 Angiogenesis

Formation of new blood vessels (angiogenesis) and lymphatic vessels (lymphangiogenesis) are thought to be essential for tumor progression and metastasis [64,65]. However, the initial enthusiasm for the anticancer potential of angiogenesis inhibitors has been tempered by controversy, highlighted at the 2002 American Association for Cancer Research meeting [66,67]. The initial encouraging results obtained with antiangiogenic agents meant that there was a rush to take this research from the bench to the clinic. The hope that antiangiogenic therapy might be the panacea for cancer has not been realized. There are many possible reasons for this, including endothelial and tumor cell heterogeneity, the presence of survival factors within the tumor micro-environment, the problem of defining the best dose and schedule and angiogenesis-independent regrowth of tumors [68]. More than 300 angiogenesis inhibitors have been discovered to date and there are currently over 80 antiangiogenic agents in clinical trials involving over 10 000 patients [64,69] but so far no therapy based on angiogenic modulation has shown sufficient clinical benefit to be approved for such an indication [64]. It has become apparent that not enough was known about the molecular mechanisms of tumor angiogenesis when trials of antiangiogenic compounds began in the 1990s, and the manner in which these drugs are administered must be changed to achieve maximum clinical efficacy [67].

After phase I studies, clinical trials were initiated before the biologically effective dose, pharmacokinetics, timing of therapy or long-term effects of the various drugs had been established. It has been argued that the traditional strategies that are used in clinical trials for anticancer therapies are not appropriate for assessing the efficacy of agents that modulate angiogenesis since most angiogenic modulators are cytostatic, slowing or stopping tumor growth, without producing an objective remission. It has also been suggested that imaging studies, for example MRI, could have a key role in assessing the efficacy of treatments [64]. For a review of the promises and pitfalls of antiangiogenic therapy in clinical trials readers should consult reference [70].

Cancer cells begin to promote angiogenesis early in tumorigenesis and this 'angiogenic switch', as it has been termed, is characterized by oncogene-driven tumor expression of proangiogenic proteins, such as vascular endothelial growth factor (VEGF), basic fibroblast growth factor (bFGF), interleukin-8 (IL-8), placenta-like growth factor (PLGF), transforming growth factor-β (TGF-β), platelet-derived enodothelial growth factor (PD-EGF), pleotrophin and others [71].

Paradoxically, tumor progression is associated with both increased microvascular density and intratumoral hypoxia. This paradox arises because the tumor vasculature is structurally and functionally abnormal, resulting in perfusion that is characterized by marked spatial and temporal heterogeneity [72]. In addition, decreased aerobic (hypoxic) conditions in tumors induce the release of cytokines that promote vascularization and thereby enhance tumor growth and metastasis [73]. Hypoxia-inducible factor 1 (HIF-1) controls oxygen delivery (via angiogenesis) and metabolic adaptation to hypoxia (via glycolosis). In xenograft models tumor growth and angiogenesis are correlated with HIF-1 expression. HIF-1 consists of a constitutively expressed HIF-1β subunit and an oxygen- and growth factor-regulated HIF-α subunit. Three members of the HIF-α family have been cloned to date: HIF-1α, HIF-2α, HIF-3α. HIF-1α has been the most extensively characterized and in human cancers it is overexpressed as a result of intratumoral hypoxia and genetic alterations affecting key oncogenes and tumor suppressor genes. HIF-1α overexpression in biopsies of brain, breast, cervical, esophageal, oropharyngeal and ovarian cancers is correlated with treatment failure and mortality. Increased HIF-1 activity promotes tumor progression, and inhibition of HIF-1 could represent a novel approach to cancer therapy [72,74].

Five mammalian vascular endothelial growth factor (VEGF) family members have been identified to date: VEGF, VEGF-B, VEGF-C, VEGF-D and placenta growth factor (PlGF). Almost all types of cancer cells express VEGF, which uses VEGF receptor 1 (VEGFR-1) and VEGFR-2 for signaling. Associations have been observed between VEGF expression, the vascular density in human tumors and patient prognosis [65]. Several studies have shown that overexpression of VEGF-C or VEGF-D induces lymphangiogenesis and promotes tumor metastasis in mouse tumor models. It has been demonstrated that VEGFR-3 signaling can be inhibited by recombinant adenoviruses expressing the VEGFR-3-Ig fusion protein (which binds VEG-C) resulting in suppression of tumor lymphangiogenesis and metastasis to regional lymph nodes, but not lung metastasis, in a mouse model [65].

Depending on their proposed mechanisms of action, angiogenic modulators can be subdivided into several categories: modulators of proteolytic enzymes; inhibitors of endothelial cell proliferation and/or survival; upstream modulators; and undefined [64]. Several antibodies are in clinical trials as angiogenic modulators including anti-VEGF (non-Hodgkin's lymphoma, breast, prostate, colorectal, renal and nonsmall-cell lung cancer); recombinant human anti-VEGF (Avastin or Avastatin) in a variety of cancers including renal cell carcinoma; and Vitaxin, an anti-α_1,β_3-integrin antibody that interferes with blood-vessel formation by inducing apoptosis in newly generated endothelial cells.

A phase I study of the molecular imaging and biological evaluation of HuMV833 (a humanized anti-VEGF antibody) in patients with progressive solid tumors illustrates one of the problems with this approach. Because of the heterogeneity with respect to antibody uptake and clearance it was not possible to identify the optimal biologically active dose. The authors concluded that either intrapatient dose escalation or larger, more precisely defined patient cohorts would be preferable to conventional strategies in the design of phase I studies with antiangiogenic compounds such as HuMV833 [75].

Although antiangiogenic therapy is a promising new approach, concerns have been raised that it will select for highly aggressive, hypoxia-adapted tumor cells. Tumor cells deficient in p53 display a diminished rate of apoptosis under hypoxic conditions, a circumstance that might reduce their reliance on vascular supply, and hence their responsiveness to antiangiogenic therapy. It has been shown that mice bearing tumors derived from $p53^{-/-}$ human colorectal cancer cells are less responsive to antiangiogenic combination therapy with anti-VEGFR2 antibody and a cytotoxic drug than mice bearing isogenic $p53^{+/+}$ tumors. Thus, although antiangiogenic therapy targets genetically stable endothelial cells in the tumor vasculature, genetic alterations that decrease the vascular dependence of tumor cells can influence the therapeutic response of tumors to this therapy [76]. In addition, the assumption that selection for endothelial cells that are resistant to the therapy is unlikely to occur has been called into question by the identification of mutations affecting proteins in apoptotic pathways in endothelial cells of patients with primary hypertension. The combination of antiangiogenic agent and an inhibitor of HIF-1 might be particularly effective, as the angiogenesis inhibitor would cut off the tumor's blood supply and the HIF-1 inhibitor would prevent the ability of the tumor to adapt to the ensuing hypoxia. Under these conditions of severe intratumoral hypoxia, a therapeutic window for inhibition of HIF-1 activity is most likely to exist. The dramatic effects of total HIF-1α deficiency on vascular development in mice also suggest that inhibition of HIF-1 could potentiate the effect of angiogenesis inhibitors on endothelial cells and reduce the potential for the development of drug resistance [72].

The blood vessels of individual tissues are biochemically distinct, and pathological lesions put their own 'signature' on the vasculature. The development of targeted pharmaceuticals necessitates the identification of specific ligand–receptor pairs and knowledge of their cellular distribution and accessibility. Using new methods, such as *in vivo* screening of phage libraries, which permits the identification of organ-specific and disease-specific proteins expressed on the endothelial surface, it is now possible to decipher the molecular signature of blood vessels in normal and diseased tissue [77]. Since in tumors, both blood and lymphatic vessels differ from normal vessels, peptides and antibodies that recognize these vascular signatures and can be used in targeted delivery therapeutic approaches are being developed [77,78]. For example, using a phage screening procedure a tumor-homing peptide has been identified which has targeting specificity for lymphatic vessels, accumulating in the nuclei of the lymphatic cells as well as in the nuclei of tumors (xenografts and spontaneous prostate and breast cancers in transgenic mice). The results show that tumor lymphatics carry specific markers and that it may be possible to specifically target therapies to tumor lymphatics in addition to tumor blood vessels [79].

A review of many of the important challenges that need to be addressed for the successful translation of angiogenesis inhibitors into clinical application highlighted that one of the most important is the lack of surrogate markers to determine efficacy in most cases [71]. For an overview of recent advances in angiogenesis, the increasing repertoire of drugs likely to manipulate it and new endothelial-specific genes with which to target vasculature, readers are referred to reference [80].

Polymorphism in the angiogenic genes/factors may in part explain the variation in tumor angiogenesis which has been observed between individuals. The establishment of a DNA repository containing samples from over 1800 breast cancer patients to identify gene polymorphisms in angiogenesis-related genes that play an important role in tumor growth and progression illustrates the intensive efforts that are underway in this area [81]. It is clear that in order to optimize antiangiogenic therapy a much greater understanding of the fundamentals of angiogenesis will be required which should lead to new approaches to attacking tumor vasculature [82].

13.4.6 Epigenetic silencing

Epigenetic inactivation of genes that are crucial for the control of normal cell growth is a hallmark of cancer cells [83,84]. These epigenetic mechanisms include crosstalk between DNA methylation, histone modification and other components of chromatin higher-order structure, and lead to the regulation of gene transcription. Unlike mutagenic events, epigenetic events can be reversed to restore the function of key control pathways in malignant and premalignant cells and re-expression of genes epigenetically inactivated can result in the suppression of tumor growth or sensitization to other anticancer therapies [84]. Small molecules that reverse epigenetic inactivation are now undergoing clinical trials. This, together with epigenomic analysis of chromatin alterations such as DNA methylation and histone acetylation, opens up the potential both to define epigenetic patterns of gene inactivation in tumors and to use drugs that target epigenetic silencing [83].

Two key changes in chromatin are associated with epigenetic transcriptional repression – DNA methylation and histone modifications. DNA methylation is the only commonly occurring modification of human DNA and results from the activity of a family of DNA methyltransferase enzymes (DNMT). DNA methylation leads to the binding of a family of proteins known as methyl-binding domain (MBD) proteins. Several of the members of this family have been shown to be associated with large protein complexes containing histone deacetylase (HDAC). To date several trials using agents that target DNMTs and HDACs have been completed or are underway [83].

13.4.7 Mitochondria

Mitochondria are intracellular organelles that play a central role in oxidative metabolism and apoptosis. The recent resurgence of interest in the study of mitochondria has been fueled in large part by the recognition that genetic and/or metabolic alterations in this organelle are causative or contributing factors in a variety of human diseases, including cancer. Point mutations, deletions or duplications of mitochondrial DNA are found in a wide range of cancers and the accumulation of mutations in mitochondrial DNA has been found to be tenfold greater than that in nuclear DNA. The many distinct differences in mitochondrial structure and function between normal cells and cancer cells provide molecular sites against which novel and selective chemotherapeutic agents might be targeted [85].

A new class of anticancer agents has already been developed that exploits the higher mitochondrial membrane potential seen in some carcinoma cells versus control epithelial cells. Collectively they are called delocalized lipophilic cations (DLCs) and they have shown some degree of efficacy in carcinoma cell killing *in vitro* and *in vivo*. Importantly, although increased membrane potential is necessary to achieve selective cytotoxicity by DLCs, it alone is not sufficient as cardiac muscle cells would also be susceptible to the cytotoxic effects of these agents, and this has not proved to be the case.

Alternative therapeutic strategies include employing mitochondrial protein-import machinery to deliver macromolecules to mitochondria and also the specific interaction of drugs with certain mitochondrial proteins. Although the use of DLCs as anticancer agents has shown promise, there is at present no real understanding of the biochemical basis for the increased mitochondrial membrane potential in carcinoma cells. The topic of drug and DNA delivery to mitochondria has been the subject of a recent issue of Advanced Drug Delivery Reviews to which the interested reader is directed for a more detailed treatment of this topic [86]. Knowledge of the specific biochemical alterations leading to the increased membrane potential will lead to a more rational approach to the choice of highly selective DLCs for clinical use in the future [85].

13.4.8 Carbohydrates

Experimental evidence directly implicates complex carbohydrates in recognition processes, including adhesion between cells, adhesion of cells to the extracellular matrix, and specific recognition of cells by one another. In addition, carbohydrates are recognized as differentiation markers and as antigenic determinants. Lectins are nonenzymatic proteins present in plants and animals, which preferentially bind to specific carbohydrate structures and play an important role in cell recognition. Modified carbohydrates and oligosaccharides have the ability to interfere with carbohydrate–protein interactions and therefore, inhibit the cell–cell recognition and adhesion processes, which play an important role in cancer growth and progression. Galectins are a family of proteins that share an affinity for β-galactoside moieties and significant sequence similarity in their carbohydrate-binding sites. Many epithelial tumors, such as colon, thyroid and breast express both galectin-1 and -3. Increased expression of galectin-1 by tumor cells is positively correlated with a metastatic phenotype and a poorly differentiated morphology. Selectins are a group of cell adhesion molecules that bind to carbohydrate ligands and play a critical role in host defense and in tumor progression and metastasis.

A modified natural polysaccharide modified citrus pectin (MCP) has been shown to have antitumor effects *in vitro* and in animal models. In phase II clinical trials on colorectal cancer patients, MCP showed clinical activity with five out of 23 patients showing tumor stabilization and one patient showing tumor shrinkage.

Interfering with normal cell recognition using a large or a small sugar molecule has been reported to block the progression of tumors by interfering with angiogenesis, cell–cell, cell–matrix interactions, tumor invasion and metastasis. So far, limited studies have been performed with modified pectin, but this molecule is undergoing clinical trials and has shown beneficial clinical effects [87].

13.4.9 Cyclooxygenase 2 (COX-2)

COX-2 is an inducible prostaglandin G/H synthase which is overexpressed in several human cancers. Oncogenes, growth factors, cytokines, chemotherapy and tumor promoters stimulate *COX-2* transcription via protein kinase C and RAS-mediated signaling. For example, the level of COX-2 is elevated in breast cancers that overexpress *HER-2/neu* as a result of increased signaling. Epidemiological evidence has shown that the use of nonsteroidal anti-inflammatory drugs (NSAIDS), which are prototypic COX-2 inhibitors, is associated with a reduced risk of several malignancies, including colorectal cancer [88 and references therein]. In a landmark clinical study, treatment with celecoxib, a selective COX-2 inhibitor, was shown to reduce the number of colorectal polyps in patients with familial adenomatous polyposis (FAP) [88].

Selective COX-2 inhibitors are being evaluated in conjunction with chemotherapy and radiotherapy in patients with cancers of the colon, lung, esophagus, pancreas, liver, breast and cervix. These studies should provide information on whether selective COX-2 inhibitors are effective in either preventing or treating cancer [89,90] and the results of these clinical trials are eagerly awaited.

13.5 Antisense RNA or oligonucleotides

Following the initial discoveries of natural antisense RNAs in prokaryotes, numerous applications of antisense RNA-mediated regulation have been demonstrated in a variety of experimental systems [91,92]. These nontranslated mRNAs directly repress gene expression by hybridizing to a target RNA, rendering it functionally inactive. Specificity of antisense RNA for a particular transcript is conferred by extensive sequence complementarity with the 'sense' or target RNA. Translation of a target mRNA is inhibited following formation of a sense–antisense RNA hybrid. In addition, the duplex molecule may become sensitive to double-strand-specific cellular nucleases. Other effects of antisense RNA may include transcriptional attenuation of the mRNA and also disruption of post-transcriptional processing events [92].

Oncogene DNA and RNA differ in nucleotide sequences from normal proto-oncogene DNA and RNA, and it is therefore theoretically possible to design specific antisense molecules to block translation of oncogene mRNA. For example, microinjection of antisense DNA sequences might bind to, and block expression of, oncogene DNA. Such 'blocking DNA' could also incorporate an anti-DNA toxin, such as an alkylating agent, to destroy the portion of the chromosome encoding the oncogene specifically. Similarly, an antibody against the oncogenic RNA sequence might bind and prevent that mRNA molecule from being used by ribosomes, or it might inhibit elongation by blocking transfer RNA from binding to the oncogenic messenger anticodon [93].

Several groups have tried to reverse the transformed phenotype by expressing large amounts of mRNA from the DNA strands complementary to the one coding an aberrant oncogene protein. In the nuclei of the cells the two complementary mRNA strands hybridize to form a double-stranded structure that effectively prevents translation of the mRNA [93]. It is now possible to design antisense oligonucleotides (ODNs), or catalytic antisense RNAs (ribozymes) which can pair with and functionally inhibit the expression of any single-stranded nucleic acid. These compounds interact with mRNA by Watson-Crick base-pairing and are therefore highly specific for the target protein. This high degree of specificity has made them attractive candidates as therapeutic agents [94] as well as valuable in the elucidation of cellular signaling pathways. Studies with these agents have probed their utility as potential therapeutic agents, especially in the realm of cancer. To give just one example out of many, ODNs directed at HER2 are in preclinical evaluation for the treatment of breast cancer [2]. With the implementation of gene therapy in early clinical trials, oligonucleotide-mediated suppression of gene expression has emerged as an important complementary strategy to gene therapy [95,96].

Several laboratories are testing retroviruses encoding antisense RNA expressed in a tissue-specific manner by use of appropriate enhancer and other control regions attached to the anticancer genes. To make this approach practical, a better understanding of retroviral tissue tropism mediated by cell surface receptors is important. It would not be ethical to infect a cancer patient with a retroviral vector carrying an antisense gene until its efficacy and more importantly its safety have been proved [93]. Despite this caveat, such an approach is already being used clinically (see Section 13.7).

Attempts to realize the promise and expectations raised by the concept of antisense blocking of specific genes involved in cancer, AIDS and a variety of other diseases have resulted in some unexpected questions arising about how these genes really work [97,98]. Even though the phosphorothioates are generally believed to represent the first generation of antisense nucleotides, they suffer from certain drawbacks like the existence of diastereoisomeric mixtures, intermediate RNA-binding affinity and certain nonspecific side effects [99]. Solutions for obtaining ODNs with improved binding properties and nuclease resistance can be realized by various chemical means [100]. Several novel approaches which can influence the biological efficacy by modulating the biodistribution and/or stability of the target mRNA by cleavage have been proposed. However, it remains to be proven whether these second-generation antisense ODNs will be useful. *In vivo* data are limited to methylphosphonates and in particular phosphorothioates which have entered clinical trials as the first generation of antisense compounds [100].

As stressed in a review of the antisense treatment of viral infection [101] many simple but critical questions remain unanswered and this is also true of its application in cancer. A review in the mid 1990s [102] focused on those aspects of chemistry and mechanism that were thought to be important and relevant for the therapeutic use of deoxynucleotide agents. The authors designated the outstanding questions as being: (1) How can oligonucleotides be delivered more efficiently to intracellular compartments? (2) How can oligonucleotides be more selectively delivered from the bloodstream to target tissues? (3) How stable must an oligonucleotide be for optimal function and low toxicity? (4) Will development of new RNase H-independent inhibitors lead to greater specificity? (5) Can new agents be developed that avoid aptameric effects and inhibit by a purely antisense mechanism?

Most of these questions, as well as the promise and the shortcomings [103,104] in the field of antisense are still relevant today.

In hematologic disorders antisense ODNs are being employed as *ex vivo* bone marrow purging agents and as potential drugs for direct *in vivo* administration to patients with leukemia [105,106]. *In vitro* data from cell culture experiments showed that an antisense ODN (G3139) designed to hybridize with the mRNA of *BCL2* can sensitize lymphoma cells to the apoptotic effects of chemotherapeutic agents. A phase I study in 21 patients with *BCL2*-positive relapsed non-Hodgkin's lymphoma patients who received an 18-mer phosphorothioate ODN complementary to the first six codons of the *BCL2* open reading frame (G3139) showed that no systemic toxicity was seen at daily doses up to 110.4 mg/m^2 and that BCL2 protein was reduced in seven of 16 assessable patients [107]. Phase I and II studies are also being undertaken to test G3139 in combination with docetaxel in patients with advanced breast cancer, hormone-refractory prostate cancer and other solid tumors [108].

Another target of antisense ODNs is protein kinase C-alpha (PKC-alpha) which belongs to a class of serine-threonine kinases. An antisense ODN directed against PKC-alpha has been evaluated in phase I and II studies in patients with low-grade lymphomas, and in combination with carboplatin and paclitaxel in patients with stage IIB or IV nonsmall-cell lung cancer. Antisense ODNs against *RAF1*, *HRAS*, *MYB*, protein kinase A and DNA methyltransferase are also undergoing preliminary clinical investigation in patients with a variety of cancers including hematologic, colorectal, breast and ovary [108].

The progress and the problems of antisense therapeutics in chronic myeloid leukemia have been recently summarized. It has become clear that antisense therapeutics is considerably more problematic than was naively initially assumed and the high expectations of the approach have yet to have a substantial impact on clinical practice, thus resulting in a waning of clinical interest in this approach. However, there is considerable evidence that antisense ODNs are effective *in vitro*. Critical analysis of the molecular and cellular behavior of antisense ODNs indicate that the clinical strategies that have been utilized so far are suboptimal for a number of reasons including unfavorable antisense chemistries, the wrong target or failure to achieve intracellular access. Considerable further basic research is required and an optimal antisense strategy is therefore some years away [106].

13.6   RNA interference/inhibition (RNAi)

RNAi is an innate cellular process which is activated when a double-stranded RNA (dsRNA) molecule of greater than 19 duplex nucleotides enters the cell, causing the degradation of not only the invading dsRNA molecule, but also single-stranded RNAs (ssRNAs) of identical sequence, including endogenous mRNAs. RNA interference methods, like antisense strategies, are based on nucleic acid technology. However, unlike the antisense approach, dsRNA activates a normal cellular process leading to a highly specific RNA degradation and to cell to cell spreading of this gene silencing effect in several RNAi models. This systemic property potentially provides great promise for therapy because the delivery problems that have plagued other nucleic acid-based therapies could be at least partly alleviated in RNAi-based gene silencing applications [109–111].

The demonstration that a single base difference in synthetic small inhibiting RNAs (siRNAs) can discriminate between mutated and wild-type (WT) p53 in cells expressing both forms and can result in the restoration of WT protein [112] indicates the potential of this approach. A better description of the systemic nature of the response in whole animals together with the ongoing improvements in *in vivo* nucleic acid delivery technologies could enable RNAi to be used therapeutically, as a single agent or in combination, sooner than is predicted at present [109,111,112].

13.7   Gene therapy

Gene therapy, the treatment or prevention of disease by gene transfer, has been alluded to several times in the preceding pages. It is regarded as a potential revolution in medicine because gene therapy is aimed at treating or eliminating causes of disease, whereas most current drugs treat the symptoms. For an erudite discussion of the problems associated with gene therapy in humans, readers are referred to Weatherall's excellent book [113]. Although this book was published over 12 years ago the issues he raised are as relevant today as they were then. Up until now, the twin Achilles heels of gene therapy have been the inability to target specifically and the inability to control exogenous gene expression [114].

One example of a successful gene therapy approach in an animal model has been the liposomal delivery of the human *APC* gene to rodent colonic epithelium where 100% of the epithelial cells which contacted the liposome–gene mixture were shown to have taken up the gene [115]. Expression was, however, only transient and did not persist beyond 3 days, thereby limiting the potential therapeutic utility of a single dose. Permanent expression of genes in colonic epithelium is not feasible with current technology, however, continuous expression can be obtained by repeated treatment with liposomal enemas every 2–3 days.

Studies such as this in animal models have revealed major impediments that need to be overcome if gene therapy is to become an everyday reality. One of these is that adenoviral vector administration commonly induces inflammation and antigen-specific cellular and humoral immune responses [116]. Immunomodulation, inducing transitory immunosuppression, by using recombinant IL-12 to block the production of IgA antibodies is one approach that has been investigated in an attempt to overcome this problem [117].

The first decade of gene therapy has been summarized up to the year 2000. Two hundred and sixteen gene transfer protocols using retroviruses had been approved or submitted to regulatory authorities in North America and Europe for the treatment of cancer, the main targets being tumor cells, antigen presenting cells, blood progenitor cells, T cells, fibroblasts and muscle cells [118]. Most clinical trials have been small phase I/II studies with the main objectives of demonstrating safety and gene transfer, and obtaining information to guide dose selection for phase II and III efficacy studies.

Most research continues to be in the area of vector development. Crucial to this is the development of efficient, multipurpose therapeutic gene delivery systems which should have the following characteristics: (a) the ability to target specific cells; (b) no limitation on the size or type of nucleic acid that can be delivered; (c) no intact viral component and therefore be safe for the recipient; (d) the ability to transduce a large number of cells regardless of their mitotic status; and (e) the potential to be completely synthetic. The development of molecular conjugates (ligands to which a nucleic acid or DNA-binding agent has been attached for the specific targeting of nucleic acids (i.e. plasmid DNA) to cells as delivery vectors) has resulted in the creation of a simple, nonviral method for the targeted delivery of nucleic acids to specific cell types. These 'synthetic viruses' can easily incorporate components that are associated with viral function [119 and references therein].

Since 1996, it has been clear that the main aspects in which vector improvement is required are: (1) specificity and efficiency of gene transfer; (2) specificity, magnitude and duration of expression; (3) immunogenicity; and (4) manufacturing. Each gene transfer system has its own combination of advantages and limitations and the reader is directed to [118] for an in-depth discussion of these.

Gene therapy in which the normal *p53* gene is introduced back into tumor cells using either physical or viral vectors (predominantly adenovirus and retrovirus) has been extensively evaluated in preclinical and clinical models. Most work has used replication-defective adenoviruses that deliver human *p53* cDNA sequences driven by strong viral promoters. Efficacy without nonspecific toxicity has been seen with the introduction of wild-type *p53* by these viral delivery systems, with reported suppression of the growth of various types of malignancies, including leukemia, prostate, head and neck, colon, cervical, glioblastoma, breast, liver, ovarian, kidney and lung tumor cell lines *in vitro* and *in vivo* in mouse xenograft models. However, responses in early human trials have been less convincing [54]. Also, it has been shown that *p53* alone, although able to partially inhibit tumor growth, is not able to eliminate established tumors in the long-term [57].

p53 replacement alone, although demonstrating some beneficial effects in head and neck cancer and in nonsmall-cell lung cancer, has not been totally successful. The emphasis has now shifted to combination therapies. Completed phase I and II studies in lung cancer with adenoviral-mediated transfer of wild-type *p53* (Ad-p53), either alone or in combination with chemotherapy or radiotherapy, have demonstrated acceptable toxicity and evidence of tumor regression with intratumoral delivery of Ad-p53. The predominant clinical effect was found to be locoregional in the area of intratumoral delivery [120].

The majority of the literature supports the premise that *p53* is a critical factor in the response of cancer cells to cytotoxic anticancer therapies. Consequently, these findings point to the immense clinical potential of *p53* gene therapy as a means to enhance conventional chemo- and radiotherapy that could lead to new more effective treatment modalities in the foreseeable future [57].

Suicide genes can be introduced into cancer cells to make them more sensitive to chemotherapeutic agents or toxins. Chemotherapeutic suicide gene therapy approaches are known as gene-directed enzyme prodrug therapy (GDEPT) or gene-prodrug activation therapy. GDEPT represents a rational technology to improve drug selectivity for cancer cells. One of the most promising approaches has been the use of retroviral vectors containing a herpes simplex virus thymidine kinase (HSVtk) gene in patients with brain tumors. In rat models it had previously been shown that it was mainly the rapidly proliferating tumor cells that integrated HSVtk and subsequent intravenous infusions with ganciclovir resulted in considerable death of cells expressing HSVtk [121 and references therein].

Another example of a suicide gene therapy-based approach is based on the bacterial enzyme caboxypeptidase G2 (CPG2), a bacterial enzyme isolated from *Pseudomonas aeruginosa*. The advantage of CP2 over HSVtk and cytosine deaminase-5-fluorocytosine (CD-5-FC), is that it activates prodrugs that are able to kill quiescent as well as proliferating cells. CPG2 was originally identified for use with the prodrug 4-[(2-chloroethyl)(2-mesyloxyethyl) amino]benzoyl-L-glutamic acid (CMDA) which releases a cytotoxic alkylating agent following activation by CPG2. The CPG2–CMDA system has been extensively evaluated *in vitro* and in preclinical xenografts in nude mice. Advantages of the CP2-CMDA system are: enhanced selectivity toward cancer cells, amplification effects, and bystander cell death [122].

All gene therapy strategies share a common component, the need for a selective delivery vehicle or vector with tumor-targeting capabilities. This need has led to the in-depth investigation of viruses as new vectors for gene therapy [120]. However, there are drawbacks to the use of viral vectors, particularly the use of replicating virus. These include the potential for immunological complications and their lack of selectivity for cancer cells. A number of avenues are being explored to overcome these limitations and several developments are now being investigated in preclinical models including improved vectors that have altered tissue tropism [54, 57]. One such interesting class of vectors is the replication-selective viruses. These can be used either to deliver therapeutic genes or as oncogenic agents in their own right.

Viruses have evolved to infect cells, replicate, induce cell death, release viral particles and, finally, to spread in human tissues. Replication of viruses in tumor tissue leads to amplification of the input dose at the tumor site, while a lack of replication in normal tissues can result in efficient clearance and reduced toxicity. Revolutionary advances in molecular biology and genetics have led to a fundamental understanding of both the replication and pathogenicity of viruses, and carcinogenesis. These advances have enabled novel agents to be engineered that enhance their safety and/or their antitumor potency. Replication-selective oncolytic viruses are being developed as a novel, targeted form of anticancer therapy (virotherapy) [122]. An example of one of these is *dl*1520. This is an adenovirus which was designed to have a deletion of a gene encoding a p53-binding protein, EIB-55kDa (*dl*1520). The hypothesis was that the virus would be selective for tumors that already had inhibited or lost p53 function.

The goal of clinical studies with *dl*1520 has been to increase sequentially systemic exposure to the virus only after safety with more localized delivery had been demonstrated. Following demonstration of safety and biological activity by the intratumoral route, trials were sequentially initiated to study intracavitary instillation, intra-arterial infusion and eventually intravenous administration. Clinical trials of combinations with chemotherapy were initiated only after safety of *dl*1520 as a single agent had been documented by the relevant route of administration. To date no objective responses have been documented with single-agent therapy in phase I or I–II trials in patients with pancreatic, colorectal or ovarian carcinomas. A favorable and potentially synergistic interaction with chemotherapy was discovered in some tumor types and by different routes of administration [122].

Replicating adenoviruses can achieve higher efficiencies of delivery compared with other vectors and can also achieve enhanced specificity in certain types of cancer. They also have the advantage of being oncolytic. There are several clinical trials ongoing with *dl*1520. Increasing the specificity and potency of this type of virus opens up many opportunities for improvement in future research [122].

One of the most important complications of cancer chemotherapy is the toxic effect that the drugs have on normal tissues – particularly the bone marrow. Several gene therapy vectors have been developed with the aim of expressing drug-resistance genes specifically in bone marrow stem cells, so protecting them from the drugs. The feasibility of this approach has been established in animal models and recent advances in the design of gene therapy vectors offer promise for future clinical applications. Gene therapy approaches have been developed to promote stable integration of drug-resistance genes in pluripotent hematopoietic stem cells. Protecting a cancer patient's hematopoietic stem cells from the toxic effects of cancer therapies involves autologous transplantation of genetically modified bone marrow cells. These cells are transduced *ex vivo* with a retroviral vector that contains a drug-resistance gene before being transplanted back into the patient. The choice of envelope protein expressed by the viral vector has proven to be an important determinant of stem-cell transduction efficiency, due to the fact that the receptors for viral envelopes are expressed at varying levels on the stem-cell surface. Mouse oncoretroviruses can only gain access to nuclear chromatin during mitosis, whereas lentiviral vectors can enter an intact nucleus directly through the nuclear pores. Various drug-resistance genes have been shown to protect hematopoietic stem cells in animal models. These include the multidrug resistance 1 (*MDR1*), dihydrofolate reductase (*DHFR*) and methylguanine methyltransferase (*MGMT*) genes. Several clinical trials have evaluated the feasibility of hematopoietic protection using *MDR1*-expressing vectors in adult cancer patients. Hematopoietic stem-cell gene therapy offers an opportunity to widen the anticancer therapeutic index [123].

In a complementary approach to expressing drug resistance genes, targeted gene delivery is being developed to damage tumor vasculature. It has been recently demonstrated that a cationic nanoparticle (NP) coupled to an integrin-targeting ligand can deliver genes selectively to angiogenic blood vessels in tumor-bearing mice. NPs conjugated to a mutant *Raf* gene, which blocks endothelial signaling in response to multiple growth factors, when injected into mice caused apoptosis of the tumor-associated endothelium, ultimately leading to tumor cell apoptosis and sustained regression of established primary and metastatic tumors [124]. This approach has great potential for the treatment of human cancer if it can be translated to the clinic [125].

The early results on the clinical efficacy of gene therapies have been generally disappointing, largely because the available gene-transfer vectors proved to be inadequate. Recently however, clinical benefit has been clearly demonstrated and great progress made in selecting and improving vectors. There is now every prospect that the second decade will see gene therapy live up to its enormous potential [118,126].

13.8 Drug discovery and drug design

The rapid pace and scale of developments in the field of genomics has resulted in a quantum leap in the way in which the pharmaceutical industry approaches the discovery and development of new drug compounds and it is estimated that this will increase the number of molecular targets from $\sim$1000 currently used by the pharmaceutical industry to as many as 10 000 [127]. Information obtained from SNPs, gene expression and array technology, proteomics and virtual (*in silico*) screening, coupled with *de novo* drug design, is being used to identify targets and structures that are potentially pharmacologically active.

It has been estimated that any two unrelated individuals differ by approximately one basepair (bp) change in every 1000 bp. Since there are approximately 3×10^9 bp in the human genome, this frequency equates to 3×10^6 differences between any two unrelated individuals. Many of these differences encode the visible differences between unrelated individuals, but others contribute to health and clinical outcome and they have the potential to identify the genetic factors contributing to complex diseases such as cancer and to medically complex phenotypes such as drug sensitivity and drug resistance [128,129].

The flow of investigation from gene sequence of potential therapeutic targets, through mRNA and protein expression, to protein structure and drug design has accelerated with automation of many of the steps and promises to revolutionize the drugs available for the treatment of cancer and other human diseases within the next few years.

13.9   Conclusions and future prospects

Much of the work on oncogene therapeutics is in a preliminary or even speculative theoretical stage. There is considerable enthusiasm for the therapeutic potential revealed by the study of oncogenes; however, it must be remembered that oncogene products are of fundamental importance to normal cell growth and differentiation, and that mutant oncogenes and their products often differ from their normal counterparts only in minor details. In theory this may be sufficient to improve the current therapeutic index, but it is possible that anti-oncogene therapies may prove as, or more toxic than the therapeutic modalities currently available. With few exceptions, the technical means available to achieve these therapeutic possibilities still range from the primitive and experimental to the currently impossible, making practical realization of many of these treatments still a good number of years away. However, as indicated above, progress towards developing effective therapies is being made.

Advances in human genome research are opening the door to a new paradigm for practicing medicine that promises to transform health care. Personalized medicine, the use of marker-assisted diagnosis and targeted therapies derived from an individual's molecular profile, will impact on the way drugs are developed and medicine is practiced. Knowledge of the molecular basis of disease will lead to novel target identification, toxicogenomic markers to screen compounds and improved selection of clinical trial patients, which will fundamentally change the pharmaceutical industry. In addition, patient care will be revolutionized through the use of novel molecular predisposition, screening, diagnostic, prognostic, phamacogenomic and monitoring markers. Although numerous challenges will need to be met to make personalized medicine a reality, with time, this approach has the potential to replace the traditional trial-and-error practice of medicine [129,130].

Personalized medicine is based on the hypothesis that diseases are heterogeneous, from their causes to rates of progression to their response to drugs. Each person's disease might be unique and therefore that person needs to be treated as an individual. With limited understanding of the molecular basis of disease, reliance has been placed traditionally on nonspecific clinical signs. As genomic tools are developed, so also will the ability to dissect disease into its component parts. Clinical phenotypes thought to be one disease will be subclassified by a new genomic taxonomy. Recent discoveries in the molecular pathology of cancer have highlighted important and clinically significant differences in the gene expression patterns of a variety of tumors, including leukemia and breast cancer. Preventive therapies have already been undertaken by the medical community as evidenced by the use of selective estrogen-receptor modulators (SERMs) for patients at risk of breast cancer.

Because tumor samples are routinely biopsied or removed, the first disease area likely to benefit from these technologies is cancer. Advances in our ability to classify disease are illustrated in a recent publication where gene-expression profiling was used to classify two related cancer types. Expression patterns of 50 genes were determined to distinguish accurately between ALL and AML [131]. Similar approaches have now been used to identify candidate prognostic markers in a number of cancers, as indicated in previous chapters.

Pharmacogenetics encompasses the involvement of genes in an individual's response to drugs with the ultimate goal of predicting a patient's genetic response to a specific drug as a means of delivering the best possible medical treatment. By predicting the drug response of an individual, it will be possible to increase the success of therapies and reduce the incidence of side effects [132,133]. SNPs are being used to classify patients whose target protein for their prescribed medicine shows sequence variation or differences in expression level.

To give just one example of the potential in relation to antiangiogenesis treatment, it has been suggested that other inhibitors of signal transduction that have potential as cocktail treatments might block different sets of angiogenic factors and should be used in combination with Herceptin (Trastuzumab). It should eventually be able to obtain angiogenic profiles of individual tumors and patients, allowing the most appropriate combination of signal-transduction inhibitors to be selected with a view to customizing this type of cancer treatment [134].

Cell line supernatants or explants from tumor tissues have already been used in large-scale expression-profiling experiments to identify new cancer markers and assays are already being used to determine eligibility for Herceptin treatment. New molecular markers might face many hurdles before they can be implemented in patient care. The issues range from regulatory approval and acceptance of these new markers, to developing assays, to resolving issues around the ethical, legal and social implications of obtaining highly sensitive genetic information.

Two factors will affect the availability of genetic testing as part of selection of drug therapy: testing technologies and test validation. For a test to be clinically valid it must also adequately predict the association of the test result with a clinical outcome. Primarily this is a function of the relationship of a gene, and sequence variants of that gene, with an expected outcome. This relationship could be difficult to establish. For example, there might be multiple genes that contribute independently to a particular drug-related response, leading to a low positive predictive value for any one of these. Also, for any gene, there might be multiple alleles that give rise to the condition, most or all of which will need to be detected. Demonstrating that a test has clinical validity can only be accomplished by careful clinical studies [132].

In addition to understanding that genetic variation defines us as individuals, it is important to understand that genetic variation also defines populations. Even if 'the right drug for the right patient' may still be some years away, the right drug for the right population may be a more achievable goal in the near future [128].

As understanding of the development of cancer and the complex signaling mechanisms involved improves, the enormous potential for intervention studies that prevent or slow down the malignant process is beginning to be appreciated. While the majority of current research is aimed at developing drugs to target key molecules it should not be forgotten that dietary constituents have potential as cancer agents and that a balanced, healthy diet may be able to reduce incidence, as is suggested by epidemiological evidence [135] and that early detection remains one of the most promising approaches to successful treatment (as for example in cervical cancer) [136].

It seems clear that our understanding of the mechanisms involved in the development of cancer and in its treatment are going to be completely revolutionized within the next 5–10 years and this will bring cancer prevention, and the ultimate goal of cancer therapy – the elimination or inactivation of each and every remaining malignant cell from the patient – considerably closer.

References

1. **Workman, P. and Kaye, S.B.** (2002) *Trends Mol. Med.*, **8**, (Suppl. *A Trends Guide to Cancer Therapeutics*), s1.
2. **Zwick, E., Bange, J. and Ullrich, A.** (2002) *Trends Mol. Med.*, **8**, 17.
3. **Fabbro, D., Parkinson, D. and Matter, A.** (2002) *Curr. Opin. Pharmacol.*, **2**, 374.
4. **de Bono, J.S. and Rowinsky, E.K.** (2002) *Trends Mol. Med.*, **8**, (Suppl.), s19.
5. **Toplay, J., Freuhauf, S., Ho, A.D. and Zeller, W.J.** (2002) *Br. J. Cancer*, **86**, 1487.
6. **Capdeville, R., Buchdunger, E., Zimmermann, J. and Matter, A.** (2002) *Nature Rev. Drug Discov.*, **1**, 493.
7. **Osborne, C.K.** (1998) *Breast Cancer Res. Treat.*, **51**, 227.
8. **Park, W-C. and Jordan, V.C.** (2002) *Trends Mol. Med.*, **8**, 82.
9. **Cauley, J.A., Norton, L., Lippman, M.E. *et al.*** (2001) *Breast Cancer Res. Treat.*, **65**, 125.
10. **Shang, Y and Brown, M.** (2002) *Science*, **295**, 2465.
11. **Zusi, F.C., Lorenzi, M.V. and Vivat-Hannah, V.** (2002) *Drug Deliv. Today*, **7**, 1165.
12. **Sporn, M.B., Suh, N. and Mangelsdorf, D.J.** (2001) *Trends Mol. Med.*, **7**, 395.
13. **Moingeon, P.** (2001) *Vaccine*, **19**, 1305.
14. **Cohen, E.P.** (2001) *Trends Mol. Med.*, **7**, 175.
15. **Zhu, D., Rice, J., Savelyeva, N. and Stevenson, F.K.** (2001) *Trends Mol. Med.*, **7**, 566.
16. **Finn, O.J. and Forni, G.** (2002) *Curr. Opin. Immunol.*, **14**, 172.
17. **Jager, E., Jager, D. and Knuth, A.** (2002) *Curr. Opin. Immunol.*, **14**, 178.
18. **Parmiani, G., Castelli, C., Dalerba, P. *et al.*** (2002) *J. Natl. Cancer Inst.*, **94**, 805.
19. **Whelan, M., Whelan, J., Russell, N. and Dalgleish, A.** (2003) *Drug Discov. Today*, **8**, 253.
20. **Coulie, P.G. and van der Bruggen, P.** (2003) *Curr. Opin. Immunol.*, **15**, 131.
21. **Schuler-Thurner, B., Schultz, E.S., Berger, T.G. *et al.*** (2002) *J. Exp. Med.*, **195**, 1279.
22. **Keilholz, U., Weber, J., Finke, J.H. *et al.*** (2002) *J. Immunother.*, **25**, 97.
23. **Egen, J.G., Kuhns, M.S. and Allison, J.P.** (2002) *Nature Immunol.*, **3**, 611.
24. **Fong, L. and Engelman, E.G.** (2000) *Annu. Rev. Immunol.*, **18**, 245.
25. **Bhardwaj, N.** (2001) *Trends Mol. Med.*, **7**, 388.
26. **Schuler, G., Schuler-Turner, B. and Steinman, R.** (2003) *Curr. Opin. Immunol.*, **15**, 138.
27. **Ye, Z., Hellstrom, I., Hayden-Ledbetter, M. *et al.*** (2002) *Nature Med.*, **8**, 343.
28. **Kohler, G. and Milstein, C.** (1975) *Nature*, **256**, 495.
29. **Trikha, M., Yan, L. and Nakada, T.** (2002) *Curr. Opin. Biotech.*, **13**, 1.
30. **Baselga, J., Albanell, J., Molina, M.A. and Arribas, J.** (2001) *Semin. Oncol.*, **28**, Suppl 16, 4.
31. **Slamon, D.J., Leyland-Jones, B., Shak, S. *et al.*** (2001) *New. Engl. J. Med.*, **344**, 783.
32. **Seidman, A.D., Fornier, M.N., Esteva, F.J. *et al.*** (2001) *J. Clin. Oncol.*, **19**, 2587.
33. **Cokgor, I., Akabani, G., Kuan, C-T. *et al.*** (2000) *J. Clin. Oncol.*, **18**, 3862.
34. **Grana, C., Chinol, M., Robertson, C. *et al.*** (2002) *Br. J. Cancer*, **86**, 207.
35. **Reiter, Y.** (2001) *Adv. Cancer Res.*, **81**, 93.
36. **Bagshawe, K.D.** (1987) *Br. J. Cancer*, **56**, 531.
37. **Tietze, L.F., Feuerstein, T., Fecher, A. *et al.*** (2002) *Angew. Chem.*, **114**, 759.
38. **Kousparou, C.A., Epenetos, A.A. and Deonarain, M.P.** (2002) *Int. J. Cancer*, **99**, 138.
39. **Casey, J.L., Napier, M.P., King, D.J. *et al.*** (2002) *Br. J. Cancer*, **86**, 1401.
40. **Lineberger, M.L., Maloney, D.G. and Bernstein, I.D.** (2002) *Trends Mol. Med.*, **8**, 69.
41. **Melief, C.J.M., Toes, R.E.M., Medema, J.P. *et al.*** (2000) *Adv. Immunol.*, **75**, 235.
42. **June, C.H.** (2001) In *Cancer Chemotherapy & Biotherapy: Principles and Practice* (Chabner, B.A. and Longo, D.L., eds), Third Edition, Chapter 32, p. 925. Lippincott, Williams & Wilkins, Philadelphia.
43. **Gottschalk, S., Heslop, H.E. and Rooney, C.M.** (2002) *Adv. Cancer Res.*, **84**, 175.
44. **Dudley, M.E., Wunderlich, J.R., Robbins, P.F. *et al.*** (2002) *Science*, **298**, 850.
45. **Yee, C. and Greenberg, P.** (2002) *Nature Rev. Cancer*, **2**, 409.
46. **Schumacher, T.N.M.** (2002) *Nature Rev. Immunol.*, **2**, 512.
47. **Wilson, M.T., Singh, A.K. and Van Kaer, L.** (2002) *Trends Mol. Med.*, **8**, 225.
48. **Demetri, G.D., Ernst, T.J., Pratt, E.S. II. *et al.*** (1990) *J. Clin. Invest.*, **86**, 1261.
49. **Lebel-Binay, S., Laguerre, B., Quintin-Colonna, F. *et al.*** (1995) *Eur. J. Immunol.*, **25**, 2340.

50. **Hurford, R.K., Dranoff, G., Mulligan, R.C. and Tepper, R.I.** (1995) *Nature Genetics*, **10**, 430.
51. **Economides, A.N., Ravetch, J.V., Yancopoulos, G.D. and Stahl, N.** (1995) *Science*, **270**, 1351.
52. **Homey, B., Muller, A. and Zlotnik, A.** (2002) *Nature Rev. Immunol.*, **2**, 175.
53. **Adams, J.** (2003) *Drug Discov. Today*, **8**, 307.
54. **Lane, P. and Lain, S.** (2002) *Trends Mol. Med.*, **8**, (Suppl. *A Trends Guide to Cancer Therapeutics*), s38.
55. **International Association of Cancer Registries.** www.iacr.fr/p53
56. **Hupp, T.R., Lane, D.P. and Ball, K.L.** (2000) *Biochem. J.*, **352**, 1.
57. **Chang, E.H., Pirollo, K.F. and Bouker, K.B.** (2000) *Mol. Med. Today*, **6**, 358.
58. **Bykov, V.J.N., Issaeva, N., Shilov, A. *et al*.** (2002) *Nature Med.*, **8**, 282.
59. **Lane, D.P. and Hupp, T.R.** (2003) *Drug Discov. Today*, **8**, 347.
60. **Ashkenazi, A.** (2002) *Nature Rev. Cancer*, **2**, 420.
61. **Kolch, W., Kotwaliwale, A., Vass, K. and Janosch, P.** (2002) *Expert Rev. Mol. Med.*, 25 April, http://www.expertreviews.org/02004386h.htm
62. **Knockaert, M., Greengard, P. and Meijer, L.** (2002) *Trends Pharmacol. Sci.*, **23**, 417.
63. **Senderowicz, A.M.** (2001) *Leukemia*, **15**, 1.
64. **Cristofanilli, M., Charnsangavej, C. and Hortobagyi, G.N.** (2002) *Nature Rev. Drug Discov.*, **1**, 415.
65. **He, Y., Kozaki, K., Karpanen, T. *et al*.** (2002) *J. Natl Cancer Inst.*, **94**, 819.
66. **American Association for Cancer Research** (2002) *Proceedings 93rd* Annual Meeting, April 6–10, 2002, San Francisco, California. Volume 43. Cadmus Professional Communications, Linthicum, MD, USA.
67. **Novak, K.** (2002) *Nature Med.*, **8**, 427.
68. **Sweeney, C.J., Miller, K.D. and Sledge, G.W.** (2003) *Trends Mol. Med.*, **9**, 24.
69. **Madhusudan, S. and Harris, A.L.** (2002) *Curr. Opin. Pharmacol.*, **2**, 403.
70. **McCarty, M.F., Liu, W., Fan, F. *et al*.** (2003) *Trends Mol. Med.*, **9**, 53.
71. **Kerbel, R. and Folkman, J.** (2002) *Nature Rev. Cancer*, **2**, 727.
72. **Semenza, G.L.** (2002) *Trends Mol. Med.*, **8**, (Suppl. *A Trends Guide to Cancer Therapeutics*), s62.
73. **Brahimi-Horn, C., Berra, E. and Pouyssegur, J.** (2001) *Trends Cell Biol.*, **11**, s32.
74. **Pili, R. and Donehower, R.C.** (2003) *J. Natl Cancer Inst.*, **95**, 498.
75. **Jayson, G.C., Zweit, J., Jackson, A. *et al*.** (2002) *J. Natl Cancer Inst.*, **94**, 1484.
76. **Yu, J.L., Rak, J.W., Coomber, B.L. *et al*.** (2002) *Science*, **295**, 1526.
77. **Pasqualini, R., Arap, W. and McDonald, D.M.** (2002) *Trends Mol. Med.*, **8**, 563.
78. **Ruoslahti, E.** (2002) *Drug Discov. Today*, **7**, 1138.
79. **Laakkonen, P., Porkka, K., Hoffman, J.A. and Ruoslahti, E.** (2002) *Nature Med.*, **8**, 751.
80. **Bikfalvi, A. and Bicknell, R.** (2002) *Trends Pharmacol. Sci.*, **23**, 576.
81. **Balasubramaniam, S.P., Brown, N.J. and Reed, M.W.R.** (2002) *Br. J. Cancer*, **87**, 1057.
82. **Munn, L.L.** (2003) *Drug Discov. Today*, **8**, 396.
83. **Brown, R. and Strathdee, G.** (2002) *Trends Mol. Med.*, **8**, (Suppl. *A Trends Guide to Cancer Therapeutics*), s43.
84. **Jones, P.A. and Baylin, S.B.** (2002) *Nature Rev. Genet.*, **3**, 415.
85. **Modica-Napolitano, J.S. and Singh, K.K.** (2002) *Expert Rev. Mol. Med.*, 11 April, http://www-ermm.cbcu.cam.ac.uk/12004453h.htm
86. **Weissig, V. and Torchilin, V.P. (eds)** (2001) *Adv. Drug Deliv. Rev.*, **49**, Nos. 1–2.
87. **Nangia-Makker, P., Conklin, J., Hogan, V. and Raz, A.** (2002) *Trends Mol. Med.*, **8**, 187.
88. **Steinbach, G., Lynch, P.M., Phillips, K.S. *et al*.** (2000) *New Engl. J. Med.*, **342**, 1946.
89. **Subbaramaiah, K. and Dannenberg, A.J.** (2003) *Trends Pharmacol. Sci.*, **24**, 96.
90. **Iñiguez, M.A., Rodríguez, A., Volpert, O.V. *et al*.** (2003) *Trends Mol. Med.*, **9**, 73.
91. **Green, P.J., Pines, O. and Inouye, M.** (1986) *Ann. Rev. Biochem.*, **55**, 569.
92. **Takayama, K.M. and Inouye, M.** (1990) *Crit. Rev. Biochem. Mol. Biol.*, **25**, 155.
93. **Buick, K.B., Liu, E.T. and Larrick, J.W.** (1988) In *Oncogenes: an Introduction to the Concept of Cancer Genes*, p. 262. Springer-Verlag, New York.
94. **Rossi, J.J.** (1995) *British Medical Bulletin*, **51**, 217.
95. **Scanlon, K.J., Ohta, Y., Ishida, H. *et al*.** (1995) *FASEB J.*, **9**, 1288.
96. **Putnam, D.A.** (1996) *Am. J. Health-Syst. Pharm.*, **53**, 151.

97. **Gura, T.** (1995) *Science,* **270**, 575.
98. **Rojanasakul, Y.** (1996) *Advanced Drug Delivery Reviews,* **18**, 115.
99. **Stein, C.A. and Cheng, Y.C.** (1993) *Science,* **261**, 1004.
100 **DeMesmaeker, A., Haner, R., Martin, P. and Moser, H.E.** (1995) *Acc. Chem. Res.,* **28**, 366.
101. **Whitton, J.L.** (1994) *Adv. Virus Res.,* **44**, 267.
102. **Heidenreich, O., Kang, S-H., Xu, X. and Nerenberg, M.** (1995) *Mol. Med. Today,* **1**, 128.
103. **Wagener, R.** (1995) *Nature Med.,* **1**, 1116.
104. **Stein, C.A.** (1995) *Nature Med.,* **1**, 1119.
105. **Agarwal, N. and Gewirtz, A.M.** (1999) *Biochim. Biophys. Acta,* **1489**, 85.
106. **Clark, R.E.** (2000) *Leukemia,* **14**, 347.
107. **Waters, J.S., Webb, A., Cunningham, D.** *et al.* (2000) *J. Clin. Oncol.,* **18**, 1812.
108. **Tamm, I., Dorken, B. and Hartmann, G.** (2001) *Lancet,* **358**, 489.
109. **Shuey, D.J., McCallus, D.E. and Giordano, T.** (2002) *Drug Deliv. Today,* **7**, 1040.
110. **Agami, R.** (2002) *Curr. Opin. Chem. Biol.,* **6**, 829.
111. **Shi, Y.** (2003) *Trends Genet.,* **19**, 9.
112. **Martinez, L.A., Naguibneva, I., Lehrmann, H.** *et al.* (2002) *Proc. Natl Acad. Sci. USA,* **99**, 14849.
113. **Weatherall, D.J.** (1991) *The New Genetics and Clinical Practice,* Chapter 10, 3rd edition. Oxford University Press, Oxford.
114. **Elledge, R.M. and Lee, W-H.** (1995) *BioEssays,* **17**, 923.
115. **Westbrook, C.A. and Arenas, R.B.** (1995) *Adv. Drug Deliv. Rev.,* **17**, 349.
116. **Wilson, C. and Kay, M.A.** (1995) *Nature Med.,* **1**, 887.
117. **Yang, Y., Trinchieri, G. and Wilson, J.M.** (1995) *Nature Med.,* **1**, 890.
118. **Mountain, A.** (2000) *Tibtech,* **18**, 119.
119. **Cristiano, R.J. and Roth, J.A.** (1995) *J. Mol. Med.,* **73**, 479.
120. **Swisher, S.G. and Roth, J.A.** (2002) *Curr. Oncol. Rep.,* **4**, 334.
121. **Culver, K.W. and Blaese, R.M.** (1994) *Trends Genetics,* **10**, 174.
122. **Kirn, D., Niculescu-Duvaz, I., Hallden, G. and Springer, C.J.** (2002) *Trends Mol. Med.,* **8**, (Suppl. *A Trends Guide to Cancer Therapeutics*), s68.
123. **Sorrentino, B.P.** (2002) *Nature Rev. Cancer,* **2**, 431.
124. **Hood, J.D., Bednarski, M., Frausto, R.** *et al.* (2002) *Science,* **296**, 2404.
125. **Reynolds, A.R., Moghimi, S.M. and Hodivala-Dilke, K.** (2003) *Trends Mol. Med.,* **9**, 2.
126. **McCormick, F.** (2001) *Nature Rev. Cancer,* **1**, 130.
127. **Dean, P.M., Zanders, E.D. and Bailey, D.S.** (2001) *Trends Biotech.,* **19**, 288.
128. **Jazwinska, E.C.** (2001) *Drug Discov. Today,* **6**, 198.
129. **Huang, Y. and Sadée, W.** (2003) *Drug Discov. Today,* **8**, 356.
130. **Ginsburg, G.S. and McCarthy, J.J.** (2001) *Trends Biotech.,* **19**, 491.
131. **Golub, T.R. Slonim, D.K., Tamayo, P.** *et al.* (1999) *Science,* **286**, 531.
132. **Spear, B.B., Heath-Chiozzi, M. and Huff, J.** (2001) *Trends Mol. Med.,* **7**, 201.
133. **Watters, J.W. and McLeod, H.L.** (2003) *Biochim. Biophys. Acta,* **1603**, 99.
134. **Izumi, Y., Xu, L., di Tomaso, E.** *et al.* (2002) *Nature,* **416**, 279.
135. **Manson, M.** (2003) *Trends Mol. Med.,* **9**, 11.
136. **Etzioni, R., Urban, N., Ramsey, S.** *et al.* (2003) *Nature Rev. Cancer,* **3**, 243.

Molecular techniques for analysis of genes

14

14.1 Introduction

An understanding of the methodology used in the analysis of genes described in earlier chapters is fundamental to understanding how these genes can be used for diagnosis and prognosis in cancer. There are many excellent molecular biology books available which describe these procedures in detail [1–3]. This chapter outlines these techniques briefly so that those unfamiliar with the technology can follow the previous chapters without immediate recourse to other books.

Genes can be studied at three levels, namely DNA, RNA or protein. These molecules can be either within the cell in prepared tissue sections (*in situ*) or in isolation. These two approaches give different information. Isolated DNA can be examined for qualitative and quantitative abnormalities. This is useful when looking for rearrangements or mutations within a particular gene and also for assessing absolute levels of a gene (gene amplification). Analysis of isolated RNA gives information about the level of transcription (gene expression). Direct analysis of protein allows the determination of protein levels or of changes in protein size.

In situ analysis provides information concerning the spatial distribution of molecules in the cell and therefore shows which cells are expressing a particular gene, RNA sequence or protein molecule, but cannot easily be used for quantitative analysis. When used on chromosome spreads this technique gives information about the chromosomal localization of a gene or, once a gene has been mapped, can be used to identify gene deletions.

In addition to the more 'traditional approaches' used in the past 20 years the analysis of genes has been revolutionized by the introduction of new technologies. Most notable of these is microarray technology which is a high-throughput method which enables information to be generated about gene expression and function. The technological advances in automation and bioinformatics have resulted in a discipline of biology termed genomics, which can be broadly defined as the generation and analysis of information about genes and genomes. The field is characterized by global, comprehensive studies that are conducted in a systematic manner. Although still in its infancy, genomics is beginning to permit the identification and characterization of individual genes and patterns of gene expression that distinguish malignant and premalignant cells from their normal counterparts [4]. In parallel, the comprehensive study of proteins and protein systems is now called proteomics. From a biomedical point of view proteomics has great potential as the bulk of phamacological interactions and diagnostic tests are directed at proteins rather than their genes [4]. Proteomics-based approaches, which enable the quantitative investigation of both cellular protein expression levels and protein–protein interactions involved in signaling networks, promise to define the molecules controlling the processes involved in cancer [5].

14.2 Analysis of DNA or RNA

14.2.1 Genetic markers

Four different types of DNA marker are used in genetic analysis. The markers segregate in Mendelian fashion and, provided they are sufficiently polymorphic, can be used in gene tracking studies, to help make genetic maps to be used in the subsequent isolation of genes and in cancer research for the identification of loss of heterozygosity (LOH).

The first markers to be developed were the restriction fragment length polymorphisms (RFLPs). Approximately every 200 bp along the length of chromosomes the sequence of DNA varies between individuals, usually by only a single base change. Where they coincide with restriction sites, the alterations can be detected by restriction enzymes. The presence or absence of any particular site is variable in the population, hence the term polymorphism, and the technique is referred to as RFLP analysis. The variations do not usually confer any phenotypic effect since they are frequently found in introns rather than coding exons (*Figure 14.1a*). Approximately 100 000 RFLPs are distributed throughout the human genome.

The second group of markers is composed of arrays of short tandem repeated sequences, each different allele being made up of a different number of repeats (*Figure 14.1b*). These markers are termed minisatellites or variable number of tandem repeats (VNTRs) and approximately 10 000 exist in the genome [7]. Many of these markers are located towards the telomeres of the chromosomes which can limit their usefulness.

The third group, which has been the one most frequently used for genetic analysis, is the microsatellites [8]. Di- (e.g. CA repeats), tri- and tetranucleotide markers are all used with CA and tetranucleotides being the most common (*Figure 14.1c*). These markers are found in the genome with approximately the same frequency as the VNTRs.

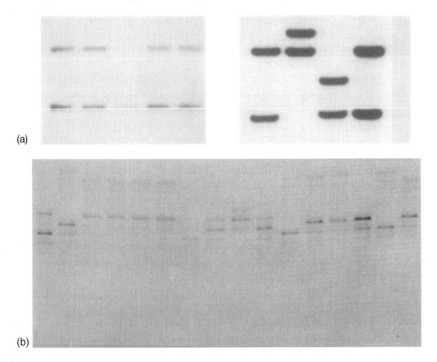

(a)

(b)

Figure 14.1

(a) Two allele RFLP linked to the *APC* gene. (b) Minisatellite marker on chromosome 16p showing several different alleles present in the population. Adapted from [6] courtesy of Steve McKay, Birmingham Heartlands Hospital.

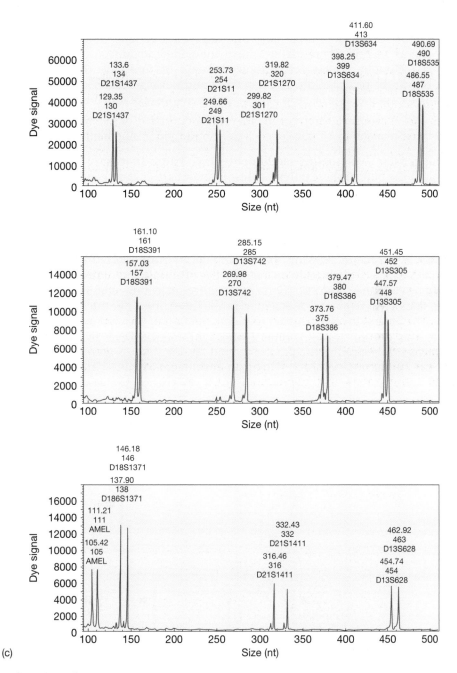

(c)

Figure 14.1 (*continued*)

(c) Twelve microsatellite markers on chromosomes 13, 18 and 21 plus a marker on the X and Y chromosomes have been analyzed together in a single multiplex PCR reaction. Primers for each marker were labeled with three different fluorochromes so that each can be resolved following gel electrophoresis and analysis on an ABI377 sequencer using Genotyper software. The name of each marker is given (e.g. D21S1437) together with the size (e.g. 125 and 133 for D21S1437) and peak area (e.g. 4921 and 4719 for D21S1437) below the alleles. Figure courtesy of Carol Hardy, West Midlands Regional Genetics Laboratory, UK.

The reasons why the latter two groups of markers have found favor over RFLPs is primarily because they are much more informative for genetic studies, as discussed below (see Section 14.2.5). Microsatellites have the added advantage of being analyzable by PCR, a technique which revolutionized much of molecular biology and which is described in the next section. The first two groups of markers are studied using Southern blotting techniques (see Section 14.2.4) and therefore take much longer to analyze.

The fourth type of marker has been identified as a result of the human genome sequencing projects. In addition to raw sequencing and mapping data, the two human genome sequencing projects have uncovered millions of single nucleotide polymorphisms (SNPs) representing additional DNA bases that vary between individuals. Although the bulk of these polymorphisms do not affect gene regulation or the functions of encoded proteins, they are being used as genetic markers to pinpoint the location of nearby genes that are responsible for a disease phenotype [4].

14.2.2 Polymerase chain reaction (PCR)

In 1985, a technique allowing specific amplification of selected sequences of DNA was described. PCR technique relies on a knowledge of at least part of the DNA sequence around the region of interest, because short specific oligonucleotides complementary to sequences either side of this region are required to prime the synthesis of the DNA sequence between them. The process involves three stages: denaturation of the double-stranded DNA, annealing of the oligonucleotide primers, and synthesis of DNA by a thermostable DNA polymerase. By repeating the process around 30 times, up to several million copies of the DNA can be made (*Figure 14.2*). The main advantages of this technique are: (1) it is rapid,

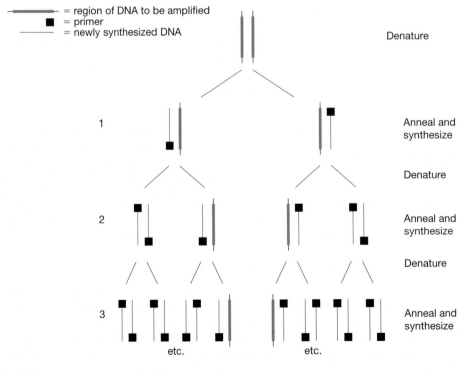

Figure 14.2

Polymerase chain reaction. Adapted from [6].

particularly as the product can often be detected directly on agarose gels (this takes several hours to perform, unlike Southern blotting which can take several days or more); (2) it is extremely sensitive (it is possible to start with a single molecule of DNA and finish with sufficient for analysis [9]); (3) DNA from unusual sources can be used (see Section 14.2.5).

PCR forms the basis of many analyses. The most important of these, with reference to the areas covered in this book, are the typing of microsatellite markers either for gene tracking studies as discussed below (see Section 14.2.5) or for the determination of loss of heterozygosity (see Section 3.2.2), for the detection of point mutations as discussed in the next section, or for the production of template DNA for sequencing.

14.2.3 Mutation detection techniques

Now that many genes have been isolated, the direct analysis of mutations has become routine practice. This procedure falls into two main groups: (1) some techniques are used where only one or two mutations are commonly found to disrupt the gene, for example the common mutations in the *RAS* gene; (2) other techniques are used as a preliminary screen to identify approximately where in a gene a mutation lies. This can then be followed up by sequencing of that region to confirm the nature of the mutation. These techniques have been used extensively in identifying mutations in the genes causing inherited cancers. In all of these genes described so far, mutations not only vary from one family to the next but in the majority of cases are relatively small abnormalities such as point mutations or small deletions and insertions and are widely distributed throughout the gene. Each of these techniques described below has different advantages such as sensitivity, ease of the procedure, cost and safety, so individuals have to balance each of these when selecting the method to use.

All of the procedures described below depend on initial PCR amplification of a region of the gene, usually the coding sequence and often a single exon, followed by analysis on a variety of different gels. Some of the more commonly used techniques are described briefly below and further details can be found in reference [10].

Allele-specific oligonucleotide (ASO) hybridization

Where only one or two common mutations, differing from the normal sequence by as little as a single base change, are known to disrupt a gene, ASOs can be used to identify the presence of the mutation. This technique involves the hybridization of the PCR product, bound to membrane filters in a denatured state, with oligonucleotide probes specific for a known mutation. These oligonucleotide probes are hybridized under specific hybridization conditions such that they only bind to the target DNA if there is perfect complementarity between the probe and the target. This technique has been used successfully to analyze the distribution of *RAS* mutations at codons 12, 13 or 61 in different tissues as described in reference [11] and in many of the preceding chapters.

Single-stranded conformation polymorphism analysis (SSCP)

SSCP is one of the most commonly used techniques for mutation detection because of its simplicity and relatively good sensitivity. Wild-type and test DNA are amplified by PCR and then denatured. The single-stranded molecules from each PCR product adopt a three-dimensional structure dependent on the sequence of the DNA. They are electophoresed through a nondenaturing gel and the migration of the fragment is dependent on the conformation adopted. If a mutation is present there will be a difference in the migration rate. Bands are visualized by the incorporation of a radioisotope or a fluorochrome during the PCR reaction or by post electrophoretic silver staining of the gel. The presence of a variation in the sequence is detected by comparison of the test sequence with a known wild-type fragment (*Figure 14.3a,b*). The sensitivity of SSCP is around 70–90% for PCR fragments less than 200 bp in length but decreases rapidly if fragments above 400 bp are analyzed. SSCP analysis can also be carried out on Hydrolink or MDE™ gels with a similar detection rate.

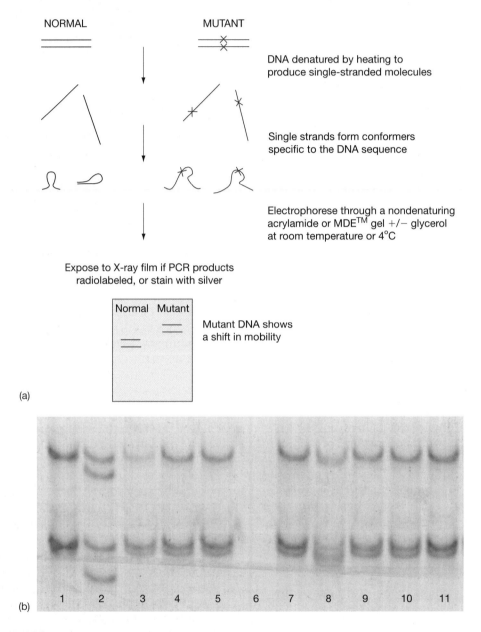

(a)

(b)

Figure 14.3

(a) SSCP analysis – schematic. (b) SSCP analysis of exon 16 of MSH2. Shifts corresponding to an alteration in the sequence of the PCR fragment are seen in lanes 2 and 8. These alterations may represent mutations or polymorphisms. Figure courtesy of Jennie Bell, West Midlands Regional Genetics Laboratory, UK.

Heteroduplex analysis

This technique depends on the fact that if both wild-type and mutant DNA are present in a PCR reaction, heteroduplex molecules are formed between the two different DNA species during the final rounds of amplification (*Figure 14.4*). Heteroduplex molecules migrate through acrylamide gels differently to homoduplex molecules, a feature which is enhanced by the use of gel matrices such as Hydrolink or MDE™ gels. The sensitivity of this technique is similar to SSCP analysis with the advantage of being extremely simple to perform. The sensitivity of the technique decreases if fragments over 300 bp are analyzed.

Denaturing gradient gel electrophoresis analysis (DGGE)

DGGE relies on the fact that a double-stranded DNA fragment exposed to a denaturant will melt in discrete domains to become single stranded and will do so in a sequence-specific manner. A single base change in the DNA is sufficient to change its melting temperature. In this technique, double-stranded DNA fragments are electophoresed through a gel containing an increasing concentration of denaturant. As the DNA migrates, individual domains of the fragment melt at a point corresponding to their melting temperature. As the fragment becomes partially melted, its migration through the gel is retarded. In order to prevent

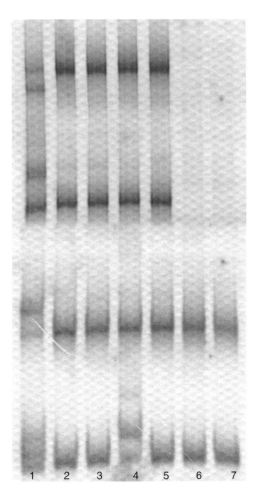

Figure 14.4

Heteroduplex analysis of four regions of *BRCA1*, exons 11A and B, exon 2 and exon 20 (in order from top of figure). Lane 2 contains a normal DNA sample and lane 1 contains positive controls for each of the four regions. Patient samples are in lanes 3–7. A shift in exon 20 of *BRCA1* is seen in the sample in lane 4 indicating the presence of a mutation or polymorphism in this fragment. Figure courtesy of Sally Rees, West Midlands Regional Genetics Laboratory, UK.

complete dissociation of the fragment, it is necessary to add a GC clamp (a sequence of approximately 40 bases, composed primarily of Cs and Gs) to one PCR primer; this region has a very high melting temperature so it will ensure that the fragment never separates into purely single-stranded molecules which would have no variation in their migration. The procedure is further enhanced by the production of heteroduplex molecules formed between one mutant and one wild-type DNA strand during the final rounds of PCR. These heteroduplexes have different melting temperatures compared with either wild-type or mutant homoduplexes and can sometimes be detected more readily on the gels. Detection of DNA fragments can be either by isotopic or nonisotopic means. DGGE is one of the most sensitive techniques as it can detect close to 100% of mutations in fragments up to 600 bp. However it is technically more difficult to set up initially and is also more expensive. The denaturing gradient can also be replaced with a temperature gradient with similar results.

DGGE analysis can also be carried out in two dimensions. For a patient under investigation, each exon of a gene is amplified and products are separated in the first dimension, based on size, on a neutral gel. The track containing the fragments is then cut out of the gel, turned through 90° and layered across a denaturing gradient and electrophoresed in the second dimension. Again any abnormal fragments will migrate differently from wild-type ones. The advantage of this technique is that a whole gene can be analyzed for a single individual in one gel.

Denaturing high-performance liquid chromatography (dHPLC)

dHPLC is a relatively new method of mutation detection which has many advantages over other techniques such as SSCP or DGGE. The method separates heteroduplex DNA molecules from homoduplex strands by using ion-paired reverse-phase high-performance liquid chromatography [12]. Partial heat denaturation decreases the retention time of mismatched heteroduplex species compared to the perfectly matched double-stranded molecules. Any PCR product containing a sequence variant will therefore be eluted earlier from the HPLC column and is detected by a UV online detector (*Figure 14.5*). Sensitivity and specificity of dHPLC

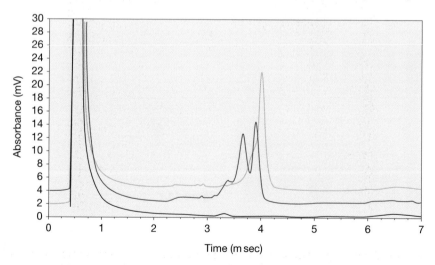

Figure 14.5

dHPLC analysis of exon 11 of *BRCA1*. The normal PCR fragment produces a single peak whereas the sample containing a mutation shows the presence of a double peak shifted to the left. Figure courtesy of Jennie Bell, Sally Rees and Marcus Allen, West Midlands Regional Genetics Laboratory, UK.

exceeds 96%. The method is rapid with each sample being analyzed in under 10 minutes. The semi-automated nature of the technique means that samples can be analyzed in a 96-well plate format without manual intervention. This means that labor costs are also lower than for other techniques [13]. dHPLC has been used widely as a rapid technique for analysis of familial cancer syndromes [14] but has also been used for many other purposes such as genotyping, determination of allelic imbalances and detection of DNA polymorphisms.

Chemical mismatch cleavage (CMC) and RNase cleavage

In this technique heteroduplexes are formed between radiolabeled wild-type DNA and mutant DNA by denaturing and reannealing. Any mismatches are chemically modified by osmium tetroxide for mismatched thymines or by hydroxylamine for mismatched cytosines. Mismatched guanines and adenines are also detected by labeling the antisense strand. The sites of chemically modified mismatches are then cleaved with piperidine, the products electrophoresed on denaturing acrylamide gels and detected by autoradiography (*Figure 14.6*). This technique is at least as sensitive as DGGE and also has the advantage that the position of the mutation can be localized. The main drawbacks to this technique used to be the hazardous nature of the chemicals used and that it is labor intensive. Use of potassium permanganate, fluorescent tags and sequencers has greatly reduced these limitations but the technique is no longer widely used.

Another technique based on the same principle is RNase A cleavage. In this method a heteroduplex molecule is formed between a wild-type riboprobe and a DNA fragment from the patient in question. Any mismatches formed because of sequence differences due to mutations can be cleaved by RNase. The main problem with this technique is its relatively low sensitivity of around 40–60%.

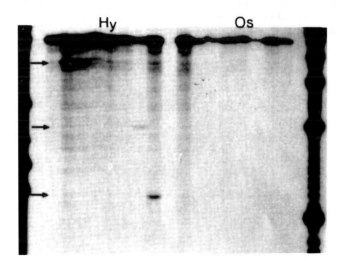

Figure 14.6

Chemical mismatch cleavage. Two groups of eight samples have been treated either with hydroxylamine (Hy) or osmium tetroxide (Os). The arrows indicate bands in two patients with similar mutations in lanes 2 and 3. Two other mismatches can be seen in lanes 7 and 8. All of these bands are seen with hydroxylamine only, indicating that they involve a cytosine. Molecular weight markers are included in the extreme right- and left-hand lanes. Photograph courtesy of Dr S. Enyat, DNA and Regional Haemostasis Laboratory, Birmingham Children's Hospital. Adapted from [6].

RT-PCR

If the sequences of intron/exon boundaries are not known it is not possible to design PCR primers either side of the exon, so it cannot be analyzed by the above techniques. One way around this problem is to produce cDNA from mRNA by reverse transcriptase PCR (RT-PCR). Primers lying within the coding sequence can then be used to amplify regions of the gene for further analysis. The technique can however only be used where the gene is expressed in accessible tissues. The analysis of cDNA is helpful for the identification both of larger deletions and of mutations affecting the splicing of mRNA which give smaller products because of the loss of several exons. In general, these differences can be identified by analysis of the cDNA from normal and affected individuals on agarose gel electrophoresis simply by the presence of small fragments in affected individuals. cDNA can also be the basis for the protein truncation test (PTT), described below, as large sections of coding sequence can be analyzed in a single reaction. For example, exons 1–14 of the *APC* gene can be analyzed together rather than in 14 separate reactions as used for SSCP or DGGE. Similarly, cDNA is useful in techniques such as chemical mismatch cleavage in which large fragments (1–2 kb) can be analyzed successfully, thereby reducing the amount of screening necessary to cover a gene.

Protein truncation test (PTT)

Many of the causative mutations in cancer genes are chain-terminating. A novel approach to mutation detection which makes use of this fact is the PTT, originally described for detection of point mutations in Duchenne muscular dystrophy [15]. The basis of this technique is again PCR but here the primers are modified to include the T7 promoter sequence as well as a eukaryotic initiation signal plus the usual sequence-specific region. Following PCR amplification of either cDNA or genomic DNA (if an exon is sufficiently large such as exon 15 of *APC* or exon 11 of *BRCA1*), the PCR products are used as templates in transcription and translation reactions resulting in the production of a peptide from the PCR fragment. In order to detect the full-length or truncated peptide, a radiolabeled amino acid is included in the translation reaction. The products of the translation reaction are separated on an SDS-polyacrylamide gel and visualized by autoradiography (*Figure 14.7*). The advantage of this technique is that whole genes can be analyzed in a few reactions each covering several kilobases [16]. However for the majority of the genes, RNA has to be used as starting material and cDNA made as the initial template for the PCR reaction. The major drawback to the technique is that it can only pick up truncating mutations. However it could in theory be adapted to use isoelectric focusing as the detection step when it could presumably also pick up missense mutations.

Multiplex ligation-dependent probe amplification (MLPA)

MLPA is a method for quantification of copy number which is relatively straightforward to perform [17,18]. The technique depends on the simultaneous amplification of up to 40 sequences, usually individual exons of a gene, in a single PCR reaction. Specific probes are designed for each of the sequences and consist of a short oligonucleotide probe and a long oligonucleotide probe. The short synthetic oligonucleotide probe contains a short target-specific sequence at the 3′ end and a common 19 nucleotide sequence at the 5′ end. The long oligonucleotide contains a target specific sequence at the 5′ end, a variable length sequence, the 'stuffer' sequence, which can be used to differentiate each of the final products, and a common region at the 3′ end. Denatured DNA is hybridized with a mixture of each of the probes, the two parts of the probe hybridize to adjacent target sequences and are ligated. All ligated products are then amplified with a single primer pair specific for the common nucleotide sequences, one of which is labeled with a fluorescent tag, to give PCR products of unique variable size, each of which corresponds to a specific exon or region of interest. These

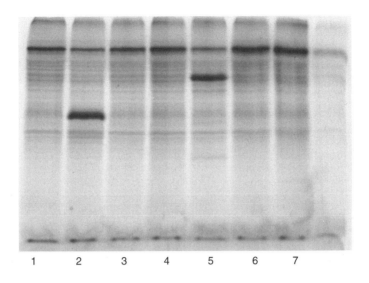

Figure 14.7

PTT analysis of a region of exon 15 of the *APC* gene. Full-length peptides are seen at the top of the gel. Shorter peptides indicating the presence of truncating mutations are seen in lanes 2 and 5. Figure courtesy of Yvonne Wallis, West Midlands Regional Genetics Laboratory, UK.

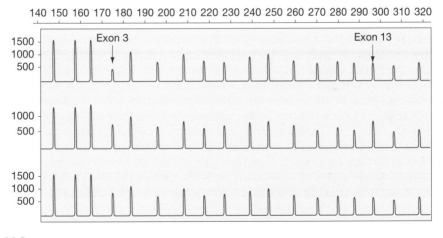

Figure 14.8

MLPA analysis of *BRCA1* in two patients (panels 1 and 2) and in a control (panel 3). Exons 3 and 13 are marked. The first panel shows a deletion of exon 3, panel two shows a duplication of exon 13. Figure courtesy of Elizabeth Redmond and Graham Taylor, Yorkshire Regional DNA Laboratory.

products can be separated by electrophoresis on a fluorescent sequencer. Relative amounts of the products reflect the copy number of target sequences. The main application of MLPA is in the detection of deletions and duplications in cancer genes and it has been shown to be of value for detection of rearrangements in *BRCA1* and *2*, *MHL1*, *MSH2* and *VHL* (*Figure 14.8*).

Direct sequencing

The 'gold standard' for mutation detection is direct sequencing of the PCR fragment. This will immediately give information about the nature of the mutation and any changes due

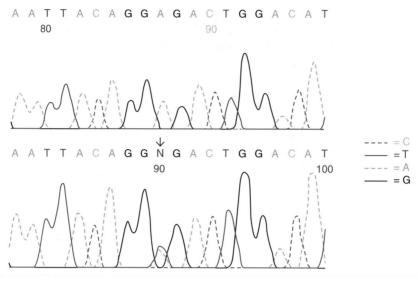

Figure 14.9

Sequence of a region of *BRCA1* indicating an A to T transition which changes the amino acid for arginine to a stop codon. Figure courtesy of Sally Rees, West Midlands Regional Genetics Laboratory, UK.

to neutral polymorphisms can be identified. Sequencing is usually based on the Sanger dideoxy chain-termination method and is traditionally carried out radioactively. Many kits are now available for cycle sequencing based on this method which makes use of PCR and means that only minute amounts of template are required. Fluorescence-based sequencing methods are now widespread. These are based either on cycle sequencing or on Sequenase™-based methods and use either fluorescently labeled primers or fluorescently labeled dideoxy terminators. These methods are the most efficient since they can be automated (*Figure 14.9*) making this a very rapid approach. In order to speed up sequencing a number of strategies have been introduced. Multiple exons of a gene can be sequenced together by an exon connection strategy in which several exons are linked together in a PCR reaction prior to sequencing [19]. A major problem to high-throughput sequencing is the time taken to read the sequences manually. This has now been overcome by the introduction of automated sequence analysis packages such as Staden or CSA [20].

14.2.4 Southern blotting

For over two decades until the development of PCR the basis of much of the analysis of genes and of gene tracking through families (see below) depended on the technique of Southern blotting, a technique named after its developer, Professor Ed Southern. All techniques subsequently developed for the analysis of RNA or protein have continued with this analogy for nomenclature hence, Northern (RNA), Western (protein) and even Southwestern (DNA/protein) blotting. Although the technique is no longer used as frequently as it once was, it still remains useful in some cases to identify large deletions or insertions, or for the analysis of some RFLPs and minisatellite markers.

The basis of Southern blotting is shown in *Figure 14.10*. DNA is extracted from the tissues or cells and digested into small fragments with restriction enzymes. These enzymes, which are isolated from bacteria, selectively cut DNA at sites dependent on specific nucleotide sequences in the genome (restriction sites; *Table 14.1*). Following digestion,

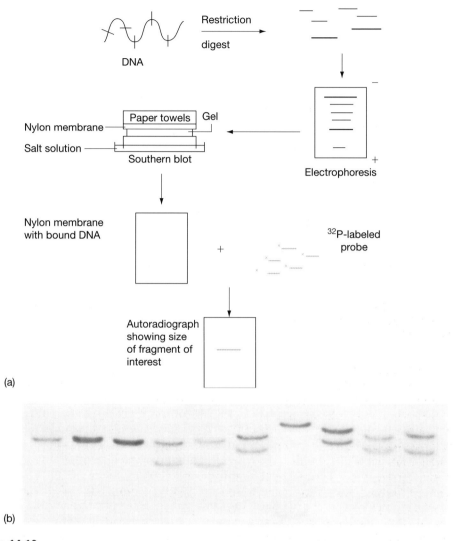

(a)

(b)

Figure 14.10

(a) Southern blotting technique. (b) Example of the end-product of Southern blotting; an autoradiograph showing bands of different sizes. Adapted from [6].

Table 14.1   Some commonly used restriction enzymes

Name	Recognition sequence	Bacterial strain of origin
*Bam*HI	GGATCC CCTAGG	*Bacillus amyloliquefaciens*
*Eco*RI	GAATTC CTTAAG	*Escherichia coli*
*Hind*III	AAGCTT TTCGAA	*Hemophilus influenzae*
*Pst*I	CTGCAG GACGTC	*Providencia stuartii*
*Taq*I	TCGA AGCT	*Thermus aquaticus*

these double-stranded DNA fragments are separated according to size by electrophoresis on an agarose gel. Treatment of the DNA in the gel with alkali produces single-stranded molecules, which are transferred or 'blotted' on to nitrocellulose or nylon membranes. The presence or absence of sequences of interest is then determined by incubating the membrane with the relevant DNA probe which has previously been labeled (with an isotope such as ^{32}P, with a molecule such as biotin or digoxigenin, or with an enzyme) and made single stranded. Hybridization between the complementary sequences of the probe and target DNA is detected by autoradiography, if ^{32}P has been used to label the probe; by an enzyme–dye complex, in the case of biotinylated or digoxigenin-labeled probes, or by chemiluminescent detection in the case of an enzyme-labeled probe, allowing recognition of the DNA sequence under investigation and assessment of alterations in its size, copy number, etc. [1].

14.2.5   Gene tracking with RFLPs and microsatellites

Gene tracking is a method by which chromosomes can be tracked through a family to determine which one of a pair segregates with a disease gene. It depends on knowing on which chromosome and roughly where along its length a gene for a particular disease is located, and also that there is a polymorphic marker situated in the same region which can be used to 'tag' that chromosomal region. Traditionally this technique relied on RFLP analysis but now usually involves the analysis of polymorphic microsatellite markers.

This technique has been used successfully for the diagnosis of many single gene disorders such as Duchenne muscular dystrophy or cystic fibrosis. In cancer research it has been used to track the defective gene in families with inherited cancers such as familial breast or colon cancer. The basis of the method is described first for RFLP analysis then compared with microsatellite analysis which is now used more frequently.

In the example shown in *Figure 14.11*, if the DNA from individuals with different DNA polymorphisms is digested with a restriction enzyme and then Southern blotted, individuals with chromosome A will have a 5-kb fragment detected by the probe, whereas individuals with chromosome B will have a 3-kb fragment detected by the same probe. The 5-kb fragment is arbitrarily called allele 1 and the 3-kb fragment allele 2.

For the detection of inherited diseases, the basis of the test is as follows: DNA from individuals in a family is digested with restriction enzymes and subjected to Southern blotting

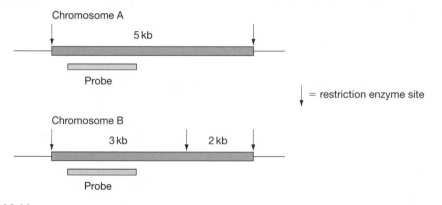

Figure 14.11

Example of a RFLP showing one chromosome with a 5 kb DNA fragment and a second chromosome with a 3 kb fragment recognized by the same probe. Arrows indicate the restriction sites. Adapted from [6].

as described earlier. Each individual in the family is then given a genotype corresponding to the alleles present on either of the two chromosomes. *Figure 14.12* shows the result for three individuals using the probe and RFLP shown in *Figure 14.11*. Individual 1 is homozygous for allele 1, that is, has two copies of chromosome A. Individual 3 is also homozygous, but this time for allele 2. The individual shown in the center is a heterozygote, that is, has one copy of chromosome A and one of chromosome B. By entering these results on to the family pedigree along with results from any other members of the family available, it is possible to determine which allele, corresponding to a particular restriction fragment length, segregates with the disease allele for that particular pedigree. In the pedigree shown in *Figure 14.13*, of an autosomal dominant disease such as FAP, the results suggest that allele 2 is segregating with the disease in this particular family. This is suggested because the affected individual (II-3) has inherited the chromosome with allele 2 from her affected mother, whereas her normal brother and sister have inherited the normal chromosome carrying allele I from their mother. By examination of this pedigree one important point can be seen; the presence or absence of a particular allele does not in itself signify that an individual is affected or normal. In *Figure 14.13* individual I-3, who has married into this family, has two copies of allele 2 and is normal. The RFLP is merely a marker which can be used to follow the pattern of inheritance of a particular chromosome through a family. The second important point to make about this type of analysis is that because many of the RFLPs are located around the disease gene rather than in it, there is the possibility during meiosis of a cross-over in the DNA between the RFLP and the gene. The further away the

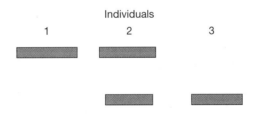

Figure 14.12

Diagram to illustrate the patterns of bands seen in three individuals following Southern blotting and hybridization with the probe described in *Figure 14.11*. Individuals 1 and 3 are homozygous for the 5 kb and 3 kb fragments, respectively, and individual 2 is a heterozygote. Adapted from [6].

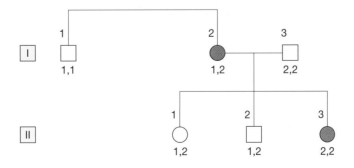

Figure 14.13

Pedigree of a family with a disease segregating in an autosomal dominant manner (see text for detailed explanation of results). The genotypes associated with the RFLPs described in *Figures 14.11* and *14.12* are shown. Adapted from [6].

RFLP, the higher the chance of such a recombination until ultimately it is so far away that it cannot be used for the analysis. All results of such a test have to carry with them an error rate to take account of the possibility of such a recombination event.

The problem with RFLP analysis is that for the majority of RFLPs only two alleles are present at any one locus. Many affected individuals will be homozygous for one or other allele, making the family uninformative. For example if individual I-2 had been homozygous for allele 2 it would not have been possible to differentiate between the high- and low-risk chromosomes. The identification of microsatellite markers has rapidly transformed linkage analysis for a number of reasons: (1) they are found widely distributed on all chromosomes; (2) the number of repeats at any one locus varies considerably in the population such that for any marker many different alleles are present, meaning that it is relatively easy to make a family informative with at least one marker; (3) microsatellites are analyzed by PCR, and results are therefore available in 1–2 days rather than the 5–7 days required for analysis by Southern blotting; (4) studies are possible even when affected individuals are dead, as archival material such as histological specimens from the time of surgery, stored as paraffin blocks, can be used for DNA extraction. Such DNA has generally been degraded by the treatments necessary for processing of the tissue so that only short fragments can be isolated. It can generally be analyzed by PCR-based techniques in which the products of the reaction are only a few hundred basepairs at most but can rarely be used for RFLP analysis in which the alleles are generally several kilobases long. The principle of the technique involves PCR amplification across the polymorphic region followed by separation of the products on acrylamide gels to identify individual alleles. Identification of the alleles is possible either by incorporation of a radionucleotide into the PCR reaction or by staining of the gels post electrophoresis. Each individual in a pedigree can then be assigned a genotype in the same way as for RFLP analysis.

It is now possible to label one of the primers used in the PCR reaction for microsatellite analysis with a fluorescent dye so that the products can be analyzed automatically. Several sets of reactions can then be analyzed together if each set of primers is labeled with a different-colored dye, speeding up the analysis by a further order of magnitude. As well as using these markers for gene-tracking studies this method is equally applicable for use in LOH studies as shown in *Figure 14.14*.

14.2.6 Single nucleotide polymorphisms (SNPs)

At the time of preparing the new edition of this book the more traditional approaches for gene tracking, as indicated in the previous section, are being replaced by high-density genetic maps of the genome based on the identification of SNPs as they are the most common form of sequence variation. In the human genome, approximately one basepair in a thousand is variant when any two chromosomes are compared. In addition to their high frequency in the genome, SNPs are very stable genetic markers and have a relatively low mutation rate in comparison with other types of genetic markers [21]. A detailed treatment of this topic and its enormous potential impact on human biology and medicine is beyond the scope of this book. A good starting point for information is the 2002 *Trends Guide to Genetic Variation and Genomic Medicine* and the references therein [22].

Over a million genetic markers in the form of SNPs are already available for use in genotype–phenotype studies in humans and novel typing strategies are emerging for using these SNPs to type populations. For SNPs to realize their full potential in genetic analysis, thousands of different SNP loci must be screened. Assay readouts on oligonucleotide arrays, in microtiter plates, gels, flow cytometers and mass spectrometers (e.g. matrix-assisted laser desorption – ionization time-of-flight (MALDI-TOF)) have all been developed but decreasing cost and increasing throughput of DNA typing are essential if high-density

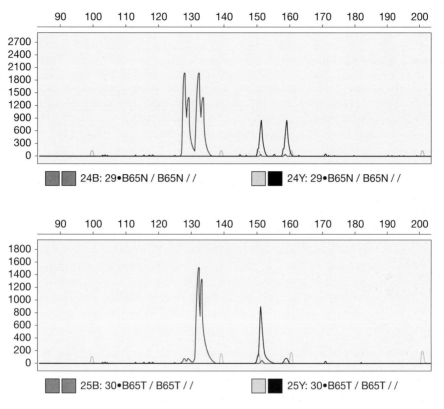

Figure 14.14

Loss of heterozygosity in a Wilms' tumor. DNA from normal tissue (upper panel) and from a Wilms' tumor (lower panel) have been compared using two microsatellite markers from chromosome 16 and a fluorescent-based PCR technique. Both markers in the tumor show only one allele compared to the two seen in normal DNA. Figure courtesy of Emma Prebble, Division of Molecular and Medical Genetics, University of Birmingham, UK.

genetic maps are to be applied on a large scale [21,23]. However, it is already possible to rapidly identify the pattern of all common genetic variation across any single candidate gene [21]. The candidate gene strategy can be approached in two ways: either by assembling a relatively large collection of potential disease genes and concentrating on scanning coding sequences for potential functional coding SNPs or by focusing on a single gene and systematically scanning all of its genomic sequence for all polymorphisms [24]. By mapping SNPs onto three-dimensional structures of proteins, problems concerning population, medical and evolutionary genetics are also being addressed [25].

In terms of identification of cancer risk, SNP validation requires proof of Mendelian inheritance and that the SNP contributes to a phenotypic change, specifically an increased risk of cancer. Although the effect of any single variant is probably small, combinations of SNPs, either as haplotypes or between distinct genes, can contribute to disease risk. It is important, therefore, that genetic association studies need to evaluate the contribution of one or more SNPs to well-defined clinical endpoints [26].

Nonetheless, SNPs have opened up a new and exciting way of studying complex diseases such as cancer and in addition to providing information on disease development, SNPs are providing an opportunity of exploiting human genetic variation in drug discovery and development [27]. It is clear that within the next 10 years our ability to understand how cancer develops and how it can be effectively treated will be revolutionized by SNP analysis.

14.2.7   Northern blotting

This technique is essentially the same as Southern blotting except that intact messenger RNA (mRNA) is separated on the gels and again detected, following blotting, by probing with a labeled DNA fragment. This allows the amount of the transcript to be determined and can also be used to detect abnormalities in the size of the mRNA [1].

14.2.8   *In situ* hybridization

This technique permits the direct analysis of sequences of DNA or RNA in tissues so that specific cells, populations of cells, or chromosomes can be examined [28]. The method involves the denaturation of DNA in a tissue section followed by the application of a labeled probe which is complementary to the sequence of interest. As with Southern blotting it is possible to use a radioactively labeled probe, but here the isotopes used are usually ^{3}H, ^{125}I or ^{35}S (*Figure 14.15*). In addition, as with Southern blotting techniques, it is now common to use probes labeled with biotin which can be detected by an enzymatic reaction. This technique has several advantages over radioisotopes, particularly speed, safety and the long-term stability of the labeled probe. The technique has been adapted to use fluorochromes and fluorescence *in situ* hybridization (FISH) probes are labeled either directly using fluorochrome-conjugated nucleotides or indirectly using reporter molecules. The advantages and limitations of FISH techniques and their applications in genetic diagnosis have been reviewed recently [29].

14.2.9   Comparative genomic hybridization

Comparative genomic hybridization is a method by which differences in copy number between tumor and normal DNA can be determined [30]. The two DNA samples are differentially labeled with two fluorochromes then hybridized to normal human chromosomes in a metaphase spread (*Figure 14.16*). The fluorescence signals are quantitated by image analysis and the ratio of one fluorochrome to the other determined along the length of the chromosome thereby allowing the determination of over- or under-representation of DNA segments in the tumor genome. Gains in DNA sequences are detected by increased signals from the fluorochrome labeling the tumor DNA whereas losses are detected by a signal from the normal DNA but not from the tumor DNA at the same site. This technique has been used to study copy number in both solid tumors and in leukemias. An adaptation of this approach, in which DNA from the tumor tissue is first amplified by degenerate oligonucleotide primed PCR (DOP-PCR), can also be used on formalin-fixed paraffin-embedded material and can give results even if only a few tumor cells are present.

14.2.10   DNA microarrays

Microarray technology is a high-throughput method which enables information to be generated about gene expression and function. The basic principle of the method is the same as for other blotting assays (e.g. Southern blotting). A detailed treatment of this technology and its applications in biomedicine is beyond the scope of this book. For an introduction to the use of microarrays in biomedical research visit [31] and good starting points for information on this technology, its applications to biomedicine, standardization and interpretation of results are the 1999 and 2002 *Nature Genetics* supplements 'The Chipping Forecast' and 'The Chipping Forecast II' and the references therein [32,33].

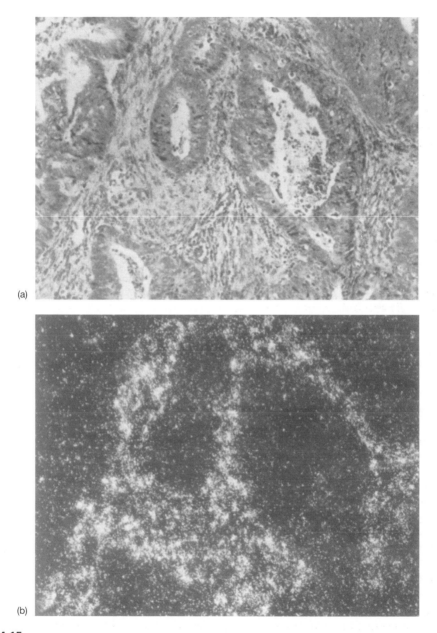

Figure 14.15

(a) Paraffin section of a colon cancer. Bright field illumination to show tissue morphology.
(b) *In situ* hybridization using a ^{35}S-labeled cDNA probe to detect type 1 collagen mRNA.
Detection is by autoradiography. Dark-field illumination shows silver grain distribution.
Photographs courtesy of D.G. Powell, G.I. Carter and R.E. Hewitt, Department of
Histopathology, University of Nottingham. Nottingham, UK. Adapted from [6].

The basic feature is that a solid support, more often glass (although other supports such as
nitrocellulose, nylon and silicon have been used) is 'spotted' robotically with oligonucleotides, cDNAs or genomic DNA defining particular genes and fluorescently or radioactively labeled mRNAs from test and control samples are hybridized to the microarrays.

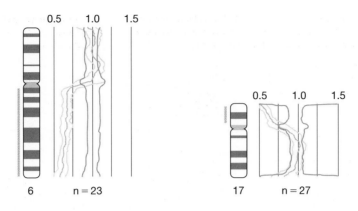

Figure 14.16

Comparative genomic hybridisation (CGH) using DNA from a pediatric brain tumor (ependymoma). This patient shows relative loss at 6q and 17p. Loses are indicated on the chromosome ideograms by a purple bar. The case was analyzed using High Resolution CGH software (Cytovision, Applied Imaging). Figure courtesy of Sara Dyer, West Midlands Regional Genetics Laboratory, U.K.

There are several variations of the technique [34] and cDNA microarrays will be used to illustrate the principle. Many gene-specific polynucleotides derived from the 3' end of RNA transcripts are individually arrayed on the solid matrix. This matrix is then simultaneously probed with fluorescently tagged cDNA representations of total RNA pools from test and reference cells, allowing determination of the relative amount of transcript present in the pool by the type of fluorescent signal generated. Relative message abundance is inherently based on a direct comparison between a 'test' cell state and a 'reference' cell state, an internal control is thus provided for each measurement. It is not possible to undertake the simultaneous analysis if a radioactive tag is used.

Microchip arrays are already being used for differential gene expression profiling in several cancers, for example brain tumors [35], ovarian cancer [36] and prostate cancer [37,38] and the National Cancer Institute has initiated the Cancer Genome Anatomy Project, one of its goals being the production and sequencing of cDNA libraries from five major human cancers (prostate, ovary, breast, lung and gastrointestinal tract) [37]. Microarrays are also being used to study carcinogenic changes seen in the livers of patients with chronic exposure to environmental toxins, such as arsenic [39].

While this technology is going to have a major impact in identifying genes involved in the progression to cancer a very important caution in the interpretation of expression profiling data has been noted [37]. These authors illustrate the problem with reference to prostate cancer and the selection of normal epithelium as a baseline control against which to compare and contrast tumor gene expression profiles. As they point out, normal epithelium in prostatic ducts ranges from atrophic to resting to hyperplastic and each has a unique pattern of gene expression. Additionally, epithelium adjacent to tumor may not be 'normal' – although phenotypically 'normal' at the light microscopic level, it may be genotypically abnormal and/or exhibit an altered gene expression profile due to its proximity to the invading tumor. In addition the degree of inflammation may also have a significant impact. For these reasons, the authors suggest that it is necessary to profile a spectrum of normal and tumor cell populations from a series of patients to distinguish between alterations that are relevant to tumorigenesis and those reflecting the biological spectrum of the 'normal' prostate, or those that have occurred for reasons unrelated to transformation [37].

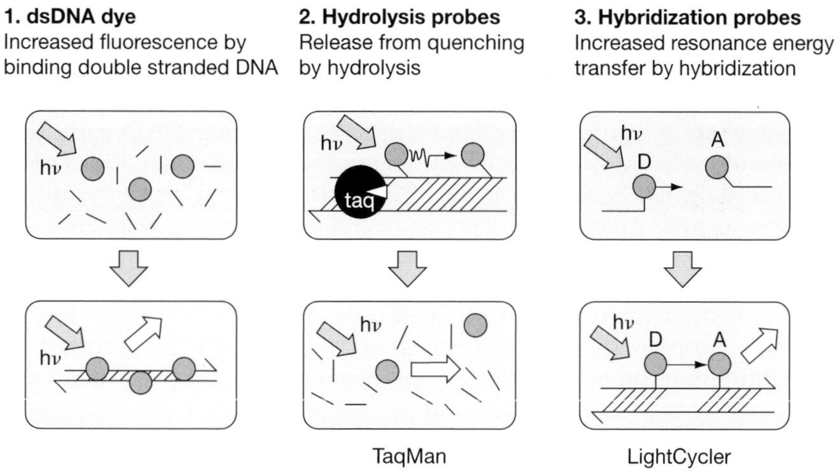

1. dsDNA dye
Increased fluorescence by
binding double stranded DNA

2. Hydrolysis probes
Release from quenching
by hydrolysis

3. Hybridization probes
Increased resonance energy
transfer by hybridization

TaqMan   LightCycler

Figure 14.17

Real-time fluorescence detection of PCR products during the amplification phase (real-time PCR). Fluorescence labeling can be performed (1) by nonspecific binding of an appropriate dye to double-stranded DNA (e.g. Cyber Green), (2) by hybridization of TaqMan probes carrying quencher and reporter dyes which are cleaved during elongation, or (3) by hybridization to a pair of probes labeled with a donor (D) and an acceptor (A) dye with increased fluorescence in juxtaposition (LightCycler method). Adapted from [42].

Despite this caveat it seems clear that microarray studies of human cancer will help to clarify the molecular events that lead to tumor development and in the identification of new diagnostic, prognostic and therapeutic targets.

14.2.11   Real-time (kinetic) RT-PCR

Currently, array technology is being used in establishing broad patterns of gene expression. Validation of expression differences is accomplished with an alternate method such as Northern blotting, *in situ* hybridization, RNAse protection assays or RT-PCR [40,41]. However, the first three of these techniques are time consuming, labor intensive, and require large amounts of RNA (>5 µg total RNA). Conventional RT-PCR utilizes 20–40 ng of RNA but quantification can be difficult. Real-time PCR evaluates product accumulation during the log-linear phase of the reaction and is currently the most accurate and reproducible approach to gene quantification. There are many commercial systems which are in current use. An example is in the detection and quantification of *BCR–ABL* fusion transcripts using the LightCycler technology, which combines rapid thermocycling with online fluorescence detection of PCR product as it occurs [41]. Fluorescence monitoring of PCR amplification is based on the concept of fluorescence resonance energy transfer (FRET) between two adjacent hybridization probes carrying donor and acceptor fluorophores. Excitation of a donor fluorophore (fluorescein) with an emission spectrum that overlaps the excitation spectrum of an acceptor fluorophore results in nonradioactive energy transfer to the acceptor (*Figure 14.17*; see also *Figure 11.6*). Once conditions are established, the amount of fluorescence resulting from the two probes is proportional to the amount of PCR product. Due to amplification in glass capillaries with a low volume/surface ratio PCR reaction times have been reduced to less than 30 min [42].

Another example is the TaqMan 5′ nuclease assay, where fluorescence is released in direct proportion to the accumulation of PCR product. Through the use of a CCD camera, fluorescence production is continually monitored in each of 384 reaction wells. The cycle number at which the fluorescence reaches a threshold value is precisely determined and used to calculate the amount of starting material by comparison to known standards (see *Figure 11.7*). Manufacturers' specifications claim a limit of detection of 50 copies of target molecule and a linear dynamic range of five orders of magnitude. Single-copy detection is possible and ten-copy detection is routine. Throughput is approximately 1500 assays per day. The versatility and flexibility of the real-time PCR method allows for application to a wide range of projects including basic gene expression studies, microarray validation, master cell bank control and FDA mandated product release testing for recombinant products. The quantity of starting material, either RNA or DNA, is determined by comparison with known standards. Comparing treated versus untreated cells or different cell types results in relative quantitation. Absolute quantitation requires the use of carefully quantitated external standards. Quantitation can be performed using two methods, cyber green or TaqMan. The cyber green method tracks the accumulation of double-stranded DNA and provides for a lower cost method of estimating gene copy number because only PCR primers are required. Higher accuracy is achieved by using the TaqMan FRET technology [43]. Although most real-time PCR assays in themselves are characterized by high precision and reproducibility, the accuracy of the data which are obtained is largely dependent on other factors such as sample preparation, quality of the standard and choice of housekeeping gene [44].

14.3 Analysis of proteins

An alternative strategy used to examine the expression of genes is to look at the proteins themselves rather than at the DNA or RNA. This method has some advantages. In particular it means that a retrospective analysis of tumor material can be made. This is made possible by the availability of paraffin blocks from surgically resected specimens. This material is not suitable for analysis of RNA although it is now possible to extract DNA from the blocks for analysis by PCR. The presence of oncoproteins in such tissue is most easily assessed by the application of antibodies. The most obvious advantage of this technique is the ability to test samples from individuals for whom the clinical outcome is known. In studies of the value of oncogenes as prognostic indicators this is invaluable as prospective studies could take years to perform. In order to study oncoprotein expression, however, antibodies have to be produced.

14.3.1 Production of antibodies to onco- or tumor suppressor proteins

In general, antibodies to the proteins have been raised against synthetic peptides rather than the native protein molecule. Short peptides, synthesized on the basis of known DNA sequences and coupled to protein carriers, have been used as immunogens. The region chosen for synthesis of the peptides is selected following the construction of a hydrophobicity plot and includes those residues thought to be exposed on the surface of the intact molecule. Both monoclonal and polyclonal antibodies have been produced in this way and the approach has found many useful applications [45].

A novel approach to producing antibodies specific for tumor cells has been to transfect the NIH-3T3 cell line with DNA fragments obtained from a human ALL and to use the transformants as immunogens to raise antibodies. Although the antigen has not been characterized in detail, at least one antibody produced in this way did not react with NIH-3T3 or with the majority of normal tissues tested but did react with certain tumors, including ALLs and sarcomas [46].

The potential for the use of these antibodies is great, not only for examining the expression of the proteins in tissues but also for monitoring the levels of oncoproteins in body fluids and as potential carriers of therapeutic agents to cells, the so-called 'magic

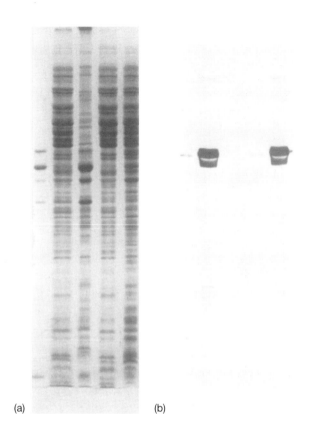

(a)   (b)

Figure 14.18

(a) Polyacrylamide gel stained for total proteins. (b) Western blot of proteins extracted from tumor cells reacted with a monoclonal antibody to a tumor marker. Adapted from [6].

bullets'. Some of their applications have been discussed in previous chapters and the techniques used in their study are described in the following sections.

14.3.2   Western blotting

An overall examination of the size and quantity of oncoproteins in cells is possible by the technique of Western blotting. Proteins are extracted from cells and separated on polyacrylamide gels on the basis of size and/or charge. They are then transferred to nitrocellulose or nylon membranes. Unlike Southern blotting, transfer of proteins by passive diffusion is inefficient; Western blotting requires the application of an electric current for several hours. Following transfer, the protein of interest can be detected by incubation of the membrane in an antibody solution followed by detection with an enzymatically labeled second antibody and direct visualization on the membrane [47] (*Figure 14.18*) or by chemiluminescence and visualization on an exposed film.

14.3.3   Immunohistochemical techniques

Immunohistochemical techniques are used widely throughout all areas of research to study expression of proteins in tissues, essentially by applying antibodies to thin sections of tissues, either frozen sections or sections of tissue which have been fixed and embedded in paraffin wax. Antibodies may be labeled directly with an enzyme such as horseradish peroxidase or alkaline phosphatase. Their binding sites can then be visualized by incubation of the tissue with a substrate which yields a colored product. More commonly,

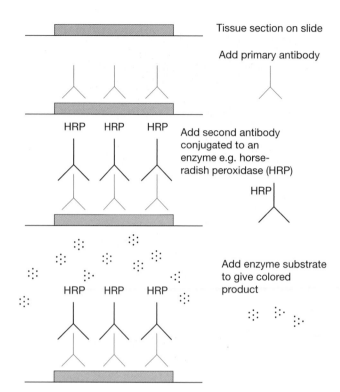

Tissue section on slide

Add primary antibody

Add second antibody conjugated to an enzyme e.g. horse-radish peroxidase (HRP)

Add enzyme substrate to give colored product

Figure 14.19

Flow diagram to illustrate the immunohistochemical technique. Adapted from [6].

a second enzymatically labeled antibody specific for the primary antibody is used; this technique is described schematically in *Figure 14.19*. A typical result showing the distribution of the MYC oncoprotein in a section of gastric carcinoma is illustrated in *Figure 14.20*.

Some care has to be taken when interpreting the results of studies using immunohistochemistry because the antigenic target for some antibodies remains controversial. A number of antibodies have been produced to the *RAS* p21 protein. Some of these antibodies are reactive with both normal and mutant forms of the protein, for example, Y13 259, whereas others have been raised to a peptide containing the amino acid sequence corresponding to one of the forms of activated *RAS* and are specific for the mutated form of p21 [48]. Another antibody (RAP 5), recognizing a peptide reflecting the amino acid sequence of one of the activated forms of *RAS*, cross-reacts with both the normal and mutated form of p21 [49]. Several antibodies (Myc1-6E10 and Myc1-9E10) to the MYC oncoprotein have been produced in a similar way, and again on further characterization it appears that one of these, Myc1-6E10, may not recognize the product of the *MYC* gene [50]. As mentioned in several of the earlier chapters, there has been some variation in results between centers using *p53* immunohistochemical staining. These are partly due to differences in technique and antibodies used and suggest that a degree of standardization is required. In addition, as mentioned in these chapters, comparisons between immunohistochemistry and detection of mutations at the DNA level have shown only a 69% correlation, leading some authors to doubt whether immunohistochemical techniques necessarily reflect *p53* mutations.

14.3.4 Flow cytometry

The development of the fluorescence-activated cell sorter (FACS) has provided an alternative method of studying oncogene expression; a method often referred to as flow cytometry.

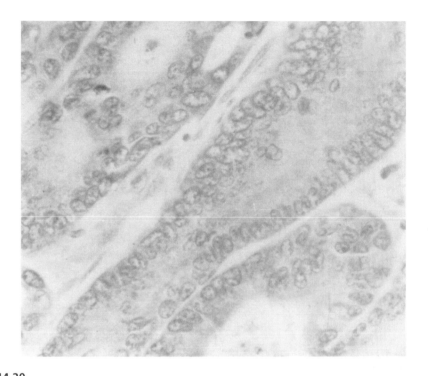

Figure 14.20

A section of gastric carcinoma reacted with an antibody to p62 (Myc1-6E10), showing cytoplasmic staining. Adapted from [6].

A wide range of particles can be analyzed using the FACS, including whole cells, nuclei, chromosomes and bacteria. Regardless of what is to be analyzed, the process is essentially the same. The particles are labeled with an appropriate fluorochrome and then passed individually through a laser beam. The emitted light for each individual particle is collected, measured and stored on computer. Other information such as particle volume and a measure of the granularity of particles can also be collected. In addition to the analysis, individual particles expressing particular preset parameters, such as fluorescence intensity, can be deflected electrostatically into collection tubes so that they can be analyzed further at the end of the experiment [51] (*Figure 14.21*). The fluorochrome used can be a fluorescent dye, such as fluorescein, coupled to an antibody specific for the protein under investigation, or it can be a nucleic acid-specific stain such as propidium iodide, ethidium bromide or DAPI. The whole procedure is extremely rapid, with thousands of particles being analyzed every second. A wide variety of parameters can subsequently be plotted either individually or against each other from the information obtained (*Figure 14.22*).

Nuclear oncogenes such as *MYC* and *FOS* have been extensively studied using flow cytometry [52]. Nuclei for analysis can be extracted from fresh tissues but can also be obtained from paraffin blocks. By dual labeling with a fluorochrome-labeled antibody plus a DNA-specific stain, it is possible to correlate the DNA content (ploidy status) with the expression of an oncogene in individual cells. The detection of aneuploidy in some tumors has been shown to correlate with a worse prognosis. The technique of combining information on oncogene expression with ploidy status may enhance its prognostic significance.

Flow cytometry is a very rapid method of determining the expression of oncoproteins on a cell-by-cell or nucleus-by-nucleus basis. It also has the advantage that, like

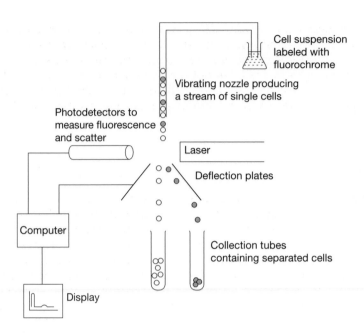

Figure 14.21

Flow cytometry schematic. Adapted from [6].

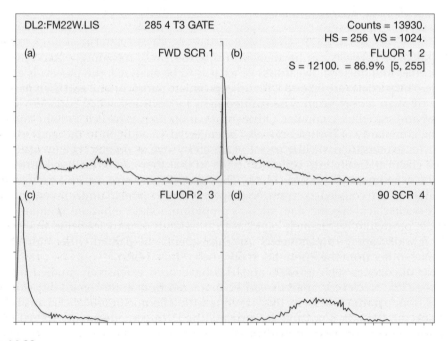

Figure 14.22

Parameters measured by flow cytometer. (a) Distribution of cell sizes; (b) fluorescence with fluorochrome 1; (c) fluorescence with fluorochrome 2; and (d) granularity of cells (90° scatter). Adapted from [6].

immunocytochemical staining techniques, archival material can be used to obtain prognostic information. One drawback in the study of nuclei obtained from archival material, in addition to the problems previously mentioned concerning the specificity of the antibodies, is the apparent loss of reactivity of the nuclear antigens because of the harsh procedures necessary to extract nuclei [53]. This may make significant differences to the levels of oncoproteins detected in the tumors and makes quantitative analysis difficult on preserved tissue.

14.3.5 Proteomics

Although genomic data and transcription profiling offer tremendous opportunities to identify and understand molecular alterations in cancer, even the complete 'genetic blueprint' has serious limitations – cellular functions are carried out by proteins and there are numerous protein modifications that are not apparent from the nucleic acid or amino acid sequence. Also, the genomic sequence does not specify which proteins interact, how proteins interact, how interactions occur or where in the cell proteins localize under various conditions.

Currently, proteomics research embraces two contrasting but complementary strategies: cell-mapping proteomics which aims to define protein–protein interactions to build a picture of the complex networks that constitute intracellular signaling pathways; and protein expression proteomics, which monitors global expression of large numbers of proteins within a cell type or tissue, and quantitatively identifies how patterns of expression change in different circumstances.

Global protein profiles can be produced for normal compared with tumor cells in a given tissue, or for cells before and after treatment with a specific drug. Currently, this is the most widely used model of proteomics and is largely dependent on 2D gel electrophoresis (2DE) for visualization of protein profiles. Expression proteomics is the protein equivalent of DNA microarray analysis that defines global patterns of RNA expression under various conditions. The main limitation of 2DE is the inability to detect proteins of medium to low abundance.

In the search for tumor progression markers or anticancer drug targets, there has been a concerted effort to define gene expression profiles at the transcript level. However, it is clear that mRNA expression data alone are insufficient to predict functional outcomes for the cell. For instance, mRNA expression data provide very little information about activation state, post-translational modification or localization of corresponding proteins. There are also many reports of the disparity between mRNA transcript and protein expression levels. At the very least, mRNA expression studies need to be supported with proteomic information to provide a complete picture of how cells are altered during malignant transformation [5].

The development of proteomics has been dependent on improvements in three enabling technologies – sample preparation, 2DE and mass spectrometry (MS)[5]. These technologies, as part of the arsenal of proteomics techniques, are also advancing the utility of proteomics in the drug discovery process [54].

Recent developments in protein microarrays have permitted analysis of enzyme–substrate, DNA–protein and different types of protein–protein interactions [55]. DNA microarrays are used for genetic analysis as well as expression analysis at the mRNA level. Protein microarrays are used for expression analysis at the protein level and in the expansive field of interaction analysis. The ultimate goal of proteomics is to characterize the information that flows through protein networks and clinical proteomics is being used to detect cancer at an earlier stage, to discover the next generation of targets and imaging biomarkers, and to tailor therapy to the individual patient [56].

The entire field of protein microarray technology shows a dynamic development driven by the increasing genomic information. New technologies such as automated protein expression and purification systems, used for the generation of capture molecules, and the need for analysis of whole 'proteomes' will undoubtedly be the driving force for rapid developments in this field in the near future [55].

14.3.6   Tissue microarrays

An offshoot of the proteomics revolution has been the recent development of tissue microarrays. These permit the comparison of hundreds of tissue sections from different tumors arranged on a single glass slide, thereby facilitating the rapid evaluation of large-scale outcome studies. Realization of this potential depends on the ability to rapidly and precisely quantify the protein expression within each tissue spot. It has been reported recently that a set of algorithms has been developed that allows the rapid, automated, continuous and quantitative analysis of tissue microarrays, including the separation of tumor from stromal elements and the subsequent localization of signals. These were validated using estrogen receptor expression in breast cancer tissue and it was demonstrated that automated analysis matched or exceeded the results obtained by conventional pathologist-based scoring [57]. This technology clearly has enormous potential in oncology.

References

1. **Sambrook, J., Fritsch, E.F. and Maniatis, T.** (1989) *Molecular Cloning: a Laboratory Manual*. Cold Spring Harbor Laboratory Press, Cold Spring Harbor; NY.
2. **Davis, L.G., Kuehl, W.M. and Batley, J.F.** (1994) *Basic Methods in Molecular Biology*. 2nd edition. Appleton & Lange, Norwalk, Connecticut.
3. **Ausubel, F.M., Brent, R., Kingston, R.E. et al. (eds)** (1994*) Current Protocols in Molecular Biology*. John Wiley & Sons, New York.
4. **Martin, D.B. and Nelson, P.S.** (2001) *Trends Cell Biol.,* **11**, S60.
5. **Simpson, R.J. and Dorow, D.S.** (2001) *Trends Biotechnol.,* **19**, S40.
6. **Macdonald, F. and Ford, C.H.J.** (1997) *Molecular Biology of Cancer*. BIOS, Oxford.
7. **Jeffreys, A.J.** (1987) *Biochem. Soc. Trans.,* **15**, 309.
8. **Weber, J.L. and May, P.E.** (1989) *Am. J. Hum. Genet.,* **44**, 388.
9. **McPherson, M.J. and Moller, S.G.** (2000) *PCR*. BIOS, Oxford.
10. **Landgren, U. (ed.)** (1996) *Laboratory Protocols for Mutation Detection*. Oxford University Press, Oxford.
11. **Bos, J.L., Fearon, E.R., Hamilton, S.R. et al.** (1987) *Nature,* **327**, 293.
12. **Xiao, W. and Oefner, P.J.** (2001) *Hum. Mut.,* **17**, 439.
13. **Jones, A.C., Sampson, J.R., Hoogendoorn, B. et al.** (2000), *Hum. Genet.,* **106**, 663.
14. **Marsh, D.J., Theodosopoulos, G., Howell, V. et al.** (2001) *Neoplasia,* **3**, 236.
15. **Roest, RA.M., Roberts, R.G., Sugino, S. et al.** (1993) *Hum. Mol. Genet.,* **2**, 1719.
16. **van der Luijt, R., Khan, PM., Vasen, H. et al.** (1994) *Genomics,* **20**, 1.
17. **Schouten, J.P., McElgunn, C.J., Waaijer, R. et al.** (2002) *Nuc. Acid Res.,* **30**, e57.
18. **Armour, J.A.L., Barton, D.E., Cockburn, D.J. and Taylor, G.R.** (2002) *Hum. Mut.,* **20**, 325.
19. **Wallace, A.J., Wu, C.L. and Elles, R.G.** (1999) *Genet. Test.,* **3**, 173.
20. **Mattocks, C., Tarpey, P., Bobrow, M. and Whittaker, J.** (2000) *Hum. Mut.,***16**, 437.
21. **Carlson, C.S., Newman, T.L. and Nickerson, D.A.** (2001) *Curr. Opin. Chem. Biol.,* **5**, 78.
22. (2002) *A Trends Guide to Genetic Variation and Medicine* (entire issue). Elsevier Science Ltd., London.
23. **Griffin, T.J. and Smith, L.M.** (2000) *Trends in Biotech.,* **18**, 77.
24. **Johnson, G.C.L. and Todd, J.A.** (2000) *Curr. Opin. Genet. Dev.,* **10**, 330.
25. **Sunyaev, S., Lathe III, W. and Bork, P.** (2001) *Curr. Opin. Struct. Biol.,* **11**, 125.
26. **Negm, R.S., Verma, M. and Srivastava, S.** (2002) *Trends Molec. Med.,* **8**, 288.

27. **Jazwinska, E.C.** (2001) *Drug Discov. Today*, **6**, 198.
28. **Buckle, V.J. and Craig, I.** (1996) In *Human Genetic Diseases: a Practical Approach* (Davies, K., ed.), p. 85. IRL Press, Oxford.
29. **Tonnies, H.** (2002) *Trends Molec. Med.*, **8**, 246.
30. **Buckle, V.J. and Kearney, L.** (1994) *Curr. Opin. Genet. Dev.*, **4**, 374.
31. http://www.frontiersingenomics.com
32. (1999) *Nature Genetics,* **21**, no.1 supplement *The Chipping Forecast* (entire issue). Nature Publishing Group, NY.
33. (2002) *Nature Genetics,* **32**, supplement *The Chipping Forecast II* (entire issue). Nature Publishing Group, NY.
34. **Duggan, D.J., Bittner, M., Chen, Y. *et al.*** (1999) *Nature Genet.*, **21**, 10.
35. **Markert, J.M., Fuller, C.M., Gillespie, G.Y. *et al.*** (2001) *Physiological Genomics*, **5**, 21.
36. **Welsh, J.B., Zarrinkar, P.P., Sapinoso, L.M. *et al.*** (2001) *Proc. Natl Acad. Sci. USA*, **98**, 1176.
37. **Cole, K.A., Krizman, D.B. and Emmert-Buck, M.R.** (1999) *Nature Genet.*, **21**, 38.
38. **Svaren, J., Ehrig, T., Abdulkadir, S.A. *et al.*** (2000) *J. Biol. Chem.*, **275**, 38524.
39. **Liu, T., Liu, J., LeCluyse, E.L. *et al.*** (2001) *Toxicological Sciences*, **59**, 185.
40. **Bustin, S.A.** (2000) *J. Molec. Endocrinol.*, **25**, 169.
41. **Rajeevean, M.S., Vernon, S.D., Taysavang, N. and Unger, E.R.** (2001) *J. Molec. Diagnostics*, **3**, 26.
42. **Hochhaus, A., Weisser,A., La Rosee, P. *et al.*** (2002) *Leukemia*, **14**, 998.
43. http://www.genelex.com/Genomicshtmls/taqman.html
44. **Klein, D.** (2002) *Trends Molec. Med.*, **8**, 257.
45. **Tanaka, T., Slamon, D.J. and Cline, M.J.** (1985) *Proc. Natl Acad. Sci. USA*, **82**, 3400.
46. **Roth, J.A., Scuderi, P., Westin, E. and Gallo, R.C.** (1985) *Surgery*, **96**, 264.
47. **Burnette, W.N.** (1981) *Anal. Biochem.*, **112**, 195.
48. **Clark, R., Wong, G., Aruheim, N. *et al.*** (1985) *Proc. Natl Acad. Sci. USA*, **82**, 280.
49. **Horan Hand, P., Thor, A., Wunderlich, D. *et al.*** (1984) *Proc. Natl. Acad. Sci. USA*, **81**, 5227.
50. **Evan, G., Lewis, C.K., Ramsey, G. and Bishop, J.M.** (1985) *Mol. Cell. Biol.*, **5**, 3610.
51. **Young, B.D.** (1986) In *Human Genetic Diseases: A Practical Approach* (Davies, K., ed.), p. 101. IRL Press, Oxford.
52. **Stewart, C.C.** (1989) *Arch. Pathol. Lab. Med.*, **113**, 634.
53. **Lincoln, S.T. and Bauer, K.D.** (1989) *Cytometry*, **10**, 456.
54. **Burbaum, J. and Tobal, G.M.** (2002) *Curr. Opin. Chem. Biol.*, **6**, 1.
55. **Templin, M.F., Stoll, D., Schrenk, M. *et al.*** (2002) *Trends Biotechnol.*, **20**, 160.
56. **Petricoin, E.F., Zoon, K.C., Kohn, E.C. *et al.*** (2002) *Nature Rev. Drug Discov.*, **1**, 683.
57. **Camp, R.L., Chung, G.G. and Rimm, D.L.** (2002) *Nature Med.*, **8**, 1323.

Index